Your Old Wiring

Your Old Wiring

DAVID E. SHAPIRO

McGraw-Hill

New York   San Francisco   Washington, D.C.   Auckland   Bogotá
Caracas   Lisbon   London   Madrid   Mexico City   Milan
Montreal   New Delhi   San Juan   Singapore
Sydney   Tokyo   Toronto

Trademarks

Caddy Clip	Lufkin	Decora
Nalox	Ideal	Tic Tracer
Penetrox	Leviton	Condulet
Wirenut	ANSI	Despard
Klein	NFPA	Circline
Channelock	CSA	NEC
Posi-Ground	UL	Wadsworth
Quadplex	Duxseal	ITE
Romex	3M	Trip-And-Light
Wiremold (and V700)	Circle F	Pushmatic
Wiggy	"Sam's Pretty Good Grocery"	Bulldog
Stakon	Hubbell	Gould
Square D	Copalum	Play-doh
QO	Twister	Duxseal
Cutler-Hammer	Ideal	FPE/Federal Pacific
Lutron	Amp	Electric
		T&B/Thomas and Betts

McGraw-Hill

A Division of The McGraw-Hill Companies

1 2 3 4 5 6 7 8 9 0   Qk/Qk   9 0 9 8 7 6 5 4 3 2 1 0 9

ISBN 0-07-135701-7

The sponsoring editor for this book was Zoe G. Foundotos, the editing supervisor was Sally Glover, and the production supervisor was Pamela Pelton. It was set in Slimbach by North Market Street Graphics.

Printed and bound by Quebecor/Kingsport

 This book is printed on recycled, acid-free paper containing a minimum of 50% recycled, de-inked fiber.

McGraw-Hill books are available at special quantity discounts to use as premiums and sales promotions, or for use in corporate training programs. For more information, please write to the Director of Special Sales, McGraw-Hill, Two Penn Plaza, New York, NY 10121-2298

CONTENTS

FOREWORD

Due to his education, training, and experience as a writer, a teacher, an electrician and an electrical contractor, David Shapiro is eminently qualified to discuss "Your Old Wiring" with you. He gives the impression of holding your hand while he accompanies you step by step through the sometimes complex subject of house wiring. Even if you do not employ any of the hands-on instructions contained in the book, for you may elect to engage a professional to do the actual work, you will, however, gain a new familiarity with the wiring in your home, and will be able to discuss it intelligently. The fact that you will have an up-to-date identification of every circuit at your panelboard will be worth the price of the book right there, the first time a fuse blows or a breaker opens up. Particularly impressive is the emphasis on safety throughout the book. Even seasoned pros get into trouble when there is a lapse of attention, as evidenced by the 59-year-old arc burn marks on my Kleins (side-cutting pliers). I started my apprenticeship 60 years ago, and soon learned about safety the hard way. You have the advantage of the many safety reminders in this book. Carefully read all of the safety tips, take them to heart, and stay alert. I leave you in the capable hands of David Shapiro with every expectation that you will enjoy reading the book (but not all at once), and with the admonition to be careful.

W. Creighton Schwan, P.E.
Hayward, California
September 2000

PREFACE

How-to books can make potentially dangerous work look too simple. This danger is especially high when what they cover doesn't exactly match what you're dealing with. "Winging it" with your wiring can be deadly. You'd have to be not just bright and informed but a full-blown genius to successfully substitute sheer reasoning power for the fruits of a century-plus of electrical development.

Going at wiring projects stone-cold is not the only way to buy trouble. I've seen mess after mess created by clever people who just were trying to match what's in their walls already; I've seen messes created even when their model, the original wiring, was installed properly by professonals. I've seen almost as many problems created by people relying on do-it-yourself books based on wiring materials and designs that did not match what was in the readers' walls.

> Homeowners clearly do their own wiring every day. Most succeed, in the sense that the installations work, and don't start fires or cause shock in the short term. How, then, can it take five years or more to become an electrician, even a residential electrician? Why isn't a book, or even several books and videos, a reasonable substitute when you're just wiring your own home? There are two answers. One is that safely doing electrical work requires more than short-term, "it works" thinking. The other is that an electrician understands wiring enough to troubleshoot and repair pretty much whatever has gone wrong. Most laypersons cannot.

If doing it yourself is so risky, why do I offer to teach you this work? There are two reasons. First, I sympathize with your desire. Whether you have trouble finding the money for a contractor, have trouble finding a professional you can trust, as the skilled trades find it increasingly difficult to recruit apprentices, or simply want to understand your own house, I know what it's like to want control over your own home. Besides, I know very well that one way or another many people do take charge of their wiring, and I'd much rather see you do it as safely as possible.

This book does three things. First, it provides the background you need in order to work safely with household electricity. Second, it shows you exactly what to do to succeed at some *relatively* simple, commonly needed electrical jobs, starting with evaluating your electrical system. Third and most important, it details the errors people commonly make as a result of ignorance or inadequate instruction. It's hard to learn from others' subtle mistakes without someone to identify them as mistakes. You will learn to see the distinctions between

This is not the book for the person just looking for quick cosmetic fixes to help unload a house, trusting that the purchaser's inspector won't look too hard. It is for the person who wants to do things right the first time, and wants them to last.

The NEC is a volume you must consult as an adjunct to using this or ANY wiring book.

"right" and "no, not like that." You'll have the chance to learn from the mistakes others have paid me to repair over the years, and avoid their frustrations.

There are three things this book doesn't try to do. First, it doesn't teach you all the laws about wiring; instead, it makes reference to the *National Electric Code (NEC)*.

The NEC is the very best source of information on what constitutes safe wiring. (This is why electricians at all levels are tested on the NEC, and why they study its changes throughout their careers.) While I will touch on commonly violated Code rules when discussing jobs such as replacing a receptable or light, the 1000-odd-page NEC is simply too much to paraphrase. (In 1881, New York City had a one-page NEC, enforced by the New York Board of Fire Underwriters; it has grown over time as consensuses have recognized improvements in safety.) You do need to go through the NEC's table of contents, especially for a sense of what the first four chapters cover. With that preparation, you can refer to it for the rules relevant to the tasks you are undertaking. Buy the current version at a local electrical distributor or from a mail-order source such as the electrical safety group IAEI (1-800-786-4234).

Second, this book doesn't dwell on background theory, physics, or formulas, except what you actually need to do the job. This book does not assume that you are an apprentice with all the training that implies, nor someone seeking knowledge that is not directly applicable.

Third, it doesn't try to teach you everything. Even my more-advanced book, *Old Electric Wiring: Maintenance and Retrofit,* which presumes a readership with a grounding in electrical work, isn't big enough for that. However, when you do have to call in a professional, this book will have made you a much better-informed consumer.

There's also one thing this book CAN'T do, despite every effort. It does not, because it cannot, remove all risk from doing your own wiring. Electricians put in a five-year apprenticeship to become journeymen—studying about as long as lawyers. A journeyman needs several more years of experience, plus an exam, to become a Master Electrician and be able to apply for electrical permits.

Even those who stick to residential work have immeasurably more training than one book can provide. Yet even Masters make deadly mistakes. Therefore, if you're not sure you understand something in this book, or if you encounter a situation not covered in it, find someone to ask.

ACKNOWLEDGMENTS

Three friends volunteered their time and intelligence to make this book clearer, and to help me avoid accidentally misleading readers. They are my sweetheart, Mary Jo Hlavaty, who brought experience as a writer and editor; my mentor, W. Creighton Schwan, a well-known code expert, consultant, and electrical writer; and my colleague Glenn Zieseniss, like Creighton a retired electrician and inspector who remains very active in electrical safety forums.

Without the support of my editor, Zoe Foundotos, McGraw-Hill might not have ended up publishing this book. I appreciate the willingness shown by Alexander Brittain, copy editor, and Sally Glover, senior editing supervisor, to accommodate my concerns about clarity and reader-friendliness.

The respect and acknowledgment I have received from members of the electrical industry for my writing over the last decade-plus, and especially in response to the publication of my earlier book, *Old Electrical Wiring*, were a significant factor in my preparation of this volume.

As I described in more detail in that first book, there are so many people who contributed to my understanding of electrical safety and to my competence as a communicator. They include my parents, Lillian Shapiro-Michael and Louis Leonard Shapiro; my teachers; editors and publishers; inspectors; and organizations such as the International Association of Electrical Inspectors. As much as or more than any of my other writing, I also owe what you will read to the customers who have called me in over the years. In the course of serving them by evaluating electrical systems, rescuing messed-up wiring, and guiding them in working with their own wiring, I too have learned.

1

Introduction

If you're not an electrician, and you want to deal with old wiring, there's never been a book to help you figure it out; never, until now. I'm both a writer and an electrician, and I've written about wiring for quite a few years. Moreover, I'm an electrician who was trained to teach. Finally, I am used to dealing with older buildings—and working with homeowners!

This book is explicitly for people other than full-fledged electricians—even for new apprentices. Because my purpose is to introduce old wiring to people who may not know anything at all about wiring, I must of necessity work—with a great deal of care—for what at first glance might appear to be limited goals. This is a first, introductory text on a subject that can be very complex.

Structure of the Book

The most important thing you will get out of this book, the thing that is most lacking in other books, is the information you will need to perform a thorough evaluation of your electrical system. I assume very little about your background, other than that you understand English. I assume very little about your electrical system, other than that it is not brand new and that it is somewhat functional. Together, these perspectives are why the first half of this book is preparation.

Ultimately, this book will teach you how to replace the three pieces of an electrical system that get the most use and get replaced most often: receptacles, switches, and lights. Since many home centers offer to teach you each of those tasks in an evening session lasting perhaps an hour, clearly I have more rigorous standards and recognize that the tasks are far more dependent on the design and condition of your electrical system.

This is why the book is carefully structured. Each chapter will build on the previous ones. The first two are an overview to introduce your electrical system. This provides a context for the following chapters. Because it's easy to hurt yourself doing even the simplest electrical job without the right precautions, the very beginning of the how-to part starts with safety tips. Even professional electricians get blown away—literally! (I've been a consultant in such a case. For further, grim reading about the horrible things that can and do happen, read Mazur's *The Electrical Accident Investigation Handbook*.)

To wire safely, you need to grasp the knowledge and begin to develop the skills covered in this book's early chapters. Do so before jumping ahead to the three basic types of repair or replacement.

Before you can wire legally, you pretty well have to at least skim various parts of the National Electrical Code (NEC), including Articles 90 and 110, and chunks of NEC Chapters Two and Three.

I urge you to take things in order. You don't want to do dangerous stuff such as working in your panel? No one says you have to. However, changing a light fixture or receptacle can be nearly as dangerous. I'll teach you why as I go through the chapters.

You don't need to read this book word-for-word. Think you already know the material that's covered in a section? Skim through; it won't hurt to confirm your impression of what you know. Wiring certainly is a context in which what you *don't* know can hurt you; moreover, what you don't *realize* you don't know can hurt you. For instance, while everyone in the building trades will be familiar with the few terms and ideas I introduce just below, the kind of electrical work on which this book focuses further on will provide even builders with a goodly amount of new information.

New Work

New work, a term that the building trades use to signify work on buildings that are not yet habitable, can be fairly simple. There are many ways to wire, but I can take new apprentices—who are essentially laypersons—and show them one and only one straightforward way to wire a job. They'll do okay, so long as I start out with some safety training and guidance in tool use. Plus lots of supervision.

Old Work

Old work, meaning work done after somebody else has already wired a building, can be far more complicated. There are at least six reasons for this difference.

- In any electrical work, once power is turned on, you *always* have to be on the lookout for live wires.

- In any old work, you run the risk of working on something that is dangerous because it was worked on by someone ignorant or indifferent, or because it suffered damage at the time of installation or later.

- On any job someone else has begun, you are at risk from any mistakes previous installers—even competent ones—overlooked.

- Once an electrical system is in place, adding to it requires some calculation. A text on new wiring should either make the equivalent calculations for you, or explain a simple way to do them. Many apprentices and journeymen fail tests required for promotion because they mess up such calculations, even the simpler ones. This is the reason a careful introductory text avoids the need for calculations, unless it is prepared to teach the applicable rules. It is why in a first text I focus on getting you to where you can safely replace devices such as switches and receptacles, rather than adding the considerable complexity associated with legally extending circuits.

- Once wiring is in place, you are faced with the problem that many different designs are safe and legal. Some match the way I might happen to wire. Others designs are equally valid and suitable. If you encounter one of the alternate designs I haven't considered, any "cookbook" approach I offer for extending you wiring, even if all the calculations already have been done, may be illegal or unsafe, or simply may not work.

- Once a building is more or less complete, even without surprises in the electrical design, the simple fact that wiring already is in place affects what you do. Suppose you want to add to wiring performed by an installer whose work was unsatisfactory, and whose work may never have been inspected. Any mistakes he or she made now become your responsibility.

Older Buildings

Sometimes an old building has been gutted; everything has been removed to the structural shell, including walls, ceilings, wiring, and plumbing. In this case, you may not need this particular book. Without the complexity resulting from existing wiring, your job is simply a variation on "new work."

"Old work" on older buildings is the opposite. It is considerably more complex than "old work" in newer buildings; it also can be far more dangerous. There are three reasons for this.

- Codes, the rules about how to wire, have changed every three years, more or less, for more than a century. The same is true of standards for equipment. These updates represent improved standards of safety, the introduction of new materials and designs, and the obsolescence or even unacceptibility of old ones. (For instance, older designs permitted far greater access to live parts of equipment.)

- Besides dealing with equipment that has become outdated and even unrecognizable or unfamiliar, you will encounter materials that have deteriorated over time. For example, insulation gets brittle and eventually crumbles; spring metal components lose their springiness.

- The longer your house has been in existence, the more uncorrected mistakes are likely to have accumulated—a danger in any old work.

Taking This Seriously

"Wiring is not a Hobby," shouts a venerable safety campaign, first sponsored by Graybar Electric in 1955. They know what they're talking about. People who fail to take wiring seriously can die, or kill others. Graybar's challenge should be directed not only at novices but also at builders, and even the occasional electrician who doesn't approach the work as a professional.

Respecting Your Limits

I urge you not to try anything you don't feel safe doing, and I strongly urge you to hold off on the tasks that I myself consider too complex for this book. It is even unwise to replace receptacles, switches, or lights when you're leery.

As I mentioned above, this book does not teach you how to add or extend circuits. Adding a circuit can be complicated because of what you find in an old panel, and extending a circuit can be even more complicated because of the way the circuit is wired. Both of these tasks pale, though, before the challenge of figuring out how your system was meant to work, and what has gone wrong. This is why in gearing you up to work safely, this book focuses on teaching you how to uncover problems, and then how to perform the three most common repairs. Troubleshooting is a prime skill learned by electricians, but it is covered briefly in the more advanced text, *Old Electrical Wiring.* It's a skill which electricians develop over their entire careers.

Even within the tasks I discuss in the greatest detail, some of the situations I describe will prompt me to say, "Leave this one alone." Some of the problems that you may run into I will explain but not teach you how to fully diagnose or to repair. These are designs, and problems, that branch out with great complexity. They present too many difficult-to-explain possibilities that are, nevertheless, essential for you to understand in order to proceed safely. It would be too easy to overload readers; I will always do my best not to put you in jeopardy.

I suggest that you absorb what's in this book and then, if you want to proceed further, invest in an intermediate-level book. If you do have the electrical background to follow sophisticated explanations, in many cases you can find the information in my other book, *Old Electrical Wiring.*

For most others, that text will be a bit advanced. Its explanations do rely on more knowledge and experience than you will have after you finish here. If just a bit more information might enable you to take on a task, the answer might be to get hold of an intermediate-level book. At this point there is no such book written with old wiring in mind. However, it is possible that you may be able to combine what you get in a more general wiring book with the information on old systems that you have garnered here.

When I come to tasks that are beyond the scope of this—and, in my experience, any—book for novices, you need to call in a pro.

Pros

When advanced-level tasks need to be handled, it's essential to call in an electrician. This leads to an important question: how do you find a trustworthy contractor?

Most parts of the country require electricians to be licensed. A license is some indication of professionalism, or at least knowledge and experience. The license certifies that the holder has demonstrated such knowledge and experience. Besides being a licensed electrician, someone offering to do your

> It would be ludicrous for an author to demand that you the reader always call in a professional electrician, whatever the fine points of the law. I myself know carpenters and handymen who can handle some electrical tasks and, more importantly, who know when tasks are beyond their competence.

work also needs a
license as an electrical
contractor, and the cor-
responding insurance. An
employee under proper
supervision by a master
electrician can do your
work.

> A *contractor* having to apply for a permit is absolutely not the same thing as a contractor having you apply for a "homeowner's permit." The former puts the responsibility and liability where it belongs, while the latter arrangement pretends you're the person doing the work.

Where there is licensure, permits and inspection usually are required. When this is the case, almost any work beyond simple equipment replacement requires that the master electrician apply for a permit and arrange for inspection by the appropriate authority. If an electrician skirts the law, he or she may well skirt safety requirements as well—or be ignorant of them. You probably have no way to recognize such indifference or ignorance.

Learning the Language

As I mentioned in the preface, it takes as long to learn electrical work as to learn other professions. And, like many other professions, electrical work has its own jargon.

I'll introduce a lot of specialized language as you go through this book. Often it is as simple as "new work" and "old work." The main reason to teach these terms is to reduce the chance that you'll misunderstand me. A second one is that if you go beyond this introductory text, it will become harder to find competent texts that pass on the background information you need, and explain the tasks you're facing, in everyday language. Besides, even at this introductory level it will be useful to make sure you're clear when you talk to professionals: electrical supply house salespersons, inspectors, and any electrician you need to call in because you recognize that a project is beyond your level of competence.

True, professionals probably know how to ask questions to figure out what you mean. However, sometimes salespersons and electricians will simply make polite noises and proceed to figure out for themselves what's going on. When this is the case, you may end up not getting what you really want. They're less likely to ignore what you have to tell them about the symptoms you've noticed or the specifications you want them to meet when you speak their language correctly.

Common Miscommunications

Uninformed persons sometimes complain, "I've got a plug with a short." It's hard for the electrician to know what they mean; therefore, the electrician often guesses. "Short" could mean "open," a broken connection. It could mean one of three types of short, briefly addressed in Chapter 5, Safety: bolted, intermittent, or arcing. A clearer description can eliminate all this guesswork.

How about "plug"? Usually, it is used to mean "cord connector"; even electricians use "plug" to mean that. One electrician knows that the other electrician means this when he or she says, "plug," but we don't trust non-electricians to mean what we would mean, or to know the lingo and use it correctly.

I've known customers to use "plug" to mean "switch" or "fuse" or "lamp socket" or "receptacle." "Socket" I've heard used both as "lamp socket" or "lampholder" and as "receptacle." I've heard "switch" for "circuit breaker" and "fuse box" for "circuit breaker panel."

So I'll use a lot of formal, technical terms, and frequently use common terms in their technical, rather than common, meanings. Should my wording ever puzzle you, check the glossary to see whether the puzzling word has a special meaning in an electrical context.

I won't go overboard, but I will eliminate ambiguity. "Bulb," for example, has a special meaning. So, even though electricians sometimes use "light bulb" the way it's normally used, in some cases I will call it a "lamp," to match the term used in the instructions that come with or are printed inside fixtures. I call the round or U-shaped openings you see when you look at the front of a modern, three-prong receptacle "the round holes," and the straight ones, "slots."

I use common terms when I can. The formal term for "light fixture" is "luminaire." I won't use it because you don't need it. Electricians use "light fixture," and there's no risk of misunderstanding.

In many cases, I'll take informality further. Some tools have become so thoroughly associated with particular brands that in normal talk, the generic name simply isn't used: "Channels" technically are "Channelock brand groove-joint pliers." Some products are referred using shorthand such as copyrighted names. "Wirenut" is a particular, branded, "twist-on solderless connector," the one invented by the Ideal Corporation in the 1920s and sold ever since. But "twist-on solderless connector," is just too much of a mouthful. People say "wirenut," and that's the term I'll use. People say "romex," not "nonmetallic sheathed cable;" "gem box," not "sectional switch box;" "BX," not "armored cable;" "breaker" at least as often as "circuit breaker." My purpose is to help you talk effectively, not fussily.

Clarity does mean that you call a breaker a breaker and not a fuse unless you are speaking generically of "overcurrent protection" rather than of what's in your particular panel; that you call a cable a "cable" and don't use "cable" for "conductor." BX is not "conduit," a usage that really can confuse an electrician or electrical distributor; and, preferably, not "the corrugated stuff." "Conduit," which means pipe, is one of a category of products that installers can pull conductors through. Because conduit is not the only member of the category you might find in your system, I use the category name, "raceway," when I am not talking specifically about conduit.

Sometimes, instead of introducing an unfamiliar global term, I will need to differentiate to an unusual extent in order to explain important concepts. The terms, "wires," "wiring," and "conductors" normally are used interchangeably. To make sure you are not confused, in some places I will assign each term a slightly different meaning. In such cases, I will use "conductor" to mean wire plus its insulation, I will use "wire" for the copper inside, and I will use "wiring" to mean one or more such conductors. Here is an even more specialized usage. When I need to talk about the skinny copper wires that a manufacturer puts together to make up a stranded conductor, I will call them "strands," not "wires."

I will do everything I can to keep myself clear. "Device" is a term that applies to "wiring devices," such as switches, and that's the only way I'll use the term.

It normally is applied to other items as well. Fuses are examples of "overcurrent devices"; twist-on wire connectors are examples of "splicing devices." I use terms other than "device" when I talk about those categories to minimize confusion.

Learning the proper terms may help you in one additional way. The Whorf/ Sapir hypothesis suggests that having the proper language can clarify your thinking. Think about it: the concept, "My circuit breaker won't reset" is a whole lot less fuzzy, and permits a much clearer approach to problem solving, than "I seem to have a bad switch in my fuse box." Incidentally, a few readers may indeed have antique electrical panels that contain fuses and switches. Any switch can die. When this seems to have been the case, "I seem to have a bad switch in my fuse box" is a dead-on description.

I

Familiarizing Yourself with Your System

2

Basic Information about Electrical Systems

Tuesday night, I got a phone call from Chris. He had seen my ad in the Yellow Pages offering electrical safety inspections, and he needed me fast. He was selling his house, and closing was scheduled for Thursday. Unfortunately, the buyer was not about to go through with the purchase unless someone checked out the basement wiring. It had been installed a few years earlier, without an electrical permit, by a handyman who apparently knew less about electricity than he thought. The man knew plumbing, but extrapolating from that knowledge had not been sufficient. Sure, the outlets worked. However, they weren't safe. Chris had not even known the wiring was flawed until his buyer showed him the home inspector's report.

The next few chapters talk about how electricity gets into your house and show you how to look for problems. They also offer you clues to the age of your electrical system, which may very well be newer than your house. Every time I describe a system element, I will describe illegalities and dangers you should be alert for when you look it over. Learning to recognize these problems is a considerable part of the "how to" in this book.

This chapter sets the groundwork for the later ones. If your electrical service is just about new, meaning that it was installed within, say, the last decade, you may be able to skim past most of the information on evaluating electrical services in Chapters 4, 6, and 7 (Service Entrance, Panel, Fuses/Circuit Breakers).

Quite frequently, part of the system has been upgraded and other parts—those more difficult to replace—have been left alone. I have seen this again and again and again. Sometimes my customer says, "The house is 60 years old, but they upgraded the entire system," yet I discover that the service cable and electrical panel are the *only* parts that have been upgraded. Other times I figure this out during the initial phone call. I've been told, "My house was built around 1950, but the basement was finished later, so its wiring should be quite a bit newer." A simple question elicits the fact that the basement wiring is rubber-insulated. Rubber insulation started making way for plastic in the 1940s. You may learn that the person from whom you bought your house gave you a very optimistic picture of the wiring (possibly out of ignorance). This is a pity, but you're better off knowing.

Regardless of whether that is the case, you will need the basic terms and concepts in this chapter.

Before you can start looking at these components or those beyond your electrical panel, you need to understand how power flows, and in particular how it flows into and through your house. At the end of this chapter, I will talk about a critically important part of your electrical system that is vulnerable to damage at any time, and thus is worth examining even if your electrical service is not very old.

Electrical Quantities

Voltage, or potential difference, is electrical pressure in relation to the zero point, essentially that of the ground itself. It is measured in volts, abbreviated "V." (In some formulas, "E" is substituted.) Electrical equipment and outlets with significant voltage are called "live" and can work; equipment with no voltage just sits there, and is referred to as "dead." Turning off the electricity feeding an outlet is called "killing" it.

There's an in-between point between dead outlets and live outlets with normal voltage. If your voltage is too low, your lights may dim, your motors (such as those in air conditioners) may suffer, and any computerized equipment may be damaged. You will look at possible causes of undervoltage shortly. Three of them are especially common in older houses.

Current is rate of flow, measured in amperes or amps, abbreviated "A" ("I" is substituted for "A" in formulas). For our purposes impedance, the aspect of a wire that impedes that flow, is its resistance, R, measured in ohms. For a given voltage, the more ohms impede current flow the fewer amps will make it around the loop—and the more the wire will heat from the amps it does carry. The amount of current a conductor can carry safely is called its ampacity (capacity in amps). The thicker a wire of a given material such as copper, the lower its resistance and hence the higher its ampacity. Too thin a wire may reduce the voltage available for utilization by the equipment it serves.

Power is measured, for our purposes, in Watts (W). Multiply the lowly watt by a thousand, and you get the kilowatt, or kW. If you install equipment designed to draw too many watts for the size of wire feeding it (an overload), many types of equipment will draw too many amps (overcurrent). Undersized wire may cause a fire by acting a bit like the element in a toaster, getting red hot and igniting materials nearby.

There's another, more familiar way of describing overloads. The rating of a circuit, of the path to the appliance, is determined by the protection (the size of the fuse or breaker) at the wire's point of origin. In some ignorant and illegal installations such as where the wrong size fuse has been installed, this rating, and thus the loads—which constitute the wattage—that the circuit might serve, exceed the ampacity of the wire.

Here is a less evident example of using a wire inadequate for the wattage it needs to handle. This one has nothing to do with improperly-wired circuits. The ongoing use of extension cords, which are intended for temporary hookups only, can be quite dangerous. These cords almost always are made up with wire smaller than the circuit can protect. This is

If you are an electrical apprentice, you know or will learn the difference between the terms I use here and the more-exact concepts and language required for precision and for more complicated work. Here, they would get in the way. A two-wire circuit does not have a neutral; however, a homeowner does not really need to recognize this fact.

Here are two concrete ways of relating watts to amps. The first concerns your pocketbook. You pay for power used over time, or kilowatt-hours, kWH. Before restructuring, most utilities charged less than a dime per kWH. A 100 watt light costs you less than a quarter to operate all day and night. 100 W / (1000 w/kW) = 0.1 kW. Then (0.1 kW) × 10 cents / kWH = 1 cent/hour. 1 cent × 24 hours = 24 cents.

The second way to relate watts and amps concerns overload. That 100 watt incandescent light fed 120 volts draws ⅚ A: 100 W ÷ 120 V. Drawing less than an amp each, quite a few such lights can burn simultaneously on a 15 amp circuit without overloading it.

one reason extension cords, so easily overloaded by feeding, say, a 1200 watt air conditioner, can ignite carpets.

How Electricity Flows

Electricity does *not* flow like water. Water can flow into your house and into your sink even if your drains are blocked. It doesn't care about the rest of its path; it may flow onto your floor.

Not so with electricity. An electrical circuit is a loop. Like a trip from home to work and back, both halves of the loop run next to each other but in opposite directions. Even though they don't make a great circle (when the wiring was done legally) but run right alongside each other, they're still a closed path. When you no longer have an unbroken circuit, the voltage maintains its pressure, like a traffic jam that ties up motorists who are eager to get moving, but the current can't flow. Phrased differently, when a wire is broken anywhere along its path, on the hot, originating side or on the "neutral" return side, the available watts can't do work such as lighting a light.

Electricity flows into your house from the utility's pole, and ultimately from their transformer, only when it simultaneously flows back. It flows out through a circuit inside your house only when the same is true within your system. The most dangerous aspect of this is that when a light is out because the loop is interrupted by, say, a broken filament or defective switch, you might assume that no voltage is available. However, it may be and often is present, waiting to do the "work" of shocking you should your body bridge the break.

One reason that I start with your electrical service, rather than beginning with your interior wiring, is to help you set priorities. Some people concerned about their old wiring start out by having their services upgraded. I think the condition of the interior wiring is much more important, and for this reason I cannot recommend that you begin by spending the money on a service upgrade unless one is needed. In the next few chapters, you will learn how to determine the need for such an upgrade as you learn how to evaluate your service.

I must emphasize that whatever procedures I describe, whatever repairs I advise you to put on your "to do" list, it is not necessary for you to do them yourself. It is perfectly reasonable, and indeed wise, for you to identify problems and then hire an electrician to perform any task you don't feel secure about. (He or she quite reasonably may be uneasy about being brought in after you've actually started repairs; see the discussion of "sweat equity" and consultation in Chapter 5, Safety.)

The Physical System

I've looked at the quantities associated with electricity, defined the circuit conceptually, and explained its rating. Now it's time for the components that can be seen and touched.

Conductors

Electricity is carried through conductors. Conductors usually consist of solid but flexible copper wires; flexible because they're skinny. We need more flexibility inside fixtures, in their attachment cords, and in extension cords—as well as in heavy wires that carry a lot of electricity. In these cases (plus others that you are less likely to encounter), each conductor almost always consists of many strands of finer wire twisted together.

Insulation

Conductors are covered with insulation—electrical insulation, which is quite different from the thermal insulation that keeps you warm in the winter and cool in the summer. Electrical insulation is some substance whose resistance, R, is extremely high. So long as this insulation protecting your wires is not damaged, electricity can't leak from one to the other, or from a wire to nearby metal or earth, or to you.

Cables

Most commonly in residential wiring, especially nowadays, conductors are in cables. Cables contain a number of insulated wires, sometimes plus filler such as kraft paper, all covered by or embedded in something that holds them together and protects them. Conduit (rigid electrical pipe), tubing, and flexible conduit are used much, much less frequently in residential electrical work nowadays, although they were employed far more frequently in the past. All these wiring systems are described in far greater detail in Chapter 10, Components.

What Type of Electrical Service Do You Have?

Your utility sends electricity into a building through its service. In modern single- or two-family houses the service usually enters in cable that encloses three conductors. Two are "hot legs," or "live" wires, and the third is the "neutral." Conduit too is used for services, mostly in businesses and in older houses. Even though metal conduit itself is grounded, it too encloses two hots and a neutral. The latter, the most important conductor, is grounded back at the utility pole and also where it comes into your system. "Grounded" means it is connected to the ground—literally, the earth. However, for an important reason that I will get to later, circuits do *not* return to the utility through the earth to any significant extent.

Saying that hot legs are 240 volts apart from each other, or 120 volts from the neutral, has a very specific meaning in practical terms. If you touch one voltmeter probe to the neutral and the other to *either* hot leg, it will read "120." Parts of your system where 240 volts are available are a bit more dangerous than parts with only 120 volts. IMPORTANT NOTE: The safe use of voltmeters is covered in Chapter 3, Testing Voltage.

System Voltage

Now it's time to return from the purely physical layout of the system to its electrical design. The next few pages explain the significance of the conductors in the service entrance cable, the voltages they bring you, and the consequences of voltage variation. I spend very little space discussing basic concepts of voltage and current such as electron flow and potential difference because I promised that this book would focus on the practical. For this reason, I will limit myself to explaining standards, rules, measurements, and calculations that will help you evaluate your system.

The Neutral

Damage to the neutral can be far more dangerous than damage to either of the other conductors, as explained below, on Page 16.

I said above that the neutral is the most critical of the three conductors. (This is true both at the service and in your interior wiring.)

The neutral is "midway between" the other two lines, in the electrical sense. The other two conductors feeding house wiring, the hots, are not grounded. In the most common arrangement, the latter conductors should be about 240 volts "apart," as defined just below; 230 or 250 is fine, but 10% variation, say 208, is definitely a problem. Each should be about 120 volts "from" the neutral. While 115 to 130 is fine; 108 certainly is not. This system is called "240/120, single-phase service." In house wiring, the term "single phase" is dropped. However, the term "two-phase," which sometimes is substituted, is quite incorrect. "Two-phase" refers to an esoteric industrial electrical system.

What Type of Electrical Service Feeds Your House?

There are variants on the 240/120 volt system. Some apartment houses and businesses are fed not at 240/120 volts, but at 208/120 volts, 4-wire, 3-phase. This too has a specific meaning. There are three hot wires; a voltmeter will measure 208 volts from each to the other two hots. There is one grounded neutral, 120 volts from each of the hot wires. Apartments in such houses, and a few stand-alone houses, usually are fed at 208/120 volts, single phase, derived from three-phase systems. For our practical purposes, all single-phase systems work essentially the same way. The hot wires in this case are 208 volts apart (with a range of acceptable variation similar to that for 240-volt systems). The grounded neutral is midway between them, but 120 volts from each one. (Why not 104 volts to the grounded neutral? This has to do with vector algebra.)

Some very old fuse boxes have 120 volt service. These fuse boxes are fed by just two wires from the utility. One is the hot, at 120 volts to ground; the other is the grounded return, equivalent for our purposes to a neutral. These very old houses (as well as some isolated houses fed from very modern solar panels) do not have 240 volts available. By the end of Chapter 4, you will be ready to determine exactly which type of system you have in even more detail.

Protect Your Appliances from Voltage Mismatches

As most houses have 240/120 volt service, that's exclusively the system I'll discuss. The wiring for 208/120 volt single-phase service is the same. Just make sure that your motorized heavy appliances are designed to run on your voltage; the same appliance usually is available in both a 208 volt and a 240 volt version.

Appliances that are more accommodating, such as electric stoves (without convection ovens), can handle either voltage, merely drawing a different amount of power at each of the two voltages. Such accommodatingly designed appliances will list both voltages on their name plates and in their instructions.

Surges

One of the most dangerous types of overvoltage is the surge, a sudden, short-lasting (but more than a microsecond) voltage rise that can be quite high. Sometimes this is the result of a lightning strike on utility lines some distance away. A surge can start a fire or destroy or degrade electrical and electronic equipment, sometimes in ways that do not show.

Surge protectors are available, and are best installed at more than one place, to "cascade" or reduce the overvoltage to successively lower levels. For instance, you could install one right at your electric meter and another at the receptacle serving your computer.

Voltage Variation

Sometimes, problems with an electrical service can cause severe voltage problems. With two-wire services, if either incoming line is disconnected, nothing will work (except, in one case, when there's some very dangerous wiring indeed). Problems with incoming lines can be worse in more modern systems. Here's one of the reasons that the neutral connection to the utility is the most critical. With either 3-wire system, problems in the connection between the house's neutral and the utility's neutral can cause the house voltage to "float." This means that instead of there being 120 volts from each hot to the neutral, the 240 or 208 volts from hot to hot can be divided any which way, such as 80 volts between one hot and the neutral and 160 volts between the other hot and the neutral in a 240/120 system, or 150 and 58 in a 208/120 system. Most equipment works poorly—and much equipment may be damaged—when the voltage available to it varies far from the voltage it is designed for. Overvoltage tends to be more dangerous than undervoltage, but either can cause harm.

In some cases of chronic service undervoltage, there is a problem with the connection between one or more of the hots from the utility source and your house's interior wiring. This too can be dangerous, albeit not quite as much so. Higher current due to voltage variations can make bad connections overheat, damaging electrical components, even starting fires.

Voltage mismatch problems, even harmful ones, also can show up during brownouts, occasions when your utility supplies lower voltage than usual. These reductions occur because their customers are using lots more electricity than the utility predicted, for instance when almost every customer is using air conditioners.

More-chronic undervoltage is rarely the result of brownouts. Neither the utility's service to your house or your neighborhood, nor the connection between your system and the utility is always the culprit. Voltage may be reduced when your house system draws a great deal of electricity, perhaps feeding an electric stove, an electric dryer, and many air conditioners simultaneously. In some cases, the wires branching out in your circuits have insufficient ampacity, given their length and the number of outlets they serve. In other cases, even more dangerous ones, there are bad connections along those circuits. Older houses are particularly prone to having too many outlets on a single circuit.

If you overload branch circuits, you can cause four problems. First, if too many appliances are used at once, that circuit may be so heavily loaded that the voltage along it is reduced significantly. This is the functional equivalent of a brownout on that circuit. Second, if an installer connects an inordinate number of outlets to a

Circuit Layout

Branch circuits can be laid out in a number of different ways, and, especially if your house is old, all types of circuits may be represented in it. The two main concepts are loop and radial, also known as daisy-chain and star, wiring. A loop or daisy-chain circuit runs from one outlet to another to another and on to the end of the line, with increasing voltage drop. The voltage drop partly is the result of the length of the circuit's path, which may be much longer than it would be if it went straight to each outlet. The voltage drop also results from the resistance of the connections along the way as the circuit passes through each outlet. A radial or star circuit spreads out from a central point, with a separate line going to each outlet it feeds. The dedicated circuit, which you might run to an outlet drawing a heavy or special load, is a pure star circuit; other circuits combine the features of the radial and the star layouts.

Parts of your layout may be pretty confusing. For instance, circuit 12 may feed from the panel to the living room, where it provides power to one of the receptacles. From there it goes to the dining room and kitchen. Even if the dining room and kitchen are laid out in a fairly straight line from the living room, the layout may vary. It may be that the line goes from the living room to the dining room and then on to the kitchen, in a pure loop. Alternately, it may run from the living room to the dining room in one loop and from the living room to the kitchen with separate wiring. This no longer is a pure daisy chain layout. The voltage drop in the kitchen may be higher in the former case than in the latter.

circuit, he or she may overload it so much that this causes nuisance tripping. "Nuisance tripping" means that the fuse or circuit breaker opens the circuit repeatedly even though there is nothing wrong with the wiring as such. (Tripping is discussed in Chapter 7, Protective Devices.) Third, the wire itself has a certain amount of resistance. Run it 50 or 75 feet (remember, *that* length gets doubled, because the electricity has to make a round trip) and you start getting a considerable amount of voltage drop even if it's feeding just one outlet at the end. (Remember the discussion of circuits and resistance above.) Fourth, every connection along the way introduces some resistance, with the same consequence. (See the discussion of "daisy-chaining" in Chapter 12, Terminations)

Grounding and Bonding

Grounding and bonding are a major part of what keeps you safe from shock. Grounding also protects your appliances from being damaged by abnormal voltage variations that frequently result from utility hiccups (some of which are themselves caused by lightning strikes on utility lines miles away). Grounding and bonding are exceedingly important throughout your electrical system, and also apply to other equipment whose electrical status concerns you, such as aluminum siding near which the lead coming down from a lighting protection array passes. (If a lightning strike passes down through your downspout, for example, you want it to travel as much as possible through the downspout right to the ground, with the least possible chance of it jumping into your house to find its way down through your wiring.)

Bonding means, for our purposes, connecting parts so that they're all grounded, and not only more-or-less grounded but connected so solidly that there can be not even a little voltage difference between them. In some cases, this latter aspect is the important one. Suppose a gas pipe is not bonded. Voltage is impressed on it from a loose wire at the furnace. This causes current to flow—in the form of a spark—from one part of the gas pipe to the other as the electricity seeks to complete its circuit. Gas plus sparks . . .

Grounding (in other parts of the world, "earthing") means bringing a part or parts to the same voltage level as the earth, our zero reference. If you stand on the earth, or on something in good contact with the earth such as concrete or a wet ground-level surface, you are grounded. When this is the case, you do *not* want to touch any ungrounded surfaces that might carry electricity because the electricity could complete its circuit by flowing through you to the ground.

Why can you be shocked by touching *one* live wire or *one* appliance that has become live? Where's the "loop" required for current to flow? The power you buy from the utility flows out along a hot wire and returns to the service through the neutral. The neutral is grounded at the service, ensuring a common earth reference for both service and premises wiring. When the hot wire accidentally touches a good ground, so much current flows from the hot wire through that ground, back to the point where the neutral is grounded at the service, that the fuse blows. When a hot wire accidentally touches a so-so ground, the current returning from the hot wire is divided between its normal return wire (unless that is interrupted) and the inadvertent ground. When you contact a live wire or surface, to the extent that you are grounded you function as such a second return path. Even though the ground itself actually is a poor conductor unless it is wet, it doesn't take a very good conductor to carry enough current from you back to the source to kill you.

Grounding and bonding start right at the service entrance. Since the part of the system that is under your control and accessible starts at the MAIN, in normal circumstances there should be a wire, your grounding electrode conductor (GEC), going from its enclosure to a clamp on your incoming cold water pipe. (I am presuming that your pipe is metallic, as it is in all systems whose plumbing was not upgraded within the last couple of decades of the 20th century.) This pipe emerging from the earth helps ground your electrical system, and thus constitutes your "grounding electrode," or, at least, part of your grounding electrode system. The wire to it may be bare, insulated, or protected in armor or conduit. It may be solid or stranded, though if it is bare, I frown on the use of a stranded GEC. I think it faces too much risk of damage. I would consider having a bare, stranded GEC replaced, unless it is absolutely out of harm's way. If you have a plastic water pipe coming in from the street (or a metal one that runs underground less than 10 feet, though I've never run into this situation), your grounding electrode system has no benefit from it, and must look elsewhere, as described below.

Modern rules require that a GEC be attached to a metal water service within five feet of where the water pipe enters. This way, if your interior pipes were ever replaced with plastic, you won't lose the grounding connection to the underground metal pipe. Until recently this was not necessarily a concern.

This GEC serves a secondary purpose as well. So long as metal interior water pipe is continuous with the entering pipe, that one bonding connection also bonds the interior water piping to the electrical system. Gas pipe is not the only interior system that requires bonding.

Suppose that a plumber is working on the interior piping, downstream from the water meter and shutoff. If the electrician's installed the GEC in the

Some people suggest disconnecting water pipes from the electrical system because small electric currents leaking to ground could cause lead from solder to leach into the water. This is a bad idea. DO NOT disconnect your water pipe ground unless your serving utility demands that you do so. (This would be quite unusual.) Every source I've read says that your risk of injury from electric shock is far greater than your risk of lead poisoning; old solder is very stable.

section the plumber is replacing, the plumber will of necessity disconnect it when getting rid of the plumbing pipe to which it is attached. If the electrician had installed the GEC upstream, the plumber would not have affected it. Sometimes the plumber has to replace the incoming water line and uses plastic in place of metal. Clearly, the GEC, or at least the part of the GEC that the water pipe constituted, is lost. In the best case, the plumber will have called in an electrician to make sure there was an adequate alternative GEC. This behavior would be rare.

Therefore, should you hire a plumber to upgrade or repair your water lines, make sure that any GEC present is not disturbed, or if it is disturbed that it is replaced. If you know they have been upgraded some time in the past, check to make sure that the GEC is still intact.

I have seen at least one case where a thoughtless plumber was replacing the pipe to which the GEC was attached, and he didn't try to remove and subsequently replace its clamp. He simply cut the grounding electrode conductor out of his way. He didn't even take the fact that it was encased in armor as an indication that it might be important! (My customer, who had ordered the plumber's work, had not been aware of this problem until I pointed it out to him and explained its significance.)

In the very worst case, disconnecting the grounding electrode system can mean that walking out of your house and touching, for instance, the metal railing could electrocute you.

In any halfway-modern wiring, there should be at least one additional grounding electrode besides that cold water pipe. "Halfway modern wiring" includes old systems that have had a service upgrade. This supplemental electrode could be a $5/8''$ diameter rod driven into the ground, or, especially in older buildings, a $3/4''$ trade size galvanized steel pipe. There are other grounding electrodes, but these two are by far the most common in residential construction, and also the easiest to examine.

⚡ CAUTION!

It is very important that you make sure the grounding electrode system is intact. Look for at least one GEC emerging from your panel, and try to follow its entire length to the clamp, to make sure the GEC in unbroken.

While you're outside examining where your service entrance conductors come through the wall (assuming that they come out the bottom of the meter base, rather than its back), look for the GEC that in modern residential wiring usually goes to a ground rod to supplement any water pipe ground. If the grounding electrode conductor is missing, you know that the service is probably

The rod electrode and the pipe electrode represent two different systems of measurement. There are measured sizes and there are trade sizes. The rod is $5/8''$ diameter as you would measure with a ruler. It is an exception. Two-by-four lumber is not 2″ by 4″, and hasn't been for many decades. Similarly, electrical pipe (conduit) and tubing and associated parts are identified in terms of trade sizes, not ruler measurements. The $3/4''$ conduit mentioned above actually is closer to an inch in diameter. There are variations even within a trade size. Any tubing or conduit of $1/2''$ trade size can be connected to a $1/2''$ trade size opening; however, the diameters and available internal volume may well differ because of their different wall thicknesses.

quite old; alternatively, it may have suffered serious abuse. With rare exceptions, no inspector would pass a modern job lacking an earth ground. One type of exception is the urban jurisdiction where authorities consider driven electrodes too risky because there are just so many cables and pipes criss-crossing underground.

If you find an outside grounding electrode, examine it. Is the GEC leading to it intact? Is it secured out of harm's way? If not, consider strapping the wire tidily against the house, or protecting it with some guard such as pressure-treated wood.

This takes care of your examination of the service's grounding and bonding, at least until you look at the grounding and bonding inside your electrical panel or panels in Chapter 6.

3

Testing Voltage

Why Should You Learn How to Perform This Specific Test This Early?

Every time you work on or even investigate wiring, you need to find out whether power is present. Testing for voltage is the very first skill you need to learn; it cannot wait until you reach Chapter 9, Tools. Voltage testers are the tools that knowledgeable people—including nonelectricians—associate most strongly with electrical work.

What Do You Need to Know About Testers?

These are the basic types of tester with their advantages and disadvantages.

The Pigtail: NOT RECOMMENDED for Basic Voltage Testing

Most frequently, you will test simply to confirm the presence or absence of power. A few decades back, many electricians would use a "pigtail," which in this case means a rubber lampholder (the socket into which you screw a light bulb) with two leads of rubber-covered flexible wire, insulated except at their ends.

A rubber pigtail is designed for temporary lighting on construction jobs, though, not for detecting voltage. What's more, it is very easy to break the light bulb. It also is very easy for the bulb's filament to break unobserved, so that the bulb doesn't light even though you are touching something live with the pigtail. The proper tools are very inexpensive, so there's no need for you to take chances with something that was not designed for the job.

There is one exception to this general condemnation. You may choose to use a pigtail for one type of voltage detection: determining that you are able to turn power on and then off and then on again at an outlet at which you already have confirmed that you can kill power. This is a sophisticated use that may come in handy later on, in the course of replacing fixtures or troubleshooting. However, it requires solid proficiency in splicing, among other things.

The Neon Tester—*One Kind Is Recommended*

The two-lead "neon" tester (the name technically is inaccurate) is a small solid "light bulb" molded to two insulated leads that have bare probes at their ends. The bulb lights when there is a suitable voltage difference between whatever the leads touch.

Neon testers' advantage over some of the more elegant (and far more expensive) testers is that they can be left plugged into a receptacle indefinitely. This can be handy when you're trying to determine which switch, or which fuse or circuit breaker, controls which receptacle. In the simplest case, you merely insert the leads; you will learn where you need to touch the leads in other cases. If you have located a voltage difference, your tester will light; by watching the tester to see when its light goes off, you will know when you have shut power off. Using the two-lead version, you can perform almost all the basic voltage tests that you can with more expensive voltage detectors.

Second, there also is a single-lead version, which is not dangerous, but probably is not worth buying. It *will* tell you when voltage is present, but it cannot be used to perform the majority of the tests you will perform beyond this basic one. It looks like a screwdriver, except that the handle contains a bulb similar to that in the two-lead version. It lights when the tip touches anything at a suitable voltage above ground level. The screwdriver-type tester is a tough, small item that easily fits in your pocket and usually costs a dollar or two. It relies on a (safe, ultra-high resistance) connection through your body to complete the circuit.

> **💡 CAUTION!**
>
> There are two-lead neon voltage testers designed for automotive wiring, which look identical to those intended for house wiring. You should not use these on house wiring. One manufacturer warned that *these have blown up when used on 120 volts.*

The Wiggy—*Most Highly Recommended*

Wiggy® is the trade name for a tester invented by a Mr. Wigginton. Colloquially, whatever the brand or variation an electrician buys, a *solenoidal* voltage tester is called a wiggy. ("Solenoidal" refers to the operating mechanism.) The wiggy is the electrician's basic voltage tester. It is sturdy and performs one single function. For these reasons, you will find it handy and unconfusing.

A wiggy, probes about to be inserted. Fingers don't need to be anywhere near metal if you use this tester.

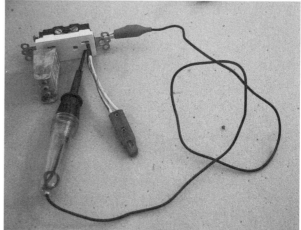

A "receptacle tester," a 2-lead "neon" tester, and **one with an alligator clip;** I can't recommend the first, and the third could be deadly if this were real, in-the-wall testing; it's for automotive use.

It has voltage markings along its length, a needle visible inside, lights, and two flexible insulated leads with probes at the ends. The probes are covered by slightly retractable, spring-loaded insulating covers. When there is a suitable voltage difference between whatever the two probes contact, the light or lights light, the needle moves to the appropriate voltage marking, and the wiggy buzzes.

This is worth emphasis: not only does it have a needle and usually a light, it also buzzes and vibrates. This lets you know when its probes are in contact with a source of significant voltage, even when the indicator is out of your field of vision or the light is insufficient, and you can't see the reading.

A second advantage of the wiggy is the retractile sheath over each lead. This means that it is not as easy to shock yourself by touching a wiggy probe that is in contact with something live as it is when you use a tester whose leads are fully exposed, such as the neon tester. The wiggy also indicates whether you're touching it to 120, 208, or 240 volts.

Finally, unlike most other voltage testers, the wiggy draws enough current that when testing voltage to ground it will trip a ground fault circuit interrupter (GFCI), an important safety device. While not officially calibrated and sanctioned for this, knowledgeable electricians trust their wiggys to perform a more meaningful test than the TEST buttons on the GFCIs themselves!

The standard wiggy has three disadvantages. First, it won't respond clearly to voltages below 90 volts. This means that if you have a very bad connection, you may get no reading. This is discussed below, under General Cautions. Similarly, if you want to test something that runs on a very low voltage such as your doorbell transformer, the wiggy (as well as the neon tester) won't register its output.

Second, the wiggy is more expensive than the neon tester, and the latter will serve most of your needs.

Third, the wiggy is not a precision tool. Like the neon tester, it certainly won't notice when your system suffers from a minor brownout. It may not indicate whether you are served by 240 volts or 208 clearly enough for absolute certainty in selecting major appliances.

Wiggy Variants

There are quite a few variations on the wiggy. Some have additional features, while some are less rugged or don't buzz.

The Multimeter—*Recommended*

The basic analog multimeter is a box with a clear window through which you can see a needle and multiple, superimposed scales. It has two flexible leads that end in rigid probes with bare ends or, more rarely, in alligator clips. The other

There is no place in a home where you can encounter a voltage higher than 208 or 240 without working *inside* an appliance. Never allow a tester (or any part of your body) to come in contact with the terminals into which you insert the ends of a fluorescent tube when power is on. Be careful not to touch the prongs attached to one end of a fluorescent tube whose other end is in its socket. Because it is unwise to touch a standard voltage tester to the terminals, you do not have a way of absolutely confirming that power to the fixture is off; this means that there is a chance high-voltage electricity could travel through the tube to shock you. (There are special testers designed for checking fluorescent sockets.)

ends of the leads normally also are rigid and bare for inserting into various holes in the tool body, depending upon which function you intend.

Here's how you use it to check for voltage. First, you plug the probe's ends into the holes marked for voltage testing. *Otherwise you may get no indication; you may even blow its internal fuse.* Then you set the scale for AC voltage of the range you will be testing. Normally one of the ranges goes from 0-300 volts. You touch the probes to, or clamp the alligator clips onto, anything that you want to test. If it detects voltage between the probes at any level within the range at which you have set it, the needle moves to a position that corresponds to the voltage—on that scale. You need to be careful not to read the wrong scale. Does using a basic multimeter sound more complicated than using a wiggy or a neon tester? It is.

Always choose the scale that encompasses the highest voltage that could be present. Even if you are testing doorbell wiring, first set the scale at 0-300 or whatever similar scale you would use to test for 240 volts. Only after you confirm that this does not show a dangerous voltage should you set the dial to a lower level.

Most versions of multimeter are not terribly rugged, and they give only visual indication. On the other hand, they can be left plugged in without damage when you need to identify circuits. Some analog versions are very, very inexpensive, though nothing approaching the price of the neon tester; digital ones cost more, as do more rugged instruments and fancier ones. Analog devices require you to approximate what number a needle moves to; digital give you fairly precise digits. You don't need the precision.

 CAUTION!

Anything digital does require a good internal battery to work, so if you leave your tester on a shelf for months at a time it may not serve you when you grab it; any tester containing an internal fuse will not register after the fuse blows; and any tester that is delicate can fail to register because it is broken.

The Receptacle Tester—*Not Recommended, Emphatically*

CAUTION!

Receptacle testers will misinform in two rare but potentially deadly cases. First, they will not recognize the presence of power where neither neutral nor ground is present; they will assert that the outlet is dead. Second, when what should be the neutral and ground are both hot, and what should be hot is connected to an actual neutral or ground, the tester will falsely indicate a correctly grounded and polarized receptacle.

(LED) "Receptacle Testers" are commonly, if rudely, called "idiot light" or "bugeye" testers. The specialized receptacle tester is a very tempting tool that purports to tell you, by combinations of colored LEDs, whether a live 15 or 20 amp, 120-volt grounded receptacle is wired correctly. A very slightly fancier version incorporates a button to test-trip GFCIs.

These are very popular with people who know very, very little about wiring, and with people who are so rushed that they cannot afford to test properly. If all you want to know is whether a receptacle is live or not, I believe that you can rely on bugeye testers.

However, you cannot use them at all to perform most of the important tests I will describe in later chapters. Even when they can perform a test, you will do better with a different tester. Some experts say that these devices can be relied upon to give a "thumbs up" when a receptacle is wired correctly, but that they may misleadingly indicate that an incorrectly wired receptacle is fine. Contrari-

Receptacle testers have told me outlets were incorrectly wired, but misdiagnosed the nature of the problems. Other times, they have misinformed users, as suggested by the engineer, that correctly-wired receptacles were miswired. After I submitted a report on a safety survey in which I evaluated the receptacles exhaustively, I got a call from the home purchaser. One of these testers had told him that the top part of

many a duplex receptacle was correctly wired, while the bottom part was wired backwards. It is not impossible to create this situation, but it would be ludicrously unlikely—to the point of impossibility—for someone to create such an arrangement. Besides, I knew from my testing that this was not the case in his home.

wise, I've been told by an engineer I trust, "They may indicate incorrect wiring, but cannot be relied upon to indicate *correct* wiring." You have no need for these so long as you have some kind of voltmeter and are not in a hurry; *take your time.*

The Tic Tracer—*Not Recommended* as a Basic Tool

The "tic tracer" or tick tracer is a noncontact tester that is supposed to register the presence of voltage by making a noise (hence the "tic" or "tick" in the name) and, usually, by glowing. I have NOT found that a basic, inexpensive tic tracer's signal or lack of signal reliably indicates the presence or absence of voltage.

The Radio—*Not Recommended* as a Basic Tool

I would not rely on a radio to indicate the absence of voltage at a particular receptacle unless I already had confirmed that it reliably recognized when voltage *was* present there. On the other hand, I have found it very useful to plug a radio into a working receptacle, turn it up high, and listen for it to go silent, as I stood at the electric panel a goodly distance away, trying to identify its circuit.

What Dangers Should You Watch Out for When Testing?

There are three very general principles that will help you stay safe when you use any electrical tester, and a few that are specific to certain tools.

Test Your Tester! Test Your Tester!

Test and retest your tester, especially when you are testing for voltage, each time *before* you begin using it. Testers die; if you rely on a defective tester, *you* could die. Go to a known source of power and make sure that your tester signals its presence.

Hold Your Tester Safely

Be very, very conscious of how you hold your tester when you insert its probes into any opening, or touch it to any surface to check for power. (This principle holds for any tool, not just your tester.) Make sure that your body is not touching parts that could become live, such as uninsulated probes.

Always hold a lead by the insulated portion, rather than by the bare steel closer to its end. Probes with retractable sheaths, such as are found on wiggies, tend to be safest. When testing a receptacle, retractable sheaths don't let the metal parts even appear until the probes are pushed into the receptacle's slots, out of finger contact. Only bugeye testers are safer to hold; as I mentioned above,

Alligator clips are by far the most dangerous types of tester contact to use, even when they have some sheathing of insulation partly over them. It is very easy to touch metal when squeezing the insulated sheathing to open the clips. Admittedly, alligator clips have one significant advantage over leads: they are self-holding. This rarely is true of leads, which can tie up both of your hands. If you are absolutely sure that you are applying them to dead metal, the clips also enable you to leave your hands off when you restore power for measurement. Still, even using unsheathed leads, such as are found on two-lead neon testers, tends to be a bit safer than attaching or touching alligator clips to surfaces that may be live, even though the surfaces *should* be dead.

though, their value as testers is very uncertain. Also, be careful that you don't hold a probe so that it bridges between something live and something grounded that touches the metal of the same probe at another point. It could arc.

Make Sure Your Tester Will Work for You at the Location You Need to Test

The best way to make sure that you can detect the *lack* of voltage at a particular location is first to *detect* voltage at that location. Do so before you attempt to kill power.

What should you do when you cannot detect voltage at a location, with all circuits and all switches on? Stop here. Do *not* assume that there is no power at a location because you do not measure any without taking more advanced steps than I cover right here. Otherwise, you may get shocked. This is where an extension cord will come in handy. When you get to Chapter 14, Replacing Receptacles, you will learn how to proceed in this case.

What Other Problems Can Impair Your Ability to Read Voltage?

There are a number of complications that can prevent you from getting a clear reading or any reading at all. (I'll offer solutions momentarily.) Readings can vary significantly based on variations in pressure. It is hard to hold leads with unvarying force; alligator clips achieve this automatically, but may not hold firmly enough and have other problems. In basic house wiring, there are few occasions where you need precise readings, but variations in readings extend far beyond imprecision. Corrosion, paint, or even the residue of old insulation can create such bad connections that you think power is absent.

You may also be misled by bad connections that are created when there is loose contact. When a screw used for an electrical connection has not been screwed down firmly, you may get no reading if you touch the wire at the screw terminal lightly. Push your probe against the wire harder, perhaps while holding the wire by its insulation, or move the switch or receptacle to which it is attached, and it may make better contact with the live part. Be careful to avoid injuring yourself by shifting the parts in such a way that you touch bare metal.

Damaged parts are another source of bad readings. Sometimes a bad connection, or some other source of overheating, causes the insulation on the wires going into a splice to melt or burn. The result can be that some wires in the splice are live and others, not quite in contact with them, are not live. You may test one bare wire coming out of the splice (or a terminal to which it is connected), and get no reading; then you disturb the splice sufficiently that it makes better contact with the other wires, putting you at risk because the part you had tested is now live.

Be extra careful when anything looks loose or looks or smells scorched. You may need to kill all power to the house, tighten and carefully position any connections you need to test, and only then restore power and test the circuit. (See Lesson: Shutting It Down in Chapter 6.)

Evaluating Your Service Entrance

Now that I've made sure some basic concepts are clear, you can start to look at your wiring. This chapter walks you from the point where the utility line reaches your house to where power enters your electrical panel. Even though you as a homeowner would not work on this, you certainly can evaluate it.

Why Is This Worth Doing?

WARNING!

I am about to start giving you "how-to" information. In the next several chapters, I will talk about opening up your electrical panel to examine various components close up. I urge you not to undo a screw, not to actually get your hands into any part of your wiring, until you have read through Chapter 5, Safety.

Checking out the lines coming from the utility to the electric system in your house is important for two reasons. There is no circuit breaker or fuse protecting this part of the system. In the event that it sustains damage, there is little to restrict the harm it may cause. Even if it does not cause harm to your house, damage to your incoming lines may interrupt power to your home in such a way that restoring power will not be a simple matter.

What Constitutes a Service?

Service Cable or Conduit

Power comes from the utility either overhead or underground. First, I'll talk about how to check out the most common overhead system. Was yours installed correctly? Is its design antique? Has it been injured? Has age caused it to deteriorate?

In some cases, power companies will take responsibility for replacing bad cable or conduit leading to the meter. They are far less likely to take responsibility for the line going from the meter into your service panel or your separate disconnect. They might charge for either type of replacement work, especially the latter, if they even take it on. Far more often, your electrician will have to buy the cable and install it. When this is the case, you may choose to invest in a service upgrade, since you'll be paying to have a chunk of that job done anyway.

Even after you have read Chapter 5, Safety, and even if you feel quite ready to work on your house wiring, it is important that you respect your "service entrance cable." Most of this part of your system remains "don't touch" material. But you don't have to be a professional to do the preliminary, hands-off evaluation.

Where Should You First Look for Damage if You Have an Overhead Service?

Start by walking outside your house. In the majority of cases, your house's wiring will be fed from overhead. When this is the case, the utility's line from the pole to your house is called the service drop. That point is the first place in your system you should examine. If you have service cable coming down the side of your house from the utility connection high on your building, has its sheath deteriorated? Before looking at these questions, you should back up a step.

What Can You Learn from an Overhead Line?

While the wires in the cable or conduit going to your meter are the beginning of your house's service entrance conductors, you should first take a good look at the point where they meet the utility lines, and even at the utility lines themselves. As you learned with regard to brownouts, unsafe conditions can hurt you whether or not the cause was your responsibility.

With overhead services, the service drop connects to a weatherhead or bracket on your building. Since about the late 1970s, depending on your area, power lines running from the pole have been triplex (three conductors twisted together), supported from the house at a single point. In older services, individual conductors are individually attached to a bracket that keeps them separate. Which design you have tells you something about how long it has been since the utility installed the drop. However, they may have replaced a deteriorated service drop from the pole without changing any part of your system itself.

Don't be fooled by changes that the utility has or has not made; their work is covered by the National Electrical Safety Code, which contains different rules than the National Electrical Code. For instance, the most certain way to determine the size of your electric service is to look at the sizes of cables and conductors. However, if you have a service upgrade, the utility may not replace *their* lines. Their calculations are based on the amount of power they expect you to draw, not the amount your service makes available. Your electric meter will tell them when you start using major new electrical appliances, and they can consider upgrading your drop at that time. Therefore, skinny—even aged—utility lines do not mean an undersized or old service. On the contrary, utilities may replace deteriorated lines at any time, so the condition of their lines, or the fact

You have no control over utility lines. Still, if they appear to be damaged, or if you notice vegetation such as tree limbs encroaching on their space, you should notify your utility. If there appears to be a location on your house from which someone could touch the lines, even by reaching out; if there appears to be a location over a driveway where a vehicle could strike a utility line; or if you have any other, similar concern—even if it is not among those listed in NEC Article 230—talk to the power company.

What you have observed may be no danger, but it could in fact be something the utility needs to fix. It also may be something that *you* rather than the utility need to correct, because the dangerous proximity is the result of carpentry done by or for you or the previous owner. Even if it's 20 years since that second floor balcony was added at a location immediately adjacent to the line feeding your house, it still may endanger your family today.

that they used triplex rather than individual conductors, does not indicate your service size. Finally, the bracket from an old service may offer them a handy place to attach a new drop, even after a previous owner upgraded your service. What this means is that utility lines suggest the age of your service, but only suggest it.

Judge Its Point of Attachment

Determining age is far less important than assessing safety. The next important question is whether the weatherhead or bracket is secured well. The weatherhead normally is held to the building by a single screw or bolt.

If the wood appears rotten or even cracked, the weatherhead could come away, putting a strain on your service entrance conductors. Reattaching a weatherhead involves working right next to the live conductors of the service drop. The one safe way for someone other than a properly trained and expensively equipped lineworker to work there is to have the utility cut off power—at their pole, not right at the house.

CAUTION!

The wood into which your weatherhead is screwed sometimes is split or rotten. Checking this, and correcting any such problem, belongs on your "to-do" list.

Examining a Mast

If the peak of the building is too low for the service drop from the utility's overhead line to have adequate clearance from the ground or from a window or other opening (as specified in NEC Sections 230-9 and 230-24), there should be a mast, meaning a length of conduit, going from the side of your house and on up above the roof line, to give the service drop sufficient clearance.

Don't confuse low-voltage lines with power lines. Power lines, as opposed to skinny, single, telephone or video cable, consist of three (or, very rarely, two) lines, either wrapped around each other or strung parallel to each other and coming to the same bracket or weatherhead high on your house.

Look at the mast carefully. Is it fairly straight and well-secured? If it's bent or leaning, it's not safe. If it's a little off vertical, a "guy wire" may make its mounting sufficiently secure. (A guy wire is a non-rusting wire secured to the mast, by a means that doesn't penetrate it, that is securely fastened to the roof on the side of the mast opposite the service drop and the utility's pole or distribution line.) Nothing other than guy wires and serviced drops should be attached to service masts—no phone or cable lines, no antennas or satellite dish conductors, not even lightning protection arrays.

It is most unusual for a mast to extend up from cable rather than continuing down in conduit to the meter, but the unusual is not necessarily bad. You certainly will run into many unusual types of installation in old wiring. This book is designed explicitly to help you differentiate those that simply are unusual from those you need to replace.

How Should You Proceed in Examining the Parts of the System That Are Under Your Control?

As you move on from the utility's responsibility to your own, you will look for equipment that has grown unsafe due to age or injury. How can you evaluate deterioration? I consider cable significantly deteriorated when the sheath (cover) no longer is intact. Wear that has progressed to the point that you can see the wires inside the sheath means that it is far gone, although not necessarily imminently hazardous. It can let water in. The water can cause electricity to leap from one of these wires to another, if its insulation is worn, or can travel down inside the cable to cause trouble at the meter or even further in. Water and wiring are a bad combination.

Your electrician pretty well can take care of small nicks in the sheath, but can do nothing to mend wear that goes right through the sheath. If the writing that originally appeared on the sheath is thoroughly worn off, or if the sheath presents a roughened texture, you have far milder deterioration. It merely indicates that at some future point you will have to replace the cable.

Old or new, service cable can be nicked, torn, cut, or otherwise damaged. These types of deterioration require that you replace the cable or, in a few cases, mend it. While your cable may have been injured in any of these ways, it is far more likely to have simply deteriorated as the result of age. In this case, mending is not a reasonable option.

The way your cable ages is a function of its design. Older service cable was built in the following way: Two rubber-insulated conductors were laid side by side, and a third, uninsulated conductor was initially spread out flat (instead of being twisted tight in a cylinder like normal stranded wires) and then wrapped in a gentle spiral around the other two. This assembly of conductors was enclosed in woven cloth, and the cloth was sealed with a weatherproof compound. This sheath of impregnated cloth was painted to further seal it and to provide a background against which printing would be legible. Manufacturers printed information on the painted surface telling us, among other things, how many conductors of what size and material they enclosed. Weathering wears away these layers more or less in order. Look for the degree of wear in order to tell whether your cable needs replacement, and how urgently.

Modern service cable contains a similar core of two insulated stranded wires and one bare conductor wrapped around them. The conductors' insulation is different, and the sheath is vinyl rather than impregnated cloth. When the insulation deteriorates, first the print becomes hard to read and then the texture of the vinyl changes, becoming less smooth. In modern cable, I have not noticed deterioration beyond this point that can be blamed strictly on aging.

The cable to the left shows slight deterioration through a "sheath" that may seem like tape that was wrapped around the original sheath after signs of deterioration were noted earlier. This is not actually the case. It is actually a protective layer found *under* the sheath; the cable is badly worn. This is not an emergency, but you need to replace such cable.

There are three other things to notice. First, an essentially round meter base strongly suggests that you have a 60-amp service. Second, the cable entering the top of the meter base not only is connected with a proper outdoor connector but also is sealed further with a malleable compound known as duct sealant or "duxseal." It indicates that the installer was particularly meticulous, a very promising sign no matter how old the service is. Third (this is a bad

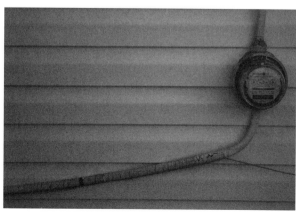

"Taped" cable sheath, 60-amp meter.

sign), straps have been removed. Replacing the missing straps is a job that any-one can do provided that the cable sheath is not so deteriorated and the con-ductor insulation so brittle that you could be exposed to the conductors inside the insulation. However, even if the straps were removed by the siding installers (and not replaced), rather than left off initially, this error should make you alert for other consequences of carelessness. The owner's choice to install new vinyl siding, but leave an out-of-date electrical service in place, suggests that he or she valued appearance over safety.

Suppose that something other than the service cable runs out through the same hole and alongside the cable; it even is secured under the same strap. Nor-mally, this is inappropriate. However, in some cases what is running down is the grounding electrode conductor (GEC), which I discussed in Chapter 2 and talk about further at the end of this chapter. The GEC is an essential part of the elec-trical service. Ideally, it runs down to the ground outdoors as close as possible to the location of your service equipment indoors; therefore, it is entirely proper to install it through the same hole, at the same time the service cable is installed.

What if you notice some other type of cable accompanying your service cable? Common types of cable are phone lines, cables for television service or satellite dishes, security lines, or cables for outdoor power or lighting. There is nothing terribly wrong about this, but you should check three things. First, make sure that your service cable was not damaged when the other cable was forced through the hole alongside it. Second, make sure that the other installer didn't leave an opening allowing moisture into your house or, worse, allowing water to reach your electrical panel. Finally, make sure that other, unassociated cables are not supported from your electric cable or conduit.

What Should You Look For if Your Service Enters in Conduit?

Overhead Service

If the service coming down the side of your house is in conduit, a bit of rust is not cause for concern; however, a spot that's rusted through is unquestionably a source of danger. You are in equal danger if the conduit shows clear damage, such as a deep dent, from having been struck hard, or if there is a hole or any missing piece in an associated fitting. (Fittings called conduit bodies are required when conduit needs to make a sharp angle, for example.)

Dented conduit is dangerous for one reason: it could pinch the insulation on the conductors within, and they could short at any time. Badly rusted conduit is dangerous for at least five reasons. First, water may be getting in, causing deterio-ration and threatening a short circuit. Second, it takes a long time to rust through thick, galvanized steel pipe; this almost certainly means your service conductors are quite old, and likely have deteriorated for that reason alone. Third, structural weakening could cause the conduit to collapse if there is significant stress on it. Such stress is especially likely if it extends above your roof to form a mast, as dis-cussed below. Finally—and this is true also of openings that did not result from rust—a person or animal could get hurt poking inside. None of this necessarily constitutes an immediate crisis, but you do need to have your service conductors replaced.

Open fitting; this "condulet" needs its missing cover replaced.

Underground Service

A service lateral is a power line that runs underground from the utility pole to your meter. It comes up to the meterbase (the box in which parts of the meter that you don't see, and the connections to the meter, are enclosed) in conduit. Therefore, look for the same problems just mentioned. Metal conduit is metal conduit, whether it comes down the side of your house or up from the ground.

In modern wiring, including some installed as a replacement for old services, you may find plastic conduit. Look for damage similar to that metal conduit suffers. You probably don't have to worry about deterioration due to aging, but keep an eye out for two common mistakes. First, clamps to hold 2″ plastic conduit against a wall are different from the clamps that should be used to hold 2″ steel conduit. If the installer used clamps that are too small, they will pull out of the wall as summer heat makes the plastic expand. If the electrician used clamps that are too large, the conduit will not be held securely. Second, if plastic conduit comes down from overhead, it may be long enough that it causes a problem as it lengthens and shortens with temperature changes. Installers have two ways of avoiding this problem. First, they can put bends in the conduit; if it is not running in a straight line, it won't pull and push on the ends as temperature changes. The second solution is to insert "expansion fittings" in the conduit. Unfortunately, neither of these solutions can be retrofitted after installation without taking the system apart. Look for distortion in the weatherhead or the meterbase, where it might have pushed them hard, and make sure that it hasn't pulled out of either, or of fittings along its length. If you find either, the conduit will have to be replaced. Whether your system uses conduit or cable, missing straps (the ones that hold it to the side of the building) usually can be replaced, missing screws likewise. Doing so is one of the least urgent tasks I will mention, but you should take care of it.

How Are Your Meter Connections?

Now look right at your meter. Where the cables enter the meterbase, were weatherproof connectors used, completely sealing the openings at the points of entry? A weatherproof connector contains a metal ring that screws down onto the body over a rubber fitting, squeezing the latter around the cable. Two-screw squeeze connectors, being intended and investigated for indoor use, technically are illegal outdoors. These merely pinch the cable between two pieces of metal. Located where rain can flow in past them, they show improper installation. Their use on the undersides of enclosures, on the other hand, is very common, and widely accepted. (It certainly does not constitute a top-flight job, though.)

Equipment that might be penetrated by moisture is supposed to be installed "so as to drain." In the past, there was such strong belief in the value of drain holes for any water that might have infiltrated or condensed and that otherwise might accumulate to damage equipment that electricians even punched holes in the bottom of enclosures that had been sealed by their manufacturers. This is undesirable. Almost any modification not intended by the manufacturer not only voids the warranty on equipment, but may explicitly violate the NEC.

If either this was done, or a two-screw connector was used, it's certainly not worth redoing the service. However, keep half an eye out for other outdated practices.

"The good old days" weren't necessarily very good. Still, some procedures that were far more common in the past have advantages over what is done today. One that is unquestionably superior to common modern practice is the use of drip loops—which the NEC still demands!

Electricians share one observation with plumbers: liquids flow downhill. Savvy installers keep this in mind when bringing a cable from a meterbase through an opening through which moisture might seek to enter by traveling along the cable. There's a great advantage to coming in from beneath the opening: water that runs along the cable will drip off below the opening rather than running into it along the cable. Cable usually leaves meterbases through openings that point downwards rather than up or straight out. I still strap service cables, among others, to the outside wall *beneath* the holes I drill, as old-timers did. If you observe this arrangement, this may add a jot to the evidence that your cable was installed quite a while back; it also will tell you that the electrician who installed it was conscientious.

Service cable attached to the meter with a weather-proof connector, but sheath worn to the point that the cable inside is exposed.

Indoor connector on underside of meter base. (The cable is starting to go.)

Electricians have never been supposed to leave openings through which rainwater can enter your house. Where the cables enter your house, check to see if the hole is at least sealed. If the parts of your system up to this point are in good enough shape to be left alone, sealing any unsealed opening goes on your list of things to do. Duct sealant can be purchased at electrical supply houses. Most caulks also are okay to use next to service cables; so is patching cement, for masonry walls. If sealant is missing, it may just be that the old sealant has come out or been pried out by some mischief-maker.

How Healthy Are the Parts of Your System That the Utility Doesn't Control?

The Disconnect

 CAUTION!

The disconnecting means is something you most certainly need to identify. Is it right there in the same box with all your fuses or breakers? Most probably, but not certainly.

At your electric service, the utility hands control over to you. The service includes everything up to and including the meter, or, when the main disconnect comes ahead of the meter, the box containing the main disconnect (colloquially the MAIN). The MAIN is where the electricity entering your house can be turned off in just one or two hand motions in an emergency.

Everything going into your property after the box containing your meter or one the load side of your disconnect, whichever comes first, is under your control, not the utility's. Therefore, it is called "premises wiring." ("After," in electrical parlance, is "downstream from"; you will encounter this term again and again.) Interior wiring is one part of your premises wiring; exterior wiring such as yard lights might be another part; garage power, a third part.

It doesn't matter whether the MAIN is on the inside or the outside of your house; in your high-rise apartment or in the basement utility area five flights down. It doesn't matter whether the disconnect is one or more switches or pairs of fuses or circuit breakers in a separate enclosure or whether the disconnect is in the "loadcenter" (fuse box or breaker panel) from which some or all your circuits branch out. (Fuses and circuit breakers are discussed in great detail in Chapter 7.) The MAIN or service disconnect even can be a switch or circuit breaker located in the enclosure containing your electric meter.

Your service includes some sort of protective device or devices that will interrupt the electrical connection automatically if too much current begins to flow from the utility. By "protective devices," I mean the familiar fuses and circuit breakers. Most often this protection is part of the manual disconnect. Evaluating the interior of your service panel will wait till Chapter 6.

Here's a place where the rules have changed in a small, but nonetheless highly significant way. In the past, disconnecting your house wiring from the utility could require up to six hand motions. You can find this design in many older homes. Modern services are safer because if something goes wrong and you don't know which circuit to turn off, using two rather than six motions saves a second or two.

If you have the older system, it is even more important that you make sure you know which fuses or which breakers need to be disconnected to kill all power. This actually may need to wait till Chapter 8, The Walk-through, or at least Chapter 6, Examining Your Panel, because far too frequently I find them mismarked.

Is a Panel with a "Main" Always Your Service Panel?

Perhaps confusingly, there's something else called a "main." If you have one or more panels downstream from the panel containing your service disconnect, there will be a circuit breaker or pair of fuses to protect it. The disconnect in your service panel protecting the "feeder" leading to each of these subpanels located further downstream normally will be labeled "Subpanel." However, the subpanel itself may contain its own MAIN.

Where Can You Find Your Disconnect?

I said that when there is nothing between your meter and a panel, that panel has to contain your service disconnect. This is not quite true. In any wiring that is even halfway modern, your service disconnect is downstream from your electric meter. There are two legal exceptions. One is where the disconnect is part of the meter base. If this is what you have, there clearly will be a door as part of the metal enclosure that is hinged and readily openable—neither locked, nor wired shut. Open the door and you will find a circuit breaker. The other is where the meters are downstream from the main disconnect. In any case the main disconnect normally is the first point in the system where you and your electrician have the right to work. The reason for this is that in the standard system, to kill power upstream from your MAIN you have to disconnect the meter, and the meter definitely is out of bounds without special authorization.

One great benefit of this standard layout where the meter comes first is that if all power to the building needs to be interrupted, removing the meter from its base—with utility permission unless it's an emergency—will do the trick. This is a job for a pro. Incidentally, if there is any possibility that wiring is so hazardous that a house should not be occupied, or if any part of a service en-trance itself is unsafe, the utility probably will open the connection at the weatherhead or pole until the local inspector gives them a "cut-in notice" certifying that it is safe to restore power.

Is Your Disconnect or Panel Mounted Properly?

The mounting location of a disconnect or panel will give you one of the first indications of modern, competent workmanship—or old or thoughtless work. Is there adequate access? The Code demands clear space in front 30 inches wide and 36 inches deep.

If the equipment is indoors, the clear space needs to be clear from the floor on up. If equipment is out-

Utility cutoff, right at the house. The line from the pole conceivably could be live.

Technically, a "feeder" serves a subpanel, and a branch circuit serves receptacles, switches, lights, and similar loads. Informally, though, a heavy line feeding a single load such as an electric stove often is called a feeder.

doors, there is no such clearance requirement above the height of the equipment itself. Indoors or out, it should be more than three feet from any gas meter, though this is not an NEC rule.

You always need to make sure there is adequate access to any component of your electrical system, such as an air conditioner or furnace disconnect or subpanel, that is likely to be worked on, or even opened and examined, with power on. Three elements are involved:

First, has the equipment itself been located so that it is easily accessible? Second, has other equipment, such as an air conditioning compressor, been installed so that access is blocked? Third, have you yourself used the space in front of it for storage?

You need to make sure there is "*ready* access," not just access, to any place you have to reach with your hand. Is the highest position at which you may need to reach a switch at least no higher than 6 feet 7 inches? This applies to your electrical panel too. The top of the panel may be higher, so long as the UP position of the topmost (usually the MAIN) breaker is no higher than 6 feet 7 inches.

What Should You Do if Your Setup Is Not Quite as It Should Be?

It doesn't make sense for you to have your disconnect moved because it only has 24 inches clearance or was mounted seven feet up. There is plenty you can do, though.

Don't block access yourself by storing goods in what should be dedicated space. If electrical equipment is replaced, have it moved to where the access is adequate. If the plumber needs to work on the furnace or the water heater, tell him or her not to block nearby electrical equipment. Because nonelectricians, even those in the other skilled trades, sometimes are not conscious of these clearance requirements, make sure that a furnace or heating-cooling installer does not block or even restrict access to the electrical shutoffs that are part of his or her own equipment.

What Is Your Service Size (and Why Should You Care)?

I have had to work on a live electrical panel crouched on a washing machine because I couldn't move the machine, and I judged, probably foolishly, that the work needed to be done more than I needed to make sure I had safe working conditions. I have leaned over boxes and squeezed between family storage items to get at a blown fuse. In the latter case, I simply wasn't going to work in the panel; the conditions were just too risky.

The next step you might take, proceeding from outside to inside your house, is to determine the size of your electric service. I say, "might take," because this step often is not essential. People with old electrical services sometimes are anxious about whether the amount of electricity available to run their appliances is adequate. This would seem to be a reasonable concern, given the difference in capacity between old services and those installed now. The smallest service the NEC allows you to put in a home nowadays is 100 amps, 3-wire (3-wire meaning that it offers 240/120 or 208/120 volts, rather than just 120). Until recently, in a few circumstances, installing a 60 amp service still was legal, but even with the 100 amp requirement, the NEC is more liberal than many jurisdictions. 150 amps is the smallest size permitted by some jurisdictions, and 200 amps is more common—though very rarely actually necessary—except where homeowners pinch a few pennies by opting for 150 amps.

How much power do you need? Relatively little, usually, compared to the amount available from a modern service. Most service changes by far are performed because the existing service panel (or, occasionally, other equipment such as the cable) is in bad shape, because it doesn't have room for additional circuits, or because it is otherwise unsatisfactory, rather than because there's insufficient power coming into the house.

One advantage to installing service conductors and busbars that are larger than your system requires is that their resistance is lower than that of smaller conductors. This helps some in reducing the chance of undervoltage should you run heavy loads.

An average house with gas serving the stove, clothes dryer, furnace, and water heater will do absolutely fine on a service providing 100 amps, 3-wire. Change some of the major appliances and systems to electric and add a heat pump or a backyard hot tub, and it may need a "heavy-up," an increase in the amount of power coming in. The term "heavy-up," incidentally, is not universal jargon. "Service upgrade" is the formal name.

I have tested the power draw of a large all-electric house served by a modern 200 amp service, while the heat pump, dryer, stove and color television were all going. I measured a power draw on each hot leg of about 40 amps (times 240 volts, thus about 9.6 kW). Why did they install a 200 amp service? Maybe worst-case calculations said they needed one. More likely, a 200 amp service sounded more impressive than a smaller size, and thus helped sell the house.

Equally likely, though, they may have been impressed not with the power, but with the fact that panels with greater ampacity provide for more circuit breakers. There's an advantage to splitting the same area up into more circuits rather than fewer. The more circuits serve various areas and loads, the less likely it is that any one circuit will be overloaded. Equally important, the fewer loads are combined the less inconvenience will be caused when one circuit trips or is turned off.

Figuring Out How Much Power Is Coming In

How much power do you have? If you have a single main circuit breaker, or one single set of cartridge fuses, the ampere rating will be marked on the side or end of the breaker handle, or on the bodies of the fuses. If your service doesn't have a single, unmistakable MAIN, the answer is harder to determine, and quite frequently miscalculated. (In equipment that lacks a single MAIN, you may very well have a split-bus panel, an outdated design that will be discussed in Chapter 6, Panels.)

Here are some features that don't tell you much about your service size. First, the panel itself will report its capacity with a label (assuming that it hasn't fallen off) saying, for instance, "150 amps."

Your electric meter's location and design are indications both of the age of your electric service and of its capacity. A meter that is behind a flat glass window set flush in a rectangular metal enclosure, or a meter protruding from a round base, is old and probably part of a service rated 60 amps or even less. A meter protruding from a square base probably is part of a 100 amp service. Meters protruding from rectangular bases probably are part of modern 150 or 200 amp services.

However, this merely represents the upper limit of possible service sizes, not necessarily the amount of power actually coming in. Some think the answer is to add up the ratings of the fuses or circuit breakers. Wrong again. These almost always add up to well over the rating of the service. The reason is load diversity: the fact that most circuits are usually dormant or lightly loaded. Good wiring design ensures a high diversity factor, meaning that a circuit almost never is drawing current at its design capacity. Certainly it is very, very rare for more than a few circuits to be heavily loaded simultaneously.

The one way to determine your service size with certainty is to examine the service cable. To determine the ampacity—and thus the amount of power available—requires judging the size of the service conductors. It is useful to know the ampacity of your system, but it is much less important to size this than to evaluate the system's condition. The latter task occupies a large part of this book.

When you don't have a single, marked disconnect, just about the only way to determine your service size, if the serving utility can't or won't tell you, is by looking up the capacity of your incoming service conductors in NEC Table 310-15 (b)(6). (Before the 1999 edition, it was Table 310-16, Note 3.) If you're lucky, the conductor size will be legibly printed on the sheath of your service cable. Be sure you read whether the cable is copper ("Cu") or aluminum ("Al"), as the two metals have different ampacities. (In the exceedingly rare and unlikely case that a cable's markings indicate that the conductors within are copper-clad aluminum, read their ampacity from the aluminum table.)

> People rightfully are concerned about the presence of aluminum branch circuit wiring. However, service conductors are different; so long as aluminum service conductors are installed correctly, nobody knowledgeable worries about their use.

Of course, if your service is in pipe or in illegibly-marked cable, you will have to judge the size of the conductors where they enter your electric panel. If there are no useful markings on the sheath, you will have to wait till you have absorbed the background necessary to consider opening your panel, which we'll discuss in Chapter 6. It may be difficult, because you need to judge the thickness of the actual wire, not the wire plus its insulation; and don't forget that the service conductors are live unless someone has temporarily removed the meter or otherwise turned off power coming from the street.

One last point: the amount of power available depends not only on conductor size but also on the number of conductors and the supplied voltage. If only two wires enter your old—in this case very old—panel from the meter, you are in the unusual situation of having 120 volt service. If four wires enter from the meter, and your breakers or fuses attach to three parallel busbar sections, you may be in the unusual situation of having 208 volt, three-phase service. I have encountered both as recently as the 1980s. In the more usual situation, where there are three wires entering your panel (normally two insulated, one bare) you almost certainly have standard, 240/120 volt service.

At this point the walk through your electrical system must be interrupted for the most important discussion in any "how-to" book: *safety.* The reason is that as you proceed inward from the service entrance, even though you still will be evaluating rather than wiring, you may start putting yourself in the proximity of live wires.

5

Safety

I'm at a meeting of the local chapter of the International Association of Electrical Inspectors. I ask the members seated around the table: "Is there anyone here who hasn't gotten a shock, one time or another?" People are amused; no hands go up. "Is there anyone here who's never taken a chance on the job he shouldn't have?" More amusement, and again no hands. "Is there anyone here who's never gotten zapped except when he knew he'd been taking chances?" Third strike.

What's so complicated about applying a screwdriver and a little intelligence to your house wiring? If a group of inspectors and electricians and consultants could suffer shocks and perhaps burns, you and your family certainly could.

What's the Rush?

This chapter will describe the major hazards associated with electrical work, explain how and when you may encounter them, and teach you how to reduce your exposure to risk. I need to provide this information *now;* two chapters from now could be too late. As we proceed beyond this chapter, I will present material that might invite you to explore potentially dangerous elements of your electrical system. This is doubly the case if yours is an older house, one whose wiring bears the fingerprints of many workers.

What Are the Basic Threats from Electricity?

The first reason for you to spend time learning about these dangers is that if you make a mistake repairs will be expensive. This is the lesser reason. The greater reason to approach wiring seriously and carefully is that it presents two potentially deadly risks: fire and shock. These risks are both immediate—they can hurt you while you work—and ongoing, meaning that they can hurt your family days or months after you do the work.

There's a third risk that frightens me even more than the first two: direct exposure to an electric arc. An arc could spatter molten copper or steel at my face. It even could blast me across the room (though *probably* not in a residential system). You can lose your eyesight gradually over the months following an electric shock; you can lose it in a moment as a result of arcing.

What Do You Need to Know About Shock?

Shock is the threat you are most likely to face as you work on your wiring. Fire, on the other hand, is an ongoing danger, but less an immediate risk to you as you work. Arcing is caused by the same problems as is shock, and threatens you simultaneously. (Protective gear is more likely to protect you from arcs than from shock.) Undetected arcing is a significant cause of electric fire. Shock—and implicitly arcing—is the danger that needs the most attention in a chapter on safety. While smoke detectors can rescue your family should your wiring start a fire, there is no equivalent mechanism to rescue you while wiring.

Shock and Your Body

Does a Tingle, a "Shock," or Only Electrocution Qualify as a Serious Danger?
Quite a small amount of voltage—much less than 120 volts—can push enough current through your body's resistance that you are unable to make your muscles let go. Eventually you collapse, electrocuted, unless you're rescued quickly. The "let-go current," the border between the current level that allows the average healthy adult to let go and the level that causes his or her muscles to lock, is astoundingly small—thousandths of an ampere.

You don't have to freeze to a live wire, though, to be injured by electric shock. For one thing, you can suffer burns, nerve damage, and many other irreversible impairments at levels well below the let-go current. For another, you may get hurt because the "sting" of shock you receive while trying to replace the porch light fixture surprises you into falling off your ladder. Don't dismiss this risk.

> Electrical work carries special risks, but you can injure yourself doing any construction or repair. This works two ways: Carpenters electrocute themselves by accidentally cutting into live wires, and electricians break their necks falling off roofs. I'll underscore this later, but keep it in mind as you read—and as you work.

If So Little Current Is So Dangerous, How Can You Protect Yourself? Use Ground Fault Circuit Interrupters (GFCIs); install them if you don't have them in place. GFCIs do a great deal to reduce the chance of electrical injury when they are wired correctly and working properly. You will learn how to install them in Chapter 14, How to Replace a Receptacle. (This delay is one of the reasons for you to read through this book in sequence before beginning any repairs.) You will find them at some, but only some, locations in modern or updated

The electric mower

The electric supply

The mowed

A cut cord, and the mower responsible for the deed; GFCIs save grave harm in these circumstances.

GFCI Peculiarities and Problems

Many GFCIs have indicator lights. In a more familiar type, when the LED is not lit, the GFCI either lacks power or has tripped. In one brand, Eagle, when the LED is not lit the GFCI either lacks power or has *not* tripped. So, unless you know the design, you have to go by the fact that the RESET button is protruding to indicate that it has tripped, and the fact that the RESET button pops out when you push the TEST button to confirm that power is present—and that the GFCI is protecting you, presuming that it was installed correctly.

wiring. These were first required in the 1974 NEC, in bathrooms. GFCI receptacles are the devices you see in many bathrooms that combine a receptacle with a TEST and a RESET button. There also is a circuit breaker version, which likewise has a TEST button.

OSHA standards now require that any circuits used to power hand tools on construction jobs be protected by GFCIs. Moreover, they classify home remodeling with construction work. They believe that you are in as much danger, and need as much protection, as the people with the hard hats.

GFCIs are set to interrupt power at about five-thousandths of an ampere, a current level *generally* low enough that the GFCIs will protect even children, if they are healthy. Even less current, though, may "shock" delicate equipment, such as a computer—or injure someone whose constitution is weak.

How to Stop Shock. Someone is being shocked! What do you do? The only way to remove the source of shock that entails no risk is to turn off the switch or circuit protective device. You will be almost as safe if you unplug the cord feeding the device that is the source of shock—but only if the cord is dry and

Standard Decora-style GFCI receptacle.

Reset-lockout GFCI receptacle.

Here Are Two Exceedingly Important Warnings About GFCIs.

First, installers can fail you. Suppose you push the TEST button on a GFCI receptacle. If the circuit has power, the RESET button should pop out and the GFCI should protect you from any power that otherwise would go through it. Therefore, when you push the TEST button and the RESET button pops out, this should demonstrate that the GFCI is functioning properly and should protect you from electrocution. However, if the GFCI was miswired, when you push in the TEST button the RESET button will pop out, but the GFCI will not actually protect you if you are at a risk of shock. A tester will not be fooled, even when the TEST button is. What gets even more complicated is that a GFCI can be wired to protect additional locations downstream; if it is wired wrong, it may not protect loads plugged into it, but will protect these additional locations. Here is a result of this complication: if you use your tester at one of the latter locations, it may trip the miswired GFCI; this has a very good chance of leading you to believe that the GFCI will protect you as needed.

Second, GFCIs fail. In a survey, an average of one out of five GFCIs did not trip when its TEST button was pushed. In areas where high levels of lightning activity regularly imposed large surges on utility systems, about half the GFCIs failed. Based on these initial data, the most likely explanation is that GFCIs are not indefinitely capable of withstanding the power surges that hit some wiring systems periodically. You may not choose to check your GFCIs regularly as the manufacturer advises, but at least push

the TEST button whenever a GFCI trips and you reset it, just to make sure that whatever tripped it did not destroy its readiness to protect you in the future.

There is a fine solution to these potential problems. A variant on the standard GFCI has a feature called "reset lockout" or some similar phrase. When tripped, it will not reset until an internal test—one that cannot be completed without power feeding into the device from the panel—has confirmed two things. First, it must detect no ground fault. (This is true of any functional GFCI.) Second, it must actively demonstrate, through the operation of internal circuitry, that it will trip as it should if called upon to do so. An added advantage to this type of GFCI compared to the older-style GFCI is that it cannot be installed incorrectly without letting you know. The "lockout" version cannot be reset if it is wired wrong.

I see two potential problems with this solution to the issue of GFCI failure. First, any technology that is new may have unanticipated flaws. Second, you may not be able to tell whether a GFCI is the old type or the new, once it is in the wall; most of the GFCIs in place, and even most of those being manufactured, still are the old type. Given these concerns, I am not ready to recommend these unreservedly. While the idea of automatic GFCI monitoring is attractive indeed, its functions are not unique. (See my discussion of wiggies in Chapter 3.) You can confirm that a GFCI continues to protect you by testing it. You can discover whether a GFCI indeed protects a downstream outlet by testing for GFCI protection *at* the outlet.

CAUTION!

I'm not telling you not to pull or push away someone who is being shocked. I'm telling you to recognize that if you do so you're being a hero, accepting an unmeasurable risk.

undamaged. Many lives are saved because someone pulls or pushes the victim away from the source of shock, or vice versa. They may take the risk out of ignorance, lack of thought, or simply courage. Unfortunately, every so often people attempting to separate victims from sources of shock become victims themselves. Often this happens even though they push or pull the victim with something they believe to be nonconductive!

What to Do in Case of Shock. How do you treat electric shock? After safely removing the source of shock, maintain vital functions as your local National Safety Council instructor teaches, perhaps at a Red Cross course. Severe shock requires treatment by a specialized trauma unit. Nerve damage and loss of various functions, including eyesight, are some of the long-term consequences specific to victims of serious electric shock.

How Can You Avoid Shock?

Consider Your Risk Factors. Voltage level is far from the only factor affecting your risk. Other factors include the resistance of your *contact with* the source, the path electricity takes through your body, the shock's duration, and your physical condition.

- Contact: how firmly are you connected? The more sensitive your touch, the less likely you will freeze to a live item, or be seriously burned. This *is* about being able to flinch, as unreliable a safeguard as that ability to jerk back is for you to rely upon.
- Path: are you conscious of what all parts of your body are touching? Safety experts don't just talk about contact with various voltages, they talk about "touch-touch potential," "touch-step potential," and "step-step potential," depending on whether electricity flows across your upper body, from upper to lower body, or from lower body to lower body. "None of the above" is ideal, but electricity flowing across your hand from thumb to pinky is much less likely to kill you than electricity flowing past your heart.
- Duration: this relates primarily to firmness of contact and to means of interrupting the source.
- Physical condition: Are you already in bad shape? Shock is a form of trauma that your body struggles to recover from. Hans Selye and many other doctors have found that stresses compound. Advanced age and illness are only two of the challenges that can make it harder for your body to fight for survival.

The old-time electrician wore rubber-soled shoes, held one hand in a pocket while working live, and wrapped insulating tape around the handles of his pliers. These steps were far from reliably effective; every electrician carried pliers with a notch in the middle of the cutting blade, created when he accidentally cut through live wires, arcing right through the pliers' steel. "Wire-stripping notches" is what we called them, and we did indeed use them for removing insulation from the ends of conductors. I used to carry a pair like that, proudly; I now recognize what foolish machismo I showed as a 22-year-old. Still these "old-timer's tricks" did save many electricians' lives.

Do You Have to Accept the Chance That You Will Get Shocked? The only way to utterly avoid being shocked is never to touch anything that conceivably could be live—and you cannot conceive of all that might become live. This book describes a great deal you can do without opening live equipment, so you certainly can skip the other parts, or read them solely to help you understand what your electrician does and why. You can survey your system and analyze what you see without lifting a screwdriver; however, you cannot open up outlets, even to examine the wiring, without exposing yourself to some risk.

How Can You Minimize Your Risk? When you are working on your wiring, there is no guarantee that everything you think is dead really is dead. You have to do your best to kill power to whatever you're working on, and then to the extent possible treat it as if it were live.

It is almost certain that you will need to deal with live wiring for the purpose of testing. Even for testing, though, try to kill power while you make ready, unless your field of operation is clear or you have no exposure to wires or terminals.

Here are two examples of what I mean. If you are testing a receptacle with its cover in place, all you need to do is make sure to use your tester safely. You have no exposure, in that your body cannot come in contact with what is behind the cover plate, except via your tester. Your field of operation is clear, in that, by the same token, your body has no way to nudge inadvertently against any of the live parts that are blocked by the cover, inside the wall.

In contrast, suppose you need to work with the cover plate removed. Before you even remove the cover plate, you will try to identify the circuit and turn it off. However, if the reason you are investigating the outlet is that the receptacle appears to be dead, you have no easy way to verify which circuit it is on. Even when you do identify the right fuse to kill the receptacle, you have no guarantee that there isn't a live circuit in the same outlet box along with the one that feeds the receptacle, which you've killed. Furthermore, because multiple hot wires can share a common return path, it may be that a live circuit returns to the panel through a common neutral it shares with the circuit you killed! This is an example of how you can be exposed to dangerous voltage, even while taking reasonable precautions. Because most receptacles are installed in boxes whose front dimensions are 3″ by 2″, you also have a limited field of operation: Your hands or tools can brush against something quite easily.

Low-voltage work is a different story. If you are replacing an in-the-wall telephone jack that is mounted in the same type of box, you still have a limited field of operation; however, it is a safe bet that you will not be exposed to 120 volts.

One approach you might take is to do the best you can to kill power, initially, then pull apart the parts you intend to test in that relative safety. This way, once you restore power and approach these parts with your voltmeter, you will have a clear path, minimizing the chance that you will hurt yourself through unintended contact.

How Do You Kill Power? *Test Your Tester!* Before you can kill power, you must confirm that you can detect power. Ensure this by "testing your tester"; I will repeat and repeat this, because it is so very important for your safety. Doing this requires understanding how to use an electrical tester safely. Read the description of how to do so in Chapter 3, Voltage Testing, before undertaking any electrical work that involves more than looking.

What More Can You Do to Make Sure That You Won't Be Exposed to Shock or Arcing? Lockout and tagout are additional techniques you can use. After testing your tester, and only after testing your tester, kill power at the source and confirm that it is off. Also kill power to any associated circuits: circuits that might share the neutral of the circuit on which you are working, and circuits that run through the box in which you will be working.

Mark the fuses or circuit breakers in a way that ensures no one will inadvertently restore power while you're working. At the least, put tape over the fuses or circuit breakers, so that anyone wanting to turn it on would have to rip off your tape to move the handle. Do your best to make sure that anyone else who may be around understands what the tape means. This marking is called "tagout." Professionals working on jobs where there may be some confusion add "lockout": They actually *padlock* the handle of the panel or of the circuit breaker itself to ensure that it stays off, as well as tagging it with an explicit warning sign.

Insidious Electrical Dangers. In some rare cases, even when you do the best you can, and test and retest to make sure everything is dead, there is some risk in working on your wiring. Either of two mistakes that someone may have made long before you started working on the system can cause a presumably dead circuit to bite. I have encountered both.

The first type of potentially deadly problem appears where some ignorant or careless installer mixed up some other circuit with the one you are working

People commonly identify multiwire circuits in one of two ways. The easy way is when you trace your circuit back to its circuit breaker and discover that it is a two-pole breaker or a pair of handle-tied breakers. The nasty way is when you get sparks as you take apart a connection that should have no power going through it. Here's what I do.

First, I identify the circuit, by which I mean the fuse or circuit breaker, feeding the device I will be working on.

Next, I remove the cover of the electrical panel.

Then I follow the conductor coming from the fuse or circuit breaker I have identified back to where it enters the panel.

Now I can see whether this wire enters with a second conductor, white or formerly white. If so, I trace this second wire back to the neutral or ground bar to confirm that it is a neutral. If it originates from a second fuse or circuit breaker, what I have found is (or should be) a 240-volt circuit. If the outlet or switch where I planned to work registered 120 volts, or if the location where I intended to work did not give me a good reading, but *should* have shown no more than 120 volts, I have detected the presence of illegal work that is messed up to the point that untangling and correcting it is beyond the scope of this book.

If I trace the hot wire back to where it enters, and I find a white or formerly white conductor plus a second conductor that is neither bare, white, nor green, I have a three-wire cable, a multiwire circuit. In this case, I trace the second nonwhite conductor to its

fuse or circuit breaker, and treat this the same way as I treat the one I originally identified, the procedure that I discuss next. (If I find it simply dead-ended, I probably have found a cable run with a spare conductor. More often than not, though, when this is the case I have found an ignorantly-installed 240-volt circuit. Be careful.)

If I trace my hot wire back to where it enters the panel, and it enters with three or more conductors, two of which are white or formerly white, I probably have found a connector or a raceway that contains conductors from more than one circuit. This is a problem that may call for bringing in a pro. If it is a connector, often I can tell which cable is associated with the conductors I am trying to trace from the angles at which the conductors enter the panel. On occasion, I have loosened the screws holding the cables in the connector and wiggled one of the cables while restraining the other, and have been able to differentiate the sets of conductors on the inside of the panel by which ones moved and which were still. With conduit, I have to trace the conductors from the location at which I will be working, usually using a continuity tester. I will not describe the latter approach in this book, mainly due to space constraints.

A second approach that you may be able to use to determine which fuse or circuit breaker feeds the other conductor passing through the box where you intend to work is to use a "tick tracer" with a sender to push a signal back. I don't quite trust this approach, so I will not go into the details.

on somewhere downstream. This usually happens at a box containing both circuits. The mixup, which uses wiring associated with one breaker to complete a circuit associated with another breaker, creates what I call an "inadvertent multiwire circuit." Work with intentionally designed, clearly identifiable multiwire circuits (as described in the sidebar) is a bit beyond an introductory text; these inadvertent ones have caught any number of professional electricians unawares. Incidentally, the danger only is present at locations where you interrupt the continuity of conductors connected either through a splice or through a device's terminals, for example, two wires on the neutral side of a receptacle. Even though you may have tested them and found no voltage while they were connected, as you disconnect them you may expose yourself to 120 volts! Therefore, every time you separate wires, treat them as though they are live, and retest them individually.

There are two ways to deal with such multiwire circuits when you don't know which breakers to flip to remove power from all possible circuits feeding into the outlet you're working on. First, and safest, kill all power to the house, and work by flashlight.

Second, if you feel certain that you can stay safe while working live, every time you remove a wire from a terminal or a splice—one that tested as dead—

retest for power between the wire you removed and the connection you removed it from. If you find no voltage, there's nothing to hurt you. Of course, if you are very, very unlucky, the conductor coming from the other circuit that (without your knowledge) is involved is the neutral coming from a light that is turned off. When someone else in the house flips its switch, you can get hurt even though you tested a minute ago. But then, that sort of risk often is present—someone could flip a breaker back on despite your warning tag and tape.

The second of these preexisting problems cannot endanger you if your home is a stand-alone single-family house. If it isn't, you face the possibility—I've run into it—that someone working on your neighbor's apartment, or on your "roofmate's" townhouse, mixed up the wires from the two dwellings. Therefore, if you are in an apartment or townhouse, and choose to kill all power to your home at your panel, you still need to test for voltage where you are about to work. The only protection from this is to kill all power in both houses. A bolted ground, discussed below, will provide equivalent, and perhaps even greater, protection—so long as the grounding connection is in place.

There are additional risks associated with undertaking any job on older wiring, beyond the risk that you will get hurt as you work. If your house's wires and splices are marginal, any work on them—even pulling them apart for evaluation—can make things worse. This is especially true when you are dealing with an old system. The best solution is to recognize your level of competence, and use the very best care you can.

Can Someone Serve You as Backup? You may benefit greatly from keeping someone around as you work. Even if the buddy system doesn't prevent your getting shocked, it might save your life. Your friend's job is to do one of two things if you seem to be getting shocked: shut off power, or separate you and the source of electricity using something nonconductive. The latter always means some risk to your buddy, as discussed above under the heading, "How to Stop Shock."

Can You Ensure That the Circuit Will Open If Something Goes Wrong? One more step you can take to protect yourself from the risk of shock or arcs is intentional grounding. Bolting to ground means making a solid connection between any conductor and a good ground. To protect yourself by bolting to ground, you need to make sure that you do have a good ground available. The first step is straightforward: confirming the presence of ground. Find a way to connect securely to the outlet box, for instance, via a screw threaded into a tapped hole in a metal box. If the box is plastic, use a suspected ground wire. Now confirm that you read line voltage between your presumed ground and a live wire, if necessary one from another location, a location known to be live and correctly polarized.

The second step is to splice or terminate all the wires you will not be working with to this source of grounding. An example of bolting to ground would be to make a temporary splice between the incoming hot and the green or bare wires while you're working in a box. For this, of course, you need to confirm which wires are the incoming hots; also that the grounds have not been interrupted upstream; and especially that there is no roundabout way that voltage could come through any of the wires from a different circuit. This is one rea-

I emphasize that you need to learn proper use of a voltmeter before you can test safely.

son why you are safer tying everything together rather than testing for voltage between each wire present and ground. When you bolt to ground, you ensure that you won't get shocked if power is restored inadvertently—from any source. Utility workers often rely on bolted grounds for safety. This should cause the fuse or breaker to immediately open the circuit if someone tries to energize it (turn it on).

Bolting to ground, and even the old-fashioned electrician's practice of flicking a presumably dead wire against a grounded surface before touching it, may seem similar to deliberate shorting. This is a method of circuit identification that I advise against, for a number of reasons, in Chapter 8, The Walk Through. The present case differs in that intentional grounding is not being used for circuit identification, you already have taken steps to make sure the circuit is dead, and, most importantly, you are taking this measure for additional safety just in case.

Even bolting to ground is no absolute guarantee of safety, for two reasons. First, if your grounding connection accidentally is interrupted, either at the box where you're working or upstream, your protection is lost. Second, you may need to undo the connection to accomplish the tasks for which you are working in the box.

Can You Guard Your Body Directly? I will discuss tool safety further in Chapter 9, Tools, but I will mention one point now: The fact that a tool has plastic on its handle does not guarantee that it is securely insulated. Most manufacturers do not certify their standard tools as insulated against electric shock. Still, it's foolish to do electrical work with tools that have metal visibly exposed in their handles. Old-timers did protect themselves by taping the handles of their pliers, though electrical tape certainly is not designed to withstand sweat.

It's wise to use only one hand, whenever possible and safe, and to keep the other clear of any source of grounding. This way, in the event that you brush against something live, you won't complete a circuit from one hand to the other, across your chest—over or through your heart and lungs. Eye protection against sparks and arcs (an arc is a longer-lasting version of a spark) always is recommended as well. To a lesser extent in house wiring, the same is true of whole body protection.

Sooner or later, you will be "sure" that some wire or terminal is dead when actually it is live. Therefore, it's good to develop the appropriate live work habits and use them as much as possible. At the very least, work dry; in damp areas, stand on dry wood. Don't wear loose clothing, keep your hair out of the way, and avoid metal jewelry. Even nails in shoes put you at slightly increased risk, depending on where you're standing. If you're working near an open, live panel, even temporarily, you may gain some safety by using a fall-in barrier, perhaps of plywood, to block live parts from inadvertent contact. Cardboard is too easy to knock out of your way, or into the panel; you're better off with something more rigid but similarly nonconductive.

Sources of Shock

How Much Voltage Will You Face? In a house, all circuits are limited to 120, 208, or 240 volts. (You may encounter higher voltages if you open up certain types of equipment such as fluorescent lights, but you will find a maximum of 240 volts in the electric circuits feeding them.) The higher the voltage, the more current will flow through a given resistance, but there is not a terribly great difference between 120 volts and 240 with regard to this danger.

More importantly, you will find no more than 120 volts *to ground* anywhere in your permanent wiring. So long as you deal with the #14 or #12 wire leading into the fixture, you cannot touch anything at a higher voltage. You will recall from Chapter 2, Basics, that while the voltage between the hot legs is 240 or 208 volts, the grounded neutral is right in the middle, 120 volts from each hot leg.

> "Outlet closers" and protective covers cannot protect children from shock if they damage appliance cords. Try to keep them from touching the cords as best you can. And recognize that there is nothing to *guarantee* that your children won't cut, crush, yank, or bite cords.

This limits the voltage you can be exposed to in your service equipment by touching a hot wire and, say, the panel itself. Whatever you touch is at the same 120 volts to ground as you're exposed to every time you remove a wall plate and stick a finger or a tool inside the box without confirming that power is off and stays off. This is also the same voltage an infant is exposed to if it chews on an appliance cord—whether or not the appliance is on!

Why Are You in Greater Danger When You Work in Your Electric Panel?

There *are* a few differences between panels and other locations that affect the degree of risk you are exposed to.

- Even though there are only 120 volts from each hot leg to ground, you do have ready access to the 240 volts between the incoming hot legs of your electric service in your panel, or between two adjacent circuit breakers or fuse holders, whereas most, but not all, other places in your electrical system only have 120 volts available between conductors or screw terminals.
- The conductors coming directly from the utility are always live.
- There are more live parts exposed in your panelboard, and the live parts directly face a wider area of your body.
- Sometimes, but not always, panels are more crowded than switches and outlets such as receptacles and lights. Crowding means you have more of a tussle to get at what you're trying to see or to move, and crowding makes you more likely to touch a finger or tool against a live part.

How Can You Get Shocked When You're *Not* Working in the Panel?

The potential for shock is available anywhere in your system. As you will learn later on in this chapter, there may well be a number of places in the ordinary branch circuits feeding lights and receptacles where the higher line-to-line voltage of 208 or 240 volts is present.

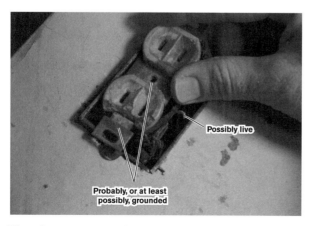

Possibly live

Probably, or at least possibly, grounded

Thumb near wire—whoops. Danger!

The Common Sources of Shock from Work on Your Branch Circuits.

You are most likely to be hurt by a 120 volt shock in one of four situations:

- you touch a hot wire and a neutral simultaneously;
- you touch a hot wire and a ground simultaneously (see below for a discussion of grounds);
- you bridge an intentional opening in a circuit, such as two terminals (usually screws, used to attach wires) of a switch in the OFF position;
- you bridge an unintentional break in a circuit, such as a terminal and a wire that has come loose from it, the two parts of an actual broken wire, or the filament and shell of a broken light bulb.

If you bridge a break, this puts you directly in series with the circuit, meaning that the 120 volts *have* to go right through you to complete its loop as it powers a light or appliance.

Simultaneously touching neutral and ground means a neutral and any kind of ground. This could be an electrical system ground such as a properly connected ground wire, conduit, cable armor, or box. It could be some other grounded surface, such as a steam radiator, a grille that is part of your forced air system, or your water tap. It could be the earth itself, or something in intimate contact with it such as a concrete slab.

A Rarer Source of 120 Volt Shock. There's another, rarer source of shock. If you touch neutral and ground simultaneously, you won't be harmed so long as no power is flowing in the circuit, which means that no light or appliance is in use. If, however, the hot wire is feeding a load, the amperes flowing in through the hot wire normally would return 100% through the neutral. If you provide an alternate path through your body, the current will divide between the two "parallel" paths because electricity has no way to "know" which path you intend it to take. Part of the current will return as it would normally, from the hot to the neutral and back to the panel; and part will take a more roundabout route, from the hot through your body to ground and through the ground back to the service where the neutral and ground are bonded to each other and to the GEC.

Almost every time this happens, most of the current will flow back through the neutral, but even a small amount of current flowing through your body can be dangerous. This is why GFCIs trip not only when there is a connection between a hot and ground, but also when there is any connection between a neutral and ground. This points to a problem discussed in Chapter 14, How to Replace Receptacles, the "bootleg ground" or "bootleg neutral." See also the discussion of downstream neutrals in Chapter 6, Examining Your Panel.

Why a "Hot" Wire or Terminal May Not Be Much More Dangerous than a Neutral. Although you can be shocked as the result of contact with the neutral, contact with the hot *is* more dangerous. The exception is when you bridge a break between two disconnected parts of the neutral. Bridging a break in hot or neutral has to be equally dangerous because the entire current is flowing through you from one side of the break to the other.

In fairly modern wiring, the hot and neutral are, respectively, a black and a white wire. Even in correct wiring, though, not all white wires serve as neutrals, for various reasons. It is harder to determine what each wire is supposed to do when you deal with old wiring; instead of black and white insulation, you may well encounter wires all of which have faded to gray. This makes it even more dangerous to assume that you are touching a neutral and therefore are relatively safe.

Why Your Fuses or Breakers May Not Protect You. Most types of shock will *not* make a fuse blow or a standard breaker trip. GFCI breakers, and to some extent AFCI breakers, both of which offer far greater protection against many kinds of shock, are the exception. Why aren't fuses or standard breakers likely to protect you?

When you simultaneously touch the hot and neutral, you simulate an appliance so far as the fuse or circuit breaker can tell. To fuses and breakers, line to ground current flow looks about the same as line to neutral current flow. The fuse or breaker won't know there's a problem, given the amount of current that is likely to flow through you.

We always face the risk that we'll touch something hot while some point on our bodies makes contact with something grounded. A utility line worker told me that a colleague, distracted for just a moment, leaned back against a live conductor and it "blew his toes off." He was speaking literally.

In theory, it is pretty safe to touch a source of almost any voltage, so long as the electricity has no chance to flow elsewhere using you as part of its path. This general rule is why birds can perch safely on power lines, even bare ones; a single power line is *all* they're touching or near. This "so long as" catch is why we cannot work around live parts; even line workers, with all their protective equipment and training, must take scrupulous care.

What You Should Make of These Warnings. Here's the most important thing to keep in mind: The source and type of shock is not the major issue. Avoid ALL shock, if only because you have no guarantee that a source will not do serious damage.

Although as a master electrician I theoretically can work at any voltage, I consider myself qualified to work at voltages up to 600 volts. This is the level referred to in the NEC as low voltage. You are qualified to work in contact with the voltage of flashlight batteries, car batteries, doorbells, thermostats, and, generally speaking, the voltages associated with transformer-fed landscape lighting. Remember, though, that some of this equipment could subject you to 120 volts. If a doorbell transformer fails in one particular way, only the high resistance of the thin bell wires will protect you, unless the line voltage side of the doorbell system shorts and trips the circuit.

Are you qualified to work with 120 volts? This is a question you have to ask before working around live equipment. Only you can answer it.

How Are You Likely to Run into These Risks?

Reasons You Would Put Yourself at Risk. You would not risk touching a live wire or screw on purpose. However, any time you work on an electrical system that hasn't been completely deenergized (had all sources of power removed), you expose yourself to some risk. Much of the work involved in evaluating wiring requires some exposure to live parts. Troubleshooting is even riskier. Ignorance, carelessness, or fatigue can cause you to touch a terminal or some other live part unintentionally.

Live work remains dangerous even for those with professional training and experience. When you plan to make or remake a connection, you may need to have the circuit on while you test the wiring. Sometimes you need to pull wires or their connections out to where you can reach them easily, or you need to undo something—for instance, partly unscrew connectors or remove tape—in order to gain easy, reliable access to the wiring. In these cases, it is far safer to take this preliminary step with power off. By following this sequence, you reduce the risk of touching live parts when you restore power for the actual testing.

The most important reason you may choose to test live wires is to make sure that you can keep yourself safe subsequently, when you make or unmake connections. There is no way to confirm that you have been successful in turning power off that is nearly so reliable as having proven that you *can* detect voltage when power is on.

Another reason you probably will test live wiring is to determine which conductor is which. For instance, as is all too common with old wiring, you may no longer be able to distinguish which conductor's insulation originally was black and which one was white. Understand very clearly that even in modern wiring, and even in correctly installed wiring, a wire's color does not determine its function reliably. With the power on, a voltage tester can be very helpful to give you a preliminary idea of each conductor's intended purpose.

Why It Makes a Difference That Your Wiring Is Not New. When you work on older—not necessarily old—wiring, you face some risk of encountering bad workmanship by previous installers. Problems that result from some previous worker's ignorance or carelessness are discouragingly common.

There are a number of such hazards that you may run into when you work on your electrical system. You will learn to avoid these practices in your own work. This will ensure that you don't put the next person—who could be you delving into your wiring in the future—at risk. Before you begin to work on your system, even before you start to access your wiring to evaluate it, you need to learn how to keep yourself safe.

Some Specific Sources of Danger

So far, the description has been relatively theoretical and abstract. Now I'll mention some specific ways electricity can sneak up and shock the unwary. You will learn how to look for these problems in Chapter 8, The Walk-Through.

- Insulation damage. Was insulation overstripped? Are there nicks or cuts in the wires' insulation? Has crowding caused insulation damage? Has the insulation dried up, even cracked, due to age or heat?

- Ineffective enclosures. Were covers left off stationary appliances? Fixed appliances?

- Ungrounded appliances. Could your appliances have shorted to their enclosures without blowing their fuses?

- "Wing it" wiring. Has someone's ignorance put you at risk? Is there bad work upstream? Do your receptacles have false grounds—has someone replaced receptacles inappropriately? Has someone installed "cheaters"? I'll explain these problems as we proceed.

Electrical Fire

How Does the Risk Arise?

Electrical fire is insidious. Fire is less of a threat to you while you work on your system than are shock and arcing. Deficiencies in your wiring threaten fire later on, when it could kill those your work is intended to serve.

The Sources of Electrical Fire. Fire requires two elements: The first is a source of heat or spark; the second is access to combustibles. High-resistance connections are very common sources of combustion heat. These include intentional connections that were done inexpertly or carelessly and unintentional connections that don't make contact quite solidly enough to blow a fuse.

I explain how to make proper connections in Chapter 12. One example of intentional connections that are inexpertly or carelessly made is bad splices. Another is loose terminations at receptacles and switches. One type of dangerous unintentional connection is off-again, on-again contact between a vagrant strand of live wire and a metal box. Another is a live screw left backed out, just barely touching a ground wire. Many a fire has started due to an imperfect but reasonably good electrical contact near thermal insulation consisting of shredded, somewhat combustible material. National Fire Protection Association research published in the early 1990s estimated that high-resistance connections cause 25,000 fires yearly. These include connections or unintended contacts that spark or arc as well as others that simply overheat.

How Heat and Sparks Gain Access to Combustibles: Incomplete Enclosure

Why Outlet Boxes May Not Seal in Any Potential Dangers. The risk of fire is greatly increased by

- boxes or panels with covers that are missing or that don't cover them completely;
- boxes with gaps (either between box and cover or between sections, in the case of multipart boxes);
- broken boxes;
- and boxes with unsealed openings.

Electric boxes, even plastic ones, are heat-resistant to the point that you could braze inside one without quickly breaching its integrity. Therefore, so long as it is closed, it provides your house with significant protection. Chapter 13 explores enclosures and cables in detail.

How Can Painting Rooms Affect Their Electrical Systems?

Painting can and often does cause electrical problems. Simply as a coating, it can

- obscure markings on switches, circuit breakers, and panel directories;
- glue cover plates, receptacles, and walls together so that electrical repair is difficult, and frequently causes damage;
- clog receptacles;
- make it impossible to tell whether a wire started out black, white, or some other color—or whether it was intentionally recolored; and
- contaminate and impair operation of electrical equipment.

I have found some circuit breaker handles very difficult to move, simply because paint had wedged them in position. I would hate to have to guess whether they were capable of tripping; they should have tripped internally anyway, regardless of what the handles did, but this is not something to bet your family's life on, in the event of an overload or short. Furthermore, if there is an electrical problem, you may need to turn the circuit breaker off manually.

Particularly when sprayed, paint can find its way into a circuit breaker panel, contaminate the contents, and cause fire or malfunction. For reasons associated with their function, residential circuit breakers are not fully sealed devices. Paint can get inside their mechanisms.

Intact, Covered Electrical Boxes May Not Provide Complete Enclosures. An even more common example of an unclosed opening is a gap between the box supporting a ceiling fixture and the edge of the fixture's canopy. (The canopy is the part of the fixture that fits against the ceiling; the fixture's wiring is connected to the house wiring in the area enclosed by the canopy and the electrical box that feeds it.) I don't worry much about igniting joists inside ceilings. I worry considerably when fixtures are mounted against wooden ceilings, such as the wainscoting used in old porches. In an old house, this is old, dry wood, and the boxes often are undersized, which means the canopy rests against a fair bit of wood beyond the box. The fiberboard used in early- to mid-20th-century ceilings is even more vulnerable—and I often find boxes recessed in it. If your fixture itself is old and in bad repair, the only thing protecting you from fire is luck. If this is the case at your house, your best choice is to turn off power to the fixture until it is repaired or rewired. You really should also insert a noncombustible barrier, such as carefully-trimmed sheet metal, over the wood of the ceiling between the edge of the box and the edge of the canopy. (Chapter 13 explains more fully how to evaluate and correct this type of hazard.) This rule, however, is very commonly ignored—by electricians as well as others. A wooden ceiling, however, does deserve the protection.

Less Obvious Causes of Electrical Fire.

Electrical Overheating. Combustibles with access to sparks and arcs such as may be found in electrical enclosures are only one source of fire. Pyrolization, literally, digestion by heat, adds to this. When wood such as structural studs and joists—or the cut edges of paneling—is heated, it is pyrolized (turned into something close to charcoal) over time. This makes it far more likely to catch fire, even without a spark.

Correctly rated, properly sized fuses and circuit breakers interrupt loads that would cause overheating. Overfusing permits such heating. Putting the wrong size fuse on a circuit violates Articles 240 and 310 of the NEC: It lets your house wiring function like the element in a toaster, treating your house like toast that has stayed in a toaster too long.

Overlamping

When a fixture is marked "60 watt maximum," using a 100 watt bulb not only can overheat the light fixture but can pyrolize nearby structural wood. When a fixture says, "Use 75 watt or smaller PAR 20 lamps," you are restricted more, not less than, by the fixture with the words, "60 watt maximum" printed near where you would screw in the bulb. A PAR lamp is a spotlight or floodlight, and PAR 20 means that the part that sticks out of the socket has a relatively narrow diameter. An "A" lamp, the standard pear-shaped light bulb, will not throw as much heat (and light) *out* of the fixture, and a bulb with a greater diameter may make direct contact with parts that it should stand clear of, when it fits in at all. Using compact fluorescent adapters in place of other light bulbs also can violate manufacturers' instructions. According to Underwriters Laboratories, Inc. representative Robert Johnson, the unplanned-for weight of a compact fluorescent adapter, in combination with the heat generated at its base, has loosened a fixture's rivets so that the compact fluorescent fell, with the socket, yanking on the fixture's wires.

Electrical insulation itself can be pyrolized. This is especially true of the very old insulation you find protecting rag (cloth-covered) wiring. Overfusing is not the only cause of charred insulation. When a careless electrician runs cable up against chimney pipe, or when the furnace people don't call in electricians to reroute cables that are in the way of heating ducts, they can create a serious hazard over time by destroying wire insulation.

Heat can create another problem for the thermoplastic-insulated wiring used from the 1940s or 1950s to the 1980s. "Thermoplastic" means "softened by heat." Elevated temperature plus pressure, such as that from being squeezed by the staples used to hold cable in place, may cause internal arcing between the wires

Pyrolization doesn't require the level of heat it takes to ignite a piece of paper; by the time you have approached the temperature of boiling water, you are well past the safe level to maintain wood's integrity.

inside the cable. I have never identified such damage by looking along a cable. All I can do once cables are in place is to avoid overheating, in large part by not over-fusing. I also avoid using more force when hammering than the amount needed to drive a staple far enough into wood so it touches my cable.

Other Ways to Reduce Your Danger While Working with Electricity

There is more to say about making your electrical installation safe. Some of the following will help you take care of yourself as you work, and some will help you ensure that what you've installed is safe or that what you have evaluated and judged to be okay really is reliable. I will talk about three things: self-preservation, seeking assistance, and handling emergencies.

Bad wiring can put you and others at risk. How can you do some of your own electrical work without endangering yourself and your family? So long as you stick to the material covered in this book, you will have the knowledge, if not the skills, to perform repairs as safely as possible. However, there are no guarantees. For one thing, good information in itself is not enough. A journey-man electrician, the lowest qualification that permits more-or-less independent work, has years of supervised training. Furthermore, nobody is an automaton.

Monitor Yourself, and "Close Up Shop" When You Need To

Contractors have tables of labor costs associated with various jobs, but they are mostly "new work" jobs and don't address the complications that are found in old wiring. Evaluation and repair are hard to fit into recipes, like specifying the number of minutes needed to bake a "perfect crust." Even if you have an idea of how long a job will take you, you may run into an unanticipated complication. Rushing is a dangerous response, as is cutting corners. If you plan to take on a job at a particular time, what do you do if you discover that you're not at your best? Respect your limits.

When you're hungry or tired or rushed or it's late in the day, you're at greater risk of messing up and getting hurt. The wisest thing to do then is to rest, not to install or repair wiring—and especially not to troubleshoot. If you're not prepared to let go of the job completely, it is at least safer to ruminate about how you will tackle the problems later, when you come back to the tasks, than to push on. When an electrical problem has you stymied, the best thing to do is to find out how to prevent imminent danger, even if this means inconveniently cutting power to some portion of your house. Solve the frustrating problem later, when you don't have the same stress impairing your judgment. You can't set a timer for yourself to evaluate or repair old wiring.

Don't Jump in Over Your Head

You are not a "master chef," and this is not a cook-

Knob-and-tube wiring—surrounding wood is pyrolized.

book. Don't jump ahead to my explanation of a job

you're eager to undertake without at least skimming the information I have laid out to take you there safely. This means looking at *all* the chapters preceding it. And don't try to handle a task that scares you.

When in Doubt, Don't Tough It Out; Call On a Pro

If you are afraid that you might do something wrong and make your wiring more dangerous, rather than safer, maybe it's time to call in an electrician to take over. Or perhaps not. You may have other options.

Stay Away from the Common Mistakes as You Seek Assistance.
Certainly there are places where it is unwise for you to turn for help.

- Salespersons at hardware stores and supply houses are not trained electricians or inspectors. Some give out bad advice not because they want to steer anybody wrong but because they're unwilling to appear ignorant. Even their best recommendations are more likely to derive from hearsay than from having proper training at the tasks that they are recommending you undertake. Generally, they don't have adequate knowledge of the NEC or of installation standards, so even when they have been "successful" at a task on frequent occasions, it is unwise for you to be guided by their experience—if you want to do things right.

    Sam wanted to run cable inside a wall going to a new outlet into an existing sectional (screwed-together) metal box. A home center clerk helpfully suggested a rather complicated operation:

    1. Clamp a connector onto the end of the cable;

    2. Knock out an opening in the back of the old box;

    3. Stick the cable up from the new box below it;

    4. Feel the way with the cable to stick it into the old box;

    5. Screw a locknut (the inside half of a connector) onto the cable to secure the connector in place;

    6. Then splice the wires in the new cable to those on the receptacle in the old box.

        There were a few problems with this.

    1. The rear knockout (embossed opening) of the old box was deep, past all the wires in the box.

    2. Even without that complication, he would have a hard time removing the knockout, without experience, because normally it is removed working from both inside *and* outside the box.

    3. Trying to force it out by hammering or pushing from within might knock the screwed-together box apart or force it back into the wall, which at the very least would damage the wall.

    4. Fishing the cable into the top of the old box without seeing what he was doing might be difficult.

    5. Finding a hole in the back of the box with the cable might prove pretty close to impossible.

    6. Even if he could wiggle the cable into the back of the old box, pointing the connector on it far enough and straight enough into the box that he

could reach 2½″ into the box with the locknut (mating with the connector on the inside) and thread it onto the connector would require awfully flexible fingers. (It even *sounds* tortuous.)

7. Tightening the locknut would require some fancy tool use.

8. Finally, adding another cable with two more insulated wires and a ground wire to what was already in the old box very likely would crowd the box well beyond what the NEC allows.

What the clerk suggested has been done, and might even need to be done to avoid wall damage (though there is at least one type of romex connector that he could have slid onto the cable from inside the box, which would allow it to be done with one part of the problem eliminated), and might possibly even be legal in Sam's case. However, it is a safe bet that the clerk had never tried it himself and had no knowledge of the issues I have listed.

• Semipros may be willing to wire with you or for you. However, there's no reason to assume that the one you meet knows much more than those hardware store clerks. Even if he or she knows what will work, at least for a while, there may be important gaps in the information, especially regarding what is legal and safe.

Myrna moved into an old house, and had a lot of ideas about how to turn it into her dream home. She got a great price from Eddie, who got a lot of her wiring done in a surprisingly short time. Eventually, though, she found serious problems with his work. She wanted him to fix it and complete it, so her wiring worked. He never came back, and soon stopped returning her calls.

Angry and frustrated, she called in a licensed electrician. After getting clear on the fact that her previous installer not only had not pulled a permit but didn't even have an electrical license, he spent a day, at premium rates, simply at the task of figuring out what functional problems and illegalities were present. The price for tearing them out, correcting them, and completing the other parts of the job was going to be much higher than if he had been called in from the start.

The contractor called the County's Chief Inspector to see what could be done about the semipro who had left Myrna high and dry. The Chief said that since Eddie did not have an electrical license, they had no power over him. While there were statutes on the book that could be used to punish unlicensed contracting, the County did not presently have the resources to bring Eddie to court or even threaten him. What about Myrna? The inspector said, "Ignorance of the law is no excuse. She should have known better."

• Professional electricians are not in the business of giving out free advice. Unscrupulous homeowners who call electrical contractors for "estimates" in order to pick their brains are not behaving ethically. They also are quite likely to encounter electricians who recognize exactly what they are attempting.

Max had no power in half his downstairs and had strung extension cords everywhere. He called three different electricians, asking them to come over, tell him what they would do to repair his problem, and give him an estimate on that work. All three told him they would come over when they had a chance; none of the three had shown as of a month later.

Stick with the Better Sources of Help

There are some places you can look that may indeed help you ensure that what you do is safe.

Seek Professional Consultation, If It Seems That Consultation May Be Enough for Your Needs

- You may be lucky enough to find the rare legitimate electrical contractor or, perhaps, retired electrician, who hires out as a consultant and can communicate at your level. *Warning: a retired pro may not have kept up with the latest materials and Code requirements.*

- A Certified Home Inspector may have a solid understanding of wiring. If so, he or she may be able to check out several of your house systems to alert you to problems, both preexisting and resulting from your efforts. However, even the American Society of Home Inspectors, ASHI, has no way to ensure that members are knowledgeable regarding the fine details, described here, or know how to perform electrical work themselves.

- Sometimes the fire marshall, or the city or county electrical inspector, will perform courtesy inspections. Approximately 55% of electrical inspectors questioned in a late-1980s survey said that they provided courtesy inspections upon request by a homeowner, especially someone who had just purchased a home. If this service is available, there may be a fee, but at least you are getting a disinterested opinion from someone who, if you're lucky, is trained and experienced. Your local electrical inspectors are in the business of protecting the public. Generally, they would much rather answer a question or two over the telephone than write up your violations or investigate your electric fire.

    Unfortunately, while a simple phone question still may get you a quick but useful answer, courtesy inspections are largely a thing of the past. Budget constraints and overwork are limiting the amount of free service inspectors can provide—even though it would enhance the safety of their communities.

- Even where they do not offer courtesy inspections, you may be able to get around the problem. There is a way to pay to have the inspectors for your local jurisdiction evaluate at least part of your system.

    Here is the key to this approach. Wherever there are inspectors and permits, anyone who adds wiring, for instance, by extending a circuit to put in a receptacle at a new location, needs to apply for a permit before even starting work. Furthermore, in many jurisdictions anyone doing a substantial amount of remodeling or replacement such as upgrading all the light fixtures needs an electrical permit.

    Almost always, you the homeowner can apply for a permit. You may have to take a simple test, consisting perhaps of 10 open-book questions, before you will be permitted to apply for a permit. Applying for any electrical permit buys you an inspection. It is that simple. Don't be shy about having the inspector out—even professionals benefit from having someone else to catch their mistakes. *Note:* Unfortunately, due to budget constraints some jurisdictions are hiring people who never have been trained in wiring to perform electrical inspections. See Afterword.

Can You Share the Work?

Usually It Is Unwise for You to Get Someone in to "Finish Up." There is some very uneasy middle ground between doing the job yourself and hiring a professional. Some homeowners say, "I'll wire everything outside the fuse box, but I want an electrician to make the final hookup." Any electrician who goes along with this is way too brave for you to want. It is very risky, both personally and legally, for a professional to finish off partly-completed amateur wiring. "I've wired everything up to the fusebox" qualifies as partly completed wiring.

When electricians—or you—take on old work, there is the chance of running into someone's dangerous errors. Remember Myrna and Eddie? When an electrician takes on work begun by an amateur who discovered that the job was beyond his skill, this hazard will feel even more likely to the pro.

One legal aspect that makes a professional nervous is that an electrician who finishes someone else's work, or who pulls a permit to cover work someone else has begun, risks ending up with liability for any damage or injury attributed to the wiring. The last person to work on it is a prime candidate for blame. This is true even when the work was begun by another, unknown, electrician. Furthermore, if a permit and inspection are required, the inspector may question the electrician's claim that the work was begun by another.

Therefore, an electrician who takes over a job ought to go over just about every inch of what has been done before—and possibly pull it out to redo. An electrician who is casual about this may possibly be casual about making sure the wiring is safe, too.

"Work-With" Can Be a Terrific Option. A very rare breed of electrician will allow you to work with him or her, allowing you to serve, essentially, as an electrician's helper. Unfortunately, there isn't a whole lot for a helper to do when it comes to troubleshooting, evaluating, or repairing and replacing devices and fixtures. This dispenses with the possibility of "sweat equity" with regard to the work covered in this book. If what you are doing is, essentially, new work, it can make sense—if you find that rare, highly competent electrician who is willing to take you on.

You Can Call in an Electrician from the Beginning, and Still Make Use of What You Have Learned. Suppose you want a helping hand. You certainly can call in a professional to take on jobs that seem too much for you. You may surprise your contractor if you say something on the order of, "It looks like Circuit 12, which has a 20 amp breaker, is overfused from the junction box by the attic stairs and on, including most of the ceiling lights. Would you check it out? If I'm right, I'd like you to derate the circuit down to 15 amps. I have a Cutler-Hammer panel."

Still, most small contractors would be happy to get a service call that sounded so straightforward, even though many jobs turn out not to be as sim-

CAUTION!

The fact that a professional will be very wary does not at all mean that you must keep going after you realize that the work you have started is beyond you. Safety First! However, starting a job and handing it off probably will cost you more than it would have to hire an electrician to begin with. Mistakes can be expensive; still, they may be necessary for learning.

ple as they sound. After all, if you understand enough to offer that clear a description of what you think you need, there is a reasonable chance that you will understand what has happened in the event that your electrician needs to explain a complication that he or she runs into.

Emergencies

In this chapter, I've covered a lot of general information about wiring safely and looking for existing hazards. (Later chapters will provide further, detailed information on safe choices and safe practices associated with specific procedures; to understand and use that information, you need additional preparation that I will provide along the way.) However careful you are, though, there is always the chance that something will go wrong in a way that is not only inconvenient or costly but dangerous. You will be far better off if you prepare for this than if you assume that it never will happen to you.

Immediate Action

If you're afraid that someone is *in danger* of shock, the first thing to do, to ensure safety, is to kill power. The second thing to do, in the event that you think the incoming power line—or your service panel or main disconnect—may be the source of danger, is to call your utility. If you think your interior wiring is putting someone in danger, call an electrician. The electrician or utility representative will tell you what to do next.

If you're afraid that you're in danger of fire, the first thing to do is to get everyone to safety. The second thing to do is to call the fire department or rescue squad. They will tell you what to do next.

If you're afraid that someone has been hurt, call for help. If your area has 911 service, this is the best number to call. If you do not have this service, do some research ahead of time to find out whom to call should you face an emergency.

Preparation

There are three ways to prepare for electricity-related emergencies.

- First, know how to shut off your circuits.
- Second, make sure there's a clear exit path that everyone knows and can navigate.
- Third, and only third, find an electrical contractor you can trust, someone who will serve as backup—on whatever terms, at whatever rates—should you get in over your head to the point that you face an emergency.

To minimize, to absolutely minimize, the chance that you will be shocked as you wire, or that someone else will be shocked as the result of your wiring, test everything carefully for voltage. Then test it again. The information you must have to do this safely and adequately is in Chapter 3, Voltage Testing.

The old-time electrician used to flick a wire against a presumably grounded surface or a ground wire (in those days, though, circuits with ground wires were rare) to make sure it was dead. Consider for a moment and you will realize that in the process of bolting to ground, you are doing just that. The same rationale holds for you as held for the old-timer: it probably will increase your safety! If you *have* made a mistake in your testing, it is better to draw an arc when you are wearing full protective gear and are alert for one as well, than to be surprised by an arc or to suffer electric shock because you are less alert to that possibility. Even a qualified electrician could make a mistake, given that the wires being tested are unlikely to be bright copper, which would allow a tester reliable contact.

6

Examining Your Panel

In this chapter, I will explain how to look into and evaluate your system from the service disconnect on to where circuits leave your loadcenter, also called a panelboard or panel. I will focus on the panel not because you will be working in it, but because evaluating and understanding it is so valuable. Your panel is important because its contents distribute incoming power out to all your circuits; in the process, its contents protect your wiring and your house. More accurately put, this equipment protects your house when, and only when, your system was installed as it should and continues to function as it should. This is why you should look at your panel, whether it is old or not so old, whether it contains fuses or circuit breakers, and whether it has been a source of worry or has seemed perfectly okay.

I will start from the general and then show you how to determine the specifics of your panel. The first thing to examine is your main disconnect, in the event that it is separate from your panel. After taking care of that, I will explain how to evaluate the mounting and the physical integrity of the panel enclosure itself. Next, I will talk about how to open a panel, and proceed to its internal design: how incoming cables are attached to the busbars that form the electrical backbone of the panel; the different ways busbars can be laid out; and how power is distributed from busbars to the fuses or circuit breakers that carry power out into the rest of your house. This will not just be familiarization; I will describe various problems or illegalities you may discover as you proceed.

The very last part of this chapter teaches an exceedingly important operation: how to interrupt power by shutting off individual circuits, or the entire house. You need to know this in order to perform any electrical work safely. Even people who never do any wiring or electrical testing should learn this; it could save the life of someone—perhaps a child—who otherwise would be electrocuted.

Separate Disconnects

In Chapter 4, Evaluating Your Service Entrance, I talked about tracing power from the utility through your meter and main disconnect to your panel. If you discovered that your service disconnect is a separate switch or circuit breaker, whether outside or inside, it's now time to examine it. Even if you don't have a separate disconnect, you will be following much the same procedure when you look at your panel. Here are some elements to evaluate:

- If it is a switch with a handle on the side, or if it is a circuit breaker, can you turn it on and off easily? (Before you try this, make sure nothing is running that will be harmed by losing power.)

- Do you have ready access, or is the path to it blocked?

- Can you stand in front of it with room to spare, or is the working space in front of it shallow or narrow, or is your leg room blocked?

- Can you get the cover off without interference, if necessary, or does a stud or drywall or some other material prevent you from getting a screwdriver in straight, pulling the cover straight out, or swinging a door open at least 90 degrees?

- If it is a fused pullout, can you pull it out and stick it back in easily?

- If it uses cartridge fuses, do the clips that hold them retain their spring tension?

- Whatever kind of disconnect it is, are there any unclosed openings or other missing parts?

- It may be all right for a disconnect mounted in the outside wall of your house to be contained in an indoor-type enclosure, one that does not have an integral cover to keep water out, provided that it is behind a door or otherwise enclosed and protected from the damp. If this was the case with yours, is it still protected? I have come across this arrangement only at houses whose electrical services were many decades old. Frequently, the enclosures no longer had doors.

- Whether or not anything is obviously missing, is there *any* access to the inside for rain, vermin, fingers, or dust?

- Is there evidence of scorching, considerable rust, or mechanical damage such as might have resulted from vandalism?

- Has its mounting gotten loose?

Just about any such problems, with a few exceptions, are grounds for calling in a pro. You may be able to clean up a modest amount of rust and apply weather-proof paint. If the disconnect is mounted to a backing (such as a quarter sheet of plywood) and the backing has gotten loose from the wall, you may be able to safely resecure the mounting to the wall surface if the backing still is in good condition.

Once you have finished with the service disconnect (if it is separate from your panel), it is time to take a look at your loadcenter.

Evaluate Your Panel's Condition from the Outside

Your panel, or panelboard, is covered in NEC Article 384. Almost all panels are found indoors, so I will talk almost exclusively about that arrangement. Unless your panel is quite new, if it is outdoors, it is extremely likely to have suffered damage from the elements and therefore to be due for replacement.

From the outside you will see a cabinet, a box that should be enclosed on all sides except the front. The front should have a metal cover (and in some designs a separate, intermediate, metal trim piece) with various openings. The cover may or may not have a blank, hinged door over it.

The first thing you should see, either on the front of the cabinet or else on the inside of the door, is the circuit directory. I'll talk about this extensively in Chapter 8, The Walk-Through. You will see how important it is at the end of this chapter, in the section called "Shutting it Down."

When you evaluate your panel, you will ask many of the same questions as you asked (or would have asked) in examining a separate disconnect, though water damage is much less likely indoors:

- Is it loosely mounted?

- Is there physical damage such as severe corrosion?

- Are there missing pieces or unclosed openings? I will talk about this a great deal more below.

It seems that installers are far more careless with indoor equipment than with outdoor, perhaps because outdoor equipment's vulnerability is so much more evident.

Is Your Panel Mounted Securely?

It is relatively rare, but far from unheard-of, for an installer to use nails or woefully-undersized screws to secure the cabinet in place. More commonly, I

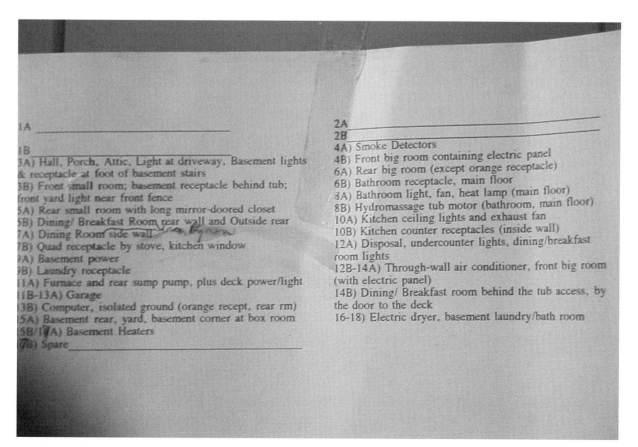

Extensive, clear circuit directory, typed rather than filled in on the form pasted in by the manufacturer.

find that the installer secured the panel to a backing that came loose over time because it couldn't handle the environment. Cut nails (cheap masonry nails) holding ½″ plywood to a brick foundation wall may do fine for a decade or two, but eventually the whole shebang may start to rot and loosen, endangering the panel secured to it. When I look at panels in damp basements, I no longer am surprised when I find them mounted on wood that looks gnawed or rotted away.

It's not physically difficult to replace mounting hardware, though anyone trying to drive screws out the back of a live panelboard needs to be careful indeed. If, however, moisture has entered the cabinet, replacing the hardware probably isn't enough. In this case the panel may well need to be replaced. If it is on a backing or a wall that is showing deterioration, it is likely that someone will have to kill incoming power in order to replace all straps and staples, repair the support surface, remount the panel, and resecure all cables and conduit. This is not a small job, but it is a necessary one. I suggest most strongly that you call in a pro if you need to replace a panel. I also urge you to call in a pro if you need to remount a panel to a wall or to new backing, unless you yourself can kill all power to the panel.

Has Your Equipment Begun to Corrode?

As with service entrance equipment, I don't worry about a little rust on or even in an indoor panel. However, I do check very carefully to make sure that there is no evidence of continued exposure to moisture. When there is any possibility that a panel is mounted on a wall that gets moist, I try to mount it on a standoff such as a piece of painted or urethane-treated marine plywood. I also check very carefully to make sure that only the cabinet has gotten wet, and not the contents. Circuit breakers or fuses that have gotten wet should be replaced. Corroded fuseholders, fuse clips, or busbars require replacement, which usually means replacing the entire panel.

Does Your Panel Itself Need Repair?

Is Your Panel Missing Pieces?

The final question to answer while looking from outside is whether all the pieces that should be protecting the panel's contents are doing so. You have no way of knowing exactly what pieces should be present, but still you can answer this question readily. Look at the panel: can you see inside, where the wires no longer are in cables or conduit, but are spliced and connected to terminals? The answer takes a fair bit of looking, but there's nothing mysterious about it. When you *do* determine that something is missing, it may not be immediately obvious just what has been lost. I find this true myself sometimes when I examine old panels.

> When you do uncover problems, only some of the solutions will be easy, or even within your capabilities.

Is the Panel Covered as It Should Be?

See whether the panel cover is properly secured to the panel. Often the panel door is missing a screw or two. Often, also, unusual screws that held the cover on were replaced with unsuitable ones when the originals were lost. The wrong ones may not hold as well, or, worse, may cut into the wires inside. If you find an electrical supply house that carries the brand of panel you own, they'll sell

Screw impinging on conductor. Even though this factory-supplied screw is suitably blunt, its threads could cut conductor insulation. Try to shift the conductors away from its path.

Panels Designed to Be Wide Open!

In very, very old panels (I've only seen this in fuse boxes), the terminals and sometimes even the busbars are not protected from access by any cover, and this apparently is just the way the ancient fuse boxes were designed. Very, very rarely, an ancient disconnect has survived that consists simply of a knife switch. This is a switch ALL of whose parts are exposed, and all of whose parts are live when it is in the ON position, except for the end of the handle. If this still is in operation in your house, upgrade! Get rid of it! Little that you will find in this book is as important to your safety as getting an electrician in to kill power to that switch and replace it. These cases are exceedingly rare.

By far the most disgusting, and very likely the most dangerous, example of contamination I have remedied was not in a home, thankfully, but in a restaurant. The panel was in the basement, immediately below plumbing that frequently backed up.

I don't present this story simply to make you shudder. The restaurant management had never noticed, never investigated, until circuits went out and they called me in. I know how crowded some people's basements are, and I know how often people never notice what's happening with their electrical panels—sometimes even forgetting where they are.

you replacements, inexpensively. (All right, expensively, but a dollar for four screws is still only a dollar.) Most panel covers are held on by washerhead machine screws, but you may have a panel with very coarse-threaded sheet metal screws lacking points. (See Chapter 11, Materials, for a description of screw types.)

Does Your Panel Have Otherwise-Proper Openings That Were Left Unclosed?

Every now and then, I come across service panels in use that are missing their covers, or in some cases missing separate inner covers. If this is your situation, get the cover replaced. Even more commonly, I find smaller unclosed openings of various sorts.

These situations are dangerous for several reasons.

- Combustibles could get in and burn or cause arcing. Besides the damage this could cause to your panel, and the inconvenience of losing power, these same unclosed openings would provide a route for fire to spread.

- If there's no cover, almost certainly there is no circuit directory. As you'll learn, circuit directories are important for both safety and efficiency.

- Moisture, dust, paint or vermin can enter to contaminate the contents and interfere with operation, even without arcing or burning.

- Depending on the opening's size and location, a member of your household could get shocked or electrocuted by accidentally touching exposed live parts.

Often, but not always, you can restore your enclosure. If a panel is fairly up-to-date (not decades old) and otherwise worth keeping, it is possible that a cover can be ordered. Unfortunately, if the cover is unavailable, you need to ask a pro to replace the panel. Visualize this situation: there is an electrical problem, perhaps even someone getting shocked. Someone hurries down to the panel to shut off the circuit. Perhaps the light is off, and perhaps obstacles have been stacked near the panel, though they shouldn't be. It has a makeshift cover or none at all, and the person reaching in to flip the circuit breaker gets electrocuted. Replacement can be worth the expense.

Do You Have Open Knockouts?

Installers bring cables or conduits into panels through "knockouts." When workers remove cables or conduit and leave the knockouts open, they expose far less of the panel than when they leave a cover off, but they do endanger you. The opening most commonly left unclosed in residential panels (as well as other electrical enclosures) is called a "half-inch knockout" or "half-inch ko" (written "k.o." or "ko" but pronounced "kay-o"). It is a circular hole, $\frac{7}{8}''$ diameter, embossed in a side of the cabinet and punched out as needed. You can close it by pressing (well, sometimes hitting) an appropriately sized ko seal in to fill the opening.

Did Someone Install Cable Without Using a Connector?

With charity, I could attribute open kos to absent-mindedness. Carelessness is dangerous when dealing with 120 volts, but I have my moments as everyone does. Not so with the next problem: a cable that enters the panel without a connector. This shows indefensibly ignorant and indifferent work. Circuits fed by such cables should be checked from beginning to end as soon as possible. In even worse cases, someone inserts several cables through a large ko without a connector. Often this leaves an even larger gap.

Your "to do" list needs to include removing all unsecured cables and, after checking their circuits for other dangers, reinserting them through connectors. When nonmetallic sheathed cables are inserted through too-large knockouts, paired "donuts" or "stamped bushings" reduce the kos to the proper size. Donuts may not, however, be used with armored cable or metal conduit. The reason for this rule is that the connection between a metal enclosure and metallic cable or conduit almost always is an essential part of the grounding. When these are stuck through openings in metal enclosures without being well secured to them, grounding is lost. This greatly increases shock hazard downstream. Donuts won't do.

There are three exceptions to this requirement for connectors on cables coming into panels. The first is Square D Company's nonmetallic "Trilliant" electrical panel, which was sold between 1990 and 1996. It is suitable for use only with romex and its cousins, and has clamps that are part of the (plastic) cabinet. Very few electricians bought these panels, but if you have one, it will be marked as a Trilliant.

The second, the one found most commonly, is not actually a cable. A grounding electrode conductor (GEC) ordinarily does not need to enter a panel through a connector. For this to be the case, it must pass through a small hole, $\frac{1}{4}$ inch or $\frac{3}{8}$ inch in diameter, rather than through a full-sized knockout. If it is enclosed in conduit or armor, or is unenclosed but enters through a full-sized ko, it needs a connector. Without a connector, it would leave too large a gap between it and the ko. If it does enter through a regular knockout without a connector, you should make sure the problem is corrected, just as you would if you saw an open ko without the GEC.

The third exception is a relatively rare one. For tidiness, with romex and its cousins, a piece of conduit coming down from the ceiling, usually large diameter conduit, may be attached to the top of the loadcenter, enclosing a number of cables. These are secured to structural elements, usually joists, up in the ceiling. The cables are not attached to the loadcenter; only the conduit is.

Does the Cover Have Dangerous Openings?

Dangerous openings are not always in the cabinet itself. You will find similar openings in panel covers, openings that also need to be closed. These represent spaces from which fuseholders or circuit breakers were removed. There are closers for these, too, at least in the case of circuit breaker panels. They are called panel blanks. They vary, because each line of panels has slightly different openings. If you come upon such openings, you probably will need to find a distributor who carries your brand of panel. You tap in panel blanks the same way you insert ko seals.

Safely Installing Panel Blanks

For the sake of your safety, I recommend that you do not try to insert a panel blank with the cover on your panel. If and only if you feel safe removing the cover, note the location of the opening to be sealed and then remove the panel cover and tap in the blank while the cover is not directly in front of the electrical panel. This way, if the blank passes through the opening instead of seating, it doesn't land inside the panel and contact a live busbar or terminal. Otherwise, you could have a problem—especially with a brand that uses metal panel blanks. *Important Note:* Before you remove the cover, always make sure that there is nothing resting on top of the panel that might fall in. Furthermore, as you unscrew and then remove the cover, make sure you don't lose control and tip a corner into the panel. If it contacts a live component, it could shock, arc, burn, or otherwise do grave harm. Be even more careful of this as you restore the cover. If you don't line it up carefully, the screws won't go in right. What's more, if a panel blank hits a circuit breaker, the blank may be dislodged.

Looking Inside Your Panel

There are more common features in the safety rules affecting fuseboxes and circuit breaker panels than there are differences. None of the dangers you learned to look for in Chapter 5, Safety, are affected by whether you have breakers or fuses or some combination of the two. The same is true of the wiring inside your panels.

The most certain way to stay safe is never to open something live. Unfortunately, it is very hard to make absolutely certain that something is dead, either in your panel, or anywhere throughout your wiring system. For all practical purposes, though, you can indeed kill all power, especially on individual circuits. Killing power to your panel, though, usually requires removing your electric meter. This requires the cooperation of an electrician and of your utility company. The power company does not grant permission merely to allow you to peek inside a panel.

I have not yet met anyone who was bold enough to wire, yet who preferred to kill all power and work by flashlight at certain jobs. One job that daunts some is removing a panel cover, even when the purpose merely is to look within and, at most, to stick a tester in. The shock potential in an electrical panel is not all that different from that to which you are exposed when working inside a live outlet, switch, or appliance. (Still, this does not mean that both are equally safe, but rather that both are similarly risky.)

The Interior

Important Note: As you prepare to look inside, you need to take the same precautions I mentioned in the context of removing the cover to insert panel blanks, and for the same reasons. Make sure that there is nothing resting on top of the panel that might fall in; make sure that you don't lose control and tip a corner of the cover into the panel.

A terminal is a place where a wire is attached to something other than a wire. A lug is a large terminal. To attach a wire to nearly any terminal found in your panel, you strip any insulation off the last half inch or so, stick that inside a hole, and tighten a setscrew against it using a slotted screwdriver. Many lugs require you to use an allen wrench instead. (These are defined in Chapter 11, Materials, and Chapter 9, Tools.) Some very old breaker panels and many fuse boxes use "screw terminals," which require that you attach wires in the manner described in Chapter 12, Connecting Conductors to Terminals. One very exotic, low-grade circuit breaker used "quickwiring," also described there.

Although it probably is not the first thing you will see when you remove the cover, the most central parts of a loadcenter are the buses. "Bus" refers to busbar, a solid, rectangular metal bar held to—but isolated from—the back of the cabinet by supports made up of good insulators. Its function is to distribute electricity, usually to fuses or circuit breakers. There needs to be one busbar (or one busbar per section, in the multisection "split bus" panel discussed later in this chapter) for each incoming hot wire and at least one for the neutrals.

The wires (most often service conductors) bringing power in to a panel or subpanel are attached to lugs that are part of the busbars, located at their ends, or to MAIN circuit breakers, or to MAIN fuseholders. Aluminum wire is common in service cable; aluminum wire is discussed in Chapter 11, Materials. If aluminum cable was used, a responsible installer would have used an antioxidant compound on the end of each incoming wire before inserting it in its lug. If you don't see even a smidgen of dark, jellylike substance where an aluminum conductor enters a lug, make sure you have an electrician check the connection.

How Does Power Get from Busbars to Branch Circuits?

Warning: Like the busbar itself, fuseholders usually are live.

If You Have Fuses

Plug Fuses.

Holders for plug fuses are screwed to the busbars. These connections to the busbars need to be very secure; if not, there will be high resistance, generating heat and causing damage. It is not hard for you to test these connections. Can you wiggle the fuseholders? They should be rock-solid. The screws can get loose but usually it is easy to tighten them. If the fuseholders are not securely mounted, and the screws holding them in place cannot be tightened—or if the fuseholders show damage such as discoloration, pitting, distortion, or broken parts—it is high time to replace the panel.

Cartridge Fuses.

Cartridge fuses are held by spring metal clips that normally are on the hidden side of blank-faced fuseholders that you can withdraw entirely from the panel to replace the fuses. The fuseholders plug into slots in the busbars.

If You Have Breakers

Branch circuit breakers for residential installation almost always are held by spring pressure against busbar "stabs," horizontal extensions from the busbars. ("Pushmatic" breakers are the exception; they are screwed to the busbars.)

One end of each circuit breaker, which is electrically isolated from its interior, is forced onto a long, more-or-less cylindrical, grounded "mounting rail"; the other end, which is connected to the internal mechanism that regulates and

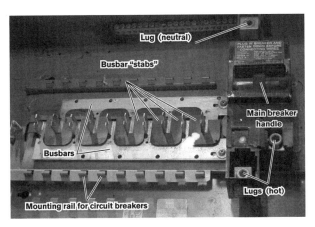

Lug (neutral)

Busbar "stabs"

Main breaker handle

Busbars

Lugs (hot)

Mounting rail for circuit breakers

Panel Interior.

passes the electricity flowing through it, is snapped onto or into the busbar. Some brands have poor reputations because their circuit breakers are not held securely by the busbars. If a single circuit breaker's busbar clip has lost its springiness, replace it. If multiple breakers are not holding well, have the panel replaced. Another reason a brand can earn a poor reputation is that its busbars are not securely mounted in the cabinet. If a loose busbar touched the cabinet you would have many of the same problems as you would if you dropped the cover into the panel, but immeasurably worse. Evidence of either of these problems means that your panel is due for replacement—whether it is 45 years old or 20.

How Are Neutral Connections Different?

Neutral wires are not protected by fuses or breakers, in even halfway modern loadcenters; therefore, the neutral bus does not need to be in the middle of the panel with the hot buses. Instead, it is a terminal strip, and because it is so unlike the hot buses, it is referred to simply as the neutral bar. You will find it off to the side, the top, or the bottom, out of the way of the fuses or breakers and the terminals to which the hot wires are attached.

Like the hot busbars, the neutral bar is a solid, rectangular metal bar held to—but isolated from—the back of the cabinet by supports made up of good insulators. Unlike the hot buses in your service panel, it is bonded to the cabinet, commonly held by a special screw provided with the panel that goes right through the neutral bar and back into the cabinet.

The GEC almost always is attached directly to the neutral bus, too. If the GEC is attached to your panel anywhere else than right on the neutral bar, a UL standard requires that the neutral bar be bonded very solidly to the cabinet, by more than just a screw passing through the bar. This reduces the chance that a serious fault will impose a large voltage between the cabinet and earth ground, which would create a grave risk of shock and of arcing.

Fused Neutrals

If you have one variety of very old fuse panel, you will not find a neutral bar. Instead, both the hot and the neutral wires will be attached to fuseholders. This is bad news, in part because it means that the panel is so old. Worse, there's a unique danger associated with this design: the fuse could blow on the neutral rather than on a hot. This would leave a nonfunctioning device that is nonetheless live and waiting to shock the unwary. This danger is similar to that of circuits msyteriously interrupted somewhere along the way, but this is even more severe.

If for reasons of economy you need to keep a fuse panel of this description, I strongly encourage you at least to have an electrician install a nonfused neutral bar and move your neutrals to it. Failing that, make sure that the fuses feeding the neutral are bypassed. One way to do this is to install 30 amp fuses on the neutrals and 15s on the hot wires. There is no reason for the neutral fuse ever to blow with this arrangement, unless there is a fault of such magnitude that the fuse on the hot side is certain to blow as well.

Most houses wired in the last several decades contain only one panel, a service panel, and it contains a single MAIN. There is no disconnect upstream from it. If you have this type, and only are interested in your system, you can skip the following material, up to "Types of Circuit."

![CAUTION!]

You should be able to kill all power to a subpanel because, by definition, it has a disconnect upstream. But confirm that it's dead; test.

Is Your Panel Wired the Way It Should Be?

All this was about what you might find in any panel. Now you will learn about the different panel layouts, so you can begin to determine whether your neutrals and your branch circuits have been installed correctly.

The first important difference is between service panels and subpanels. Neutrals should not be installed the same way in the two, so you need to check out which type of panel your circuits branch out from. Any panel that does not contain your service disconnect is a *subpanel,* even if it is the only panel from which branch circuits originate, with nothing ahead of it but your disconnect. (Usually, a subpanel is installed to add circuits when a service panel is full, or to bring multiple circuits to a location that is distant from your service panel conveniently.)

A subpanel may be a "Main Lug Only" (MLO) panel, or (though this is less common) it may have one or more MAIN disconnects inside it to shut off power to the branch circuits originating in it. It's still a subpanel. The cable feeding it is a *feeder,* rather than a service cable. NEC Article 215, which applies to feeders,

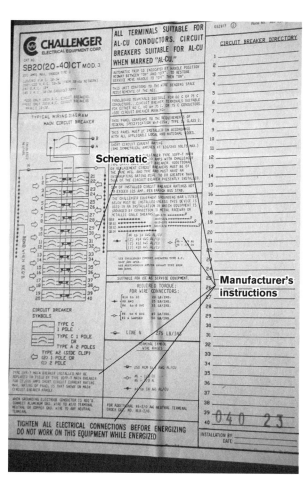

Directory and schematic.

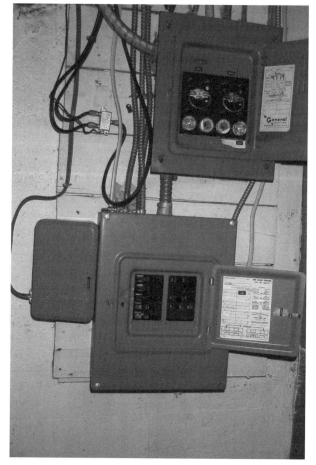

A fused service panel feeding a circuit breaker sub-panel. This approach can add circuits and convenience, inexpensively.

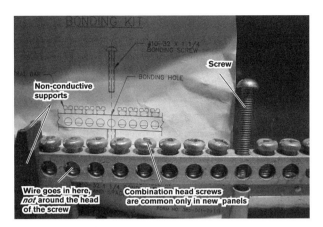

Neutral bar with bonding kit for installation when used in a service panel. The bonding screw should go through the bar into the back of the cabinet. Otherwise, the non-conductive supports would isolate the bar, as required for a *neutral* bar in a *subpanel.*

Add-in grounding bar, needed especially in subpanels.

has different rules than Article 230, which applies to service entrance cables and service panels. You could have nothing but your service panel, or have circuits in a service panelboard plus more in adjacent or remote subpanels. All these possibilities can be legal and safe. There is one important difference between how service panels and subpanels should be wired.

Floating Neutrals

If you have a subpanel or subpanels, ground wires from branch circuits should be attached to the ground bar—or to the enclosure, to which the ground bar is bonded—but never to the neutral bar. In these same subpanels, neutral conductors are attached to the neutral bar, which floats, meaning that it is insulated from the panel itself because the bonding screw or its equivalent is not installed. A neutral in a subpanel should never be attached to the ground bar.

A very common mistake made by amateurs is connecting grounds and neutrals in subpanels. A separate, insulated, neutral wire always must be run between the service equipment and the subpanel.

If ANY ground wires enter the subpanel, the installer must provide a separate means of attaching them to the cabinet. This almost always is a grounding bar. The panel comes with its neutral bar installed, just like the hot buses. The grounding bar generally has to be purchased separately and added to the subpanel. To repeat this important distinction: a bonding means such as a screw normally is provided with the neutral bar but the neutral is NOT to be bonded to the cabinet in a subpanel. Many laypersons—including handymen and "Home Center" salespersons—are unaware of these requirements.

Should You Also Be Concerned About Grounding the Wires Leaving Your Subpanel?

Another problem I have encountered is that sometimes nonelectricians buy grounding bars that fit their subpanels but whose openings are too small to accom-

Downstream Grounds

At one point and only one point—in the service equipment, near where the GEC is connected—the neutral is harmless to the touch because it is at ground level. In the rest of your electrical system, it generally carries electricity, completing the loop from the hot back to the panel, and thus can pose a hazard. The reason for this involves concepts that already have been presented—applied in a new way.

The feeders to a subpanel, including its neutral, use large-diameter wire, which means that they have relatively low resistance. Because low voltage differences at a particular level of current mean low resistance, it also is true that low resistance means low voltage differences. Therefore, with large diameter wire, since the neutral is at ground level at its origin in the service panel, and thus would be quite safe to touch (if you didn't have to stick your fingers so near live terminals to do so), it seems that it should be equally safe to touch at its other end.

However, another principle also applies. While at a given level of voltage, low current indicates that there is high resistance, it also is true that at a particular level of resistance high current means relatively high voltage differences between the two ends of a conductor. By high voltage differences I mean that the neutral at the subpanel no longer is really at ground level as represented by the service panel and the GEC. This means that if you touch the neutral at a subpanel and simultaneously touch a good ground you can get shocked.

Another way of looking at this is that current flowing back, completing the circuit from the hot wires coming from the service to the subpanel, will take every path it can. When the neutral and ground are illegally bonded at your subpanel, the main return path for the circuit will be through the neutral, but another path will be through the ground. Like the neutral, the grounding connection from the subpanel to the service has some resistance. And because of the same principle just described, the current flowing through it will create some voltage difference between it and the GEC. Therefore, if you bond the neutral and the ground at a subpanel, anything and anyone touching the "ground," whether at the cabinet or on a branch circuit fed from that cabinet, can get shocked.

Still, the chance of injury at a subpanel probably is low. If an installer mingles ground and neutral yet further downstream, the risk is greater because the resistance of the return path is higher. There are two historically sanctified exceptions to the rule forbidding downstream grounds. Electric stoves and dryers fed in a certain specified way directly from the service panel can be bonded to the neutral. However, many safety experts believe the exception should be retired, and, indeed, it is on its way out.

Before leaving the subject of downstream grounds, I will mention that some illegal designs can cause them to show up along the branch circuits serving various outlets in your house. I cover this later.

modate the ground wires that need to enter. This may or may not be a cause for concern. If the installer accommodated the difference by cutting away or bending back a significant portion of the strands, or if the wire does not enter the opening far enough for the setscrew to hold against it securely, the wire needs to be removed. It should be moved to another opening that is large enough for it, if one is free. If none is available, it should be attached to a lug that is screwed to the cabinet with a machine screw, through a tapped hole (drilled and threaded by your electrician). Sheet metal screws won't do.

Downstream bonding of ground and neutral is unfortunately frequent. Far more rarely, installers will forget to bond the neutral in—meaning to—the service panel with the panel's "main bonding jumper." Most commonly, this omission simply is the result of forgetting to install a green-headed screw provided by the manufacturer through the neutral bar and into the box behind it. Look for such a screw. This task may not be easy, but it is important. Don't confuse one of the two screws holding the neutral bar to the panel with this (nowadays green-headed) screw.

If you don't find such a screw going through the bar in your service panel, look for a small unpainted metal bar or strap that is stuck through one of the holes in the neutral bar, and at the other end is screwed to the back of the cabinet. This type of jumper can be even more effective. Make sure that it is attached by what looks like a machine screw; sometimes a very careless installer loses the screw and substitutes a flat head metal or drywall screw. If you don't find proper bonding, bring this to the attention of your electrician.

If he or she doesn't find a jumper in your service panel, have one installed. Make this a high priority. An unbonded neutral could result in significant voltage between neutral and ground throughout the system, and even result in grounded parts of the system—exposed parts that you might touch—being isolated from the grounding electrode system. The earlier discussion of "floating" will cue you in to possible consequences.

> This bonding looks different from water-pipe bonding for the grounding electrode system. However, bonding means the same thing in both contexts—making a solid electrical path.

Look very carefully for neutral bar bonding. If you find any of these problems—especially the lack of service bonding—put correcting them near the top of your "to do" list.

Busbar Layout

Modern loadcenters have ONE main breaker (or, in very rare circumstances, ONE MAIN fuse block) bolted onto the hot busbars. The hot (or "live") service conductors enter its terminal lugs, and it feeds power into the busbars for each hot wire. The incoming service neutral attaches directly to the neutral bar, in new as well as in old panels.

> (A reminder: there are some truly antique panels which did fuse the neutral. Almost all of those that are still in use have had that neutral fuse bypassed. This is the only instance where it is appropriate—indeed far, far safer—to do so, whether by literally putting a penny behind a plug fuse, putting a piece of copper plumbing tubing in the place of a cartridge fuse, or employing some tidier alternative.)

A second possibility, still quite legal in one variant, is that a circuit breaker panel could be backfed. This means that instead of the MAIN being bolted to the busbars, usually at the top of the panel, it could be a breaker plugged onto them just like the branch circuit breakers. The incoming hot legs will be connected to it, and instead of power coming out from the busbars through it to feed cables going to a downstream load, power will come through it into the busbars. The way you recognize this design, if it has been installed correctly, is a MAIN legend next to one breaker. This is different from a MAIN LIGHTING legend by a breaker, or a MAIN marking alongside a whole group of breakers.

I haven't seen backfed panels installed in years. (They *are* still around.) The major disadvantage to them is that it is easier for the service or feeder cables to get loose from a breaker plugged into the side of the busbars than from lugs at the top. The second disadvantage, and one that has caused a new rule to be added to the NEC, is that the breaker itself could come loose if it is only plugged in place. Therefore, should anyone use this design nowadays, a special clip designed by the manufacturer must be added to ensure that the MAIN could NOT come loose and risk arcing. If I found this design in a customer's house, I'd keep my eye peeled for other problems that would justify replacing the service panel.

In many older panels (including almost all old fused loadcenters), the busbar for each hot wire had two or (rarely) more sections. This is why they are called "split-bus" panels. One section is called the MAIN section, and the hot service conductors entered connectors bolted directly to it. In a split-bus fuse box, the MAIN section usually has room for two or four pullout fuseholders, each of which could hold two fuses; rarely, you will come across one with six pullouts. In a split-bus breaker panel, this section has spaces for six double-pole (240 volt) circuit breakers. There is no backup, no line-side coordinated protection, for circuits fed from the MAIN section. Ideally, nothing smaller than a 30 amp, 240 volt circuit should originate from it. At the least, it should hold only double-pole breakers, so that there are no more than six handles to throw if you need to kill all power.

The other section in one of these split-bus panels is the branch circuit section. In the MAIN section of one of these old loadcenters, there is one circuit breaker, generally rated at 60 amps, or one fuseholder, generally with 60 amp cartridge fuses, identified as the "Lighting Main."

The NEC uses the term "lighting circuits" to refer to those serving general-duty loads of lights and receptacles. These are distinguished from "small appliance circuits," circuits dedicated either to receptacles in kitchens or dining rooms or to a receptacle or two dedicated to laundry appliances and perhaps ironing. The only other circuits that are not spoken of as lighting loads are those that feed hard-wired appliances such as water heaters and hydromassage tubs; hence the term, "Lighting Main." This may reduce your confusion when you look items up in the NEC.

Cables attached to the Lighting Main are brazed directly to the other (normally, physically lower) sections of the hot busbars. This allows the Lighting Main to serve as backup for the branch circuits originating from that lower section to feed lights, receptacles, and small appliances.

What should you do when that lower section is full, and you want to add more circuits? The correct answer is to upgrade your system, or at least to get an auxiliary panel (unless your panel allows you to "double-up" branch circuit breakers, as described in Chapter 7, Protective Devices). Put another large breaker or fuseholder in the MAIN section of the old split-bus panel, and use it to feed a subpanel.

One wrong, but all too common, way to add another branch circuit is to put the 15-amp, 120 volt circuit for the new range hood in the main section, ignoring its need for backup. This approach is dangerous, but very common. Check for this violation if you have a split-bus panel.

Types of Circuits.   This is a good place to clarify that a "circuit" is whatever is fed by one or more fuses or circuit breakers. Therefore, if you have two cables leaving a panel separately to feed different loads, and they originate from two different fuses, they represent two circuits. If those two cables are spliced together and pigtailed to the same fuse, they simply are one circuit despite the fact that they head in different directions and may serve quite different types of loads.

There are a few other possibilities. If two nonwhite wires, or a white and a nonwhite wire, head out of the panel from two different fuses or breakers, they are a single 240 volt circuit (unless you have an ancient panel with fused neutrals, such as I discussed earlier). If two nonwhite and one white wire head out together, they could represent a combination 240/120 volt circuit, or simply two 120 volt circuits sharing a neutral, an arrangement called a multiwire circuit. A multiwire circuit has a few disadvantages compared to two separate 120 volt circuits with their own neutrals, but it saves on materials. So long as its two hots are fed from the two busbars rather than from a single busbar, the neutral carries the *difference* between the loads the two hots are feeding. This is because the voltages on the two busbars are 180 degrees out of phase. The voltage on one is −120 with respect to ground at the moment when the other is +120. The most common mistake I find in multiwire circuits is that the two circuits have been fed from the same leg of the service. This has the potential of overloading the neutral, because now it will carry the *sum* of the current in the two hots. What makes this overloading especially dangerous is that the neutral is not protected by a fuse or circuit breaker.

Circuit Capacity and Overfusing

The "ampere rating" of a circuit is the number on the fuse or circuit breaker feeding it. All conductors attached to the circuit need to be suitable for its rating. If any of them are otherwise safe, but not large enough, the circuit needs to be "derated." Derating a circuit means that the circuit breaker is replaced with a smaller one or the fuse is replaced with a smaller one and the fuseholder modified so that overfusing no longer is possible. ("Smaller" refers to the number marked, not the fuse's or breaker's physical size.)

Overfusing

"Overfusing" (the term is loosely applied to circuit breakers as well) is a serious fire hazard. It is one of the first things I look for in evaluating someone's wiring.

> The place where wires attach to a branch circuit fuse or circuit breaker is referred to as a terminal rather than a lug.

To uncover this danger, you need to get hold of a sample of #14 wire, a sample of #12, and a sample of #10. If any wire coming from a fuse or breaker whose rating is larger than 15 amps looks more like the #14 than the #12 or #10, it may be overfused. Similarly, if any wire coming from a fuse or breaker larger than 20 ampere looks more like the #14 or #12 than the #10, it may be overfused. The principle can be extended to other sizes, but the most commonly overfused wire is #14. Do note that these sizes refer to the wire itself. What you compare is *not* the diameter of the insulation around the wires.

> Outlets include receptacles, lights, and directly ("hard-") wired appliances. Switches, fuses, and circuit breakers are not outlets—they control electricity but don't draw any.

This completes your introduction to the complex subject of panels. In the next chapter, you will begin to evaluate your system downstream from your electrical panel, improving the panel directory in the process. First, though, comes an exceedingly important lesson.

Lesson: Shutting It Down

One of the first things to learn in dealing with your wiring system is how to shut it off. Shutting a circuit off will mean different actions, depending on the type of panel you have. You may have to

- flip the switch of a switch-type circuit breaker;
- punch the rare, "Pushmatic" type of circuit breaker so the "OFF" legend appears;

- unscrew a screw-in fuse (this is also known as a plug fuse, with subcategories called "Edison-base" and "fustat" or "no-tamp"; or a minibreaker, which looks somewhat like an Edison-base fuse with a pushbutton in the middle), backing it out at least one full turn;

- withdraw a fused pull-out;

- or throw the switch controlling the switched fuse. (Panelboards with switched fuses are old and rare.)

Do take care to read any legends present on the equipment, just in case a switch has been installed upside-down. The NEC allows circuit breakers to be installed such that punching them, flipping them sideways, or flipping them down shuts them off. Having to flip them from down to up to turn them off is the only forbidden design. Especially in older houses, though, someone may have made that very mistake. You don't want to turn something on when you think you're turning it off. Always read and test.

Sequence
It's slightly preferable to shut off individual loads downstream first before turning off the circuit at the panel. So when you can do so without too much trouble, turn off lights and appliances served by a branch circuit before flipping its breaker, and turn off individual circuits before turning off the main. Why? It's easier on the electrical contacts; the breaker will last longer. The same principle holds for disconnecting an appliance. If it contains a switch or control, turn that off rather than just unplugging the cord. This is not terribly important, except in the case of some electronic equipment such as computers and stereos. Certainly, if something is going wrong with an electrical device, yank the plug if that is the fastest way to kill it. And if killing all power in the house is the fastest way to avoid or withdraw power from a shock or fire, PLEASE do it! Killing all power should take an absolute maximum of six hand motions. That principle is discussed next.

Dead Circuits, Deadly Busbars

Service Hazards

In a correctly-wired house, it's easy to make a subpanel safe: identify the breaker or, occasionally, the fused pullout feeding the cable that brings power over to the subpanel, and shut off that main breaker or switch, or pull that fuseholder.

Services are different. Don't make the unthinking assumption that because you shut off your main, you've made it totally safe to work in a service panel. The conductors coming in, and the terminals where they attach, still are deadly. Even removing the cover needs to be done carefully. If you tip a corner of the cover into the panel to where it contacts something live, the result can be very bad. Boom!

In a split-bus panel, shutting off all the (maximum of six) mains leaves the busbars in the main section still ready to bite you. Still, after you shut down your main breakers or throw your main switches or pull your main fuse blocks, everything *else* in the house beyond those disconnects should be dead. Should be.

This assumes that the installation you're dealing with was done legally. On at least one occasion, I have shut off all power in a row house, cut into a cable in the wall, and discovered that it was live. It was fed off the next-door neighbor's system. Far more commonly, I encounter this type of hazard where a larger house has been broken into apartments with separate electric services. Test everything you can test, even when it should be dead.

There is one way to work entirely safely inside your service panel, provided that power first enters the house after passing through your electric meter. This is to remove the electric meter from its base. (This only applies to the modern meters that have a glass cylinder, containing the meter itself, sticking out.) I recommend that, at this point in your training, you leave this approach to a licensed electrician, so I will leave its description to a more advanced volume.

To add circuits in the future, you need to do two things. You need to attach the wires bringing power out and returning it, and you need to add more circuit breakers, if this can be done, or, though it very rarely is possible, to add more fuseholders.

Now that we have a physical description of panels, it is almost time to look at their place in electrical systems. From a loadcenter, various branch circuits head out through walls and ceilings to feed outlets, meaning places where electricity can be drawn for use. The next chapter completes your introduction to your electrical panel. The following one walks you along the outlets, primarily so that you can perform a very important task: labeling your circuits.

7

Protective Devices

Roy sounded perplexed when he called me. "Our family friend Bob has an older panel and there's this strange thing with not one but two fuses in it. Big cylinders. He measured 230 volts. He was able to run an extra 120 circuit, he tells me, by sticking a wire under the screw next to one of the fuses, and he thinks he can get another 120 circuit from the screw by the second fuse. I'm not so sure about this. How can Bob get anything but 230 volts here?"

I helped Roy understand how fuseholders connect to busbars, but I also expressed some concern about the safety of Bob and his family. As I questioned Roy, it began to sound as though Bob might have unknowingly put his family in grave jeopardy by connecting his new circuit in a way that would result in its wires melting and starting a fire, rather than blowing a fuse, should anything go wrong on the circuit.

You will learn three things in this chapter. First, you will learn the difference between the way a fuse and a circuit breaker protects you, and the functional significance of the difference. Second, you will learn to read your fuses or breakers; "blown" or "tripped" may not always be an obvious condition. Third, you will learn to detect two serious but often overlooked mistakes installers make in selecting fuses and circuit breakers for main and subpanels.

Understanding fuses and circuit breakers will be necessary for you to learn about the circuits branching out from them in Chapter 8, The Walk Through.

Relatively modern houses, including just about all houses built today, use circuit breakers. This does not mean that you should swap over to breakers if your panel still is a fusebox, has adequate power and room for your needs, and has not deteriorated. Understanding the operational differences between fuses and circuit breakers will help you decide whether to be satisfied with fuses.

What Protective Devices Protect You From

Fuses and circuit breakers protect against two dangers: overloads and shorts. Overloads occur when normal types of load draw too much current and, for our purposes, not just momentarily. Short circuits occur when leakage—not appliance use—causes significant current flow. Shorts are a type of "fault," or problem, with wires or insulation. The more serious type of ground fault was discussed in Chapter 5, Safety.

What You Can Expect from Fuses and Breakers

Here's a simple but important generality about protective equipment that is easy to forget: newer varieties do not mean newer house wiring. I am called in again and again by homeowners who are dumbfounded by learning this about their houses, which sport spanking-new panels; that's why I emphasize it. Panelboards cannot protect you from all electrical dangers. Older house wiring presents more dangers than old fuse boxes, including more of the dangers that standard protective equipment—old or new—cannot protect you from because it responds too slowly.

Even within its limits, protective equipment does not eliminate electrical problems; it protects you from some of the most serious *consequences* of electrical problems. Whichever protective system you use, you won't be able to restore power if you have a continuous short or severe overload. Moreover, if you do manage to restore power but don't know why your fuse blew or breaker tripped, you may be putting yourself in grave danger. A problem with your wiring may start a fire without again opening the circuit. Be thankful that the protective device operated—and investigate. See Chapter 5, Safety, for a beginning—and don't be shy about calling an electrician.

Characteristics

Interrupting Capacity

There are limits to the protection offered by any equipment designed to shut out danger, whether it is the door shutting out intruders or the fuse shutting off power. The width of your front doorway is equivalent to the size of your service; its ability to withstand a strong intruder corresponds to your MAIN's ability to cut off high current flow. The relationship between your MAIN and the utility is quite unfamiliar and rarely called upon, but essential. Ampere Interrupting Capacity or Rating, AIC or AIR, protects against fire, arcing, and explosion. The rating indicates whether your breaker (AIR) or fuse (AIC) can interrupt the flow of current before the utility can supply enough energy to blow up your house, or at least your panel and everything in its vicinity. Yes, blow it up.

As utilities serve more and more central air conditioners, electric furnaces, hot tubs, and convection ovens, they have to increase the power capacity available for houses in each neighborhood to draw. This means higher-rated transformers and low-resistance power lines. And this means more "available fault current": the ability to pour more and more power in when your house asks for it—or seems to be asking for it.

An older circuit breaker has an Ampere Interrupting Rating of 5,000 amperes, or 5kAIR. In the event of a very bad short in your wiring, or a lightning strike somewhere upstream on the utility line, as many as five thousand amps could begin to flow into your house. A circuit breaker with a 5kAIR rating will quickly interrupt that current—without exploding. If the available fault current is more than 5kA, your breaker or fuse may *not* withstand it, and the energy could do grave harm, so you need a breaker with a higher AIR.

MAIN breakers generally will have higher AIR ratings than branch circuit breakers. This allows the branch circuit breakers to have lower AIR capacities without the risk of their blowing up in the face of a very bad fault. Similarly, cartridge fuses, used as MAINs in fuse boxes, have intrinsically higher AIC ratings than do plug fuses. Nowadays it is quite common for a utility to present residences with 10kA of potential fault current, and 25kA is far from uncommon. If the utility has upgraded its system, you may need to determine the AIC rating of your MAIN or, in the "split-bus" design discussed in Chapter 6, Examining Your Panel, your MAINs.

There are a couple of ways to determine or at least estimate available fault current. First, you can ask your utility. They may be reluctant to calculate it, but they may be willing to give you a ballpark figure. Second, you can simply look at the wiring configuration. If you live in an old residential neighborhood, and the only work the utility has done in the last decade or two is to trim trees away from power lines, change the connections from service drops to services, and replace meters, your available fault current probably has not changed. If, however, they were working on your neighborhood's poles and may have changed transformers, or if you live near a substation, or if there are large commercial or industrial establishments near your house, there is a good chance your available fault current has increased. You *could* get someone to do an engineering analysis based on the length of your service conductors and the utility's drop. I would call this overkill, though, in an unchanged residential neighborhood; I have the formula, but have never gone through the calculations even for my own house.

Installation Instructions

Most fuse-holders and circuit breakers are designed to have only one wire attached. Either the instructions found inside the panel (discussed later in this chapter, under "schematics") or, in the case of a circuit breaker, a legend right on the breaker, will tell you if an exception applies. The schematic is almost certainly the place where you will learn the number and sizes of conductors permitted in each opening of the neutral bar. The breaker itself, on the other hand, most likely is where you will find restrictions on the sizes and number of conductors that may be attached to it. This very often will be in tiny print. Violations of these rules do affect reliable protection; however, they are not hard to remedy.

Usually these rules are violated because there seemed to be no place to add another wire. It often is acceptable to bring another cable into a full panel, and connect it to a lightly-loaded existing circuit. However, unless you have explicit permission to double up, the existing and new wires should be spliced to a third wire, and that third wire alone be fed directly by the circuit. It's called a *pigtail*. (Splicing will be discussed at length in Chapter 12.) To repeat: normally you will attach a single conductor such as a pigtail, not multiple conductors, to the fuse-holder or circuit breaker.

This is more than fussiness. Respecting this requirement is important both to ensure that the wire is held tight and that the protective device (fuse or breaker) operates as calibrated. To make a tight connection, you need to use the connector the way it is designed; to reverse a saying, a round peg may slide out of a square hole. The concern about calibration has to do with the fact that attaching more than one wire affects dissipation of the heat associated with overload—the heat whose build-up the circuit breaker or fuse is designed to detect.

What to Read

If your house is protected by circuit breakers and you expect never to deal with fuses, you can skip the section on fuses and read only the one on circuit breakers. Similarly, if you have fuse protection and never expect to deal with circuit breakers, you can read the section on fuses and skip the one on circuit breakers. However, I recommend that you skim through both. There are factors common to both types of systems that I have chosen not to repeat in both sections for the sake of efficiency. Coordinated protection is an example covered only in the next section. It is a requirement that far more often is violated in the case of circuit breaker panels, and I have not touched on it in this introduction, nor in the section on fuses.

Changing Standards and Grandfathering

Codes and standards are addressed in Chapter 11, Materials. However, I want to explain one principle at this point because shortly I will begin comparing more modern and older designs to a much greater extent than I have until now. Some older equipment designs and older methods—by older I mean even just a few years old!—no longer may legally be installed. The industry comes together to improve its standards every three years. Anything that was installed legally, however long ago, is legal to keep and usually to repair, but whenever you have something not up to today's Code standards you probably have something that's not up to today's *safety* standards. When this turns out to be the case, you might consider upgrading that part of your system when it needs work, even though generally you are not required to do so.

Provided that an older system was installed properly, it remains quite legal so long as it continues to function. Considering that the NEC is changed every three years, this has to be the case, or no wiring would remain legal; an impossible situation.

There are four exceptions to the principle (it's not a formal rule) of grand-fathering. First, "not okay" doesn't become okay merely because it's old and hasn't started a fire yet; grandfathering is not a statute of limitations for bad installations. Second, if something is falling apart and really dangerous, obviously it has to be fixed, and grandfathering doesn't change this. Third—and this is especially relevant to any reader who plans to make changes to a wiring system—once you've made significant changes, what you've worked on no longer is grandfathered but must comply with the standards your local inspector uses to evaluate brand-new wiring. However, in many jurisdictions inspectors are more flexible when dealing with older systems than with new. Fourth is GFCI rules, discussed in Chapter 5.

The principle of "Once okay, forever acceptable" is known as "grandfathering." However, "acceptable" can be far from optimal.

Circuit Breakers

What Exactly Does "Circuit Breaker" Mean and How Does One Work?

Circuit breakers are designed to turn power off and on at the flick of a handle or, in rare cases, the push of a square button. Besides this manual mode, circuit breakers have an automatic mode, protecting you from overloads and shorts. You don't replace a tripped breaker when it trips; you use the handle to reset it.

Other Special Characteristics

A manufacturer designs a breaker marked "SWD," SWitching Duty, not just to protect a circuit but specifically to turn equipment on and off, and to stand up to this use indefinitely, serving in lieu of any other switching device. Breakers without the marking also will serve this function, but manufacturers do not really design them to serve as switches on a daily basis. Similarly, manufacturers design breakers marked "HACR," for Heating, Air Conditioning, Refrigeration, specifically for loads involving the compressors associated with such equipment, in addition to more-ordinary loads. (This does not mean that you need an HACR-rated breaker for your refrigerator, window air conditioner, or furnace.) SWD and HACR markings usually are not on the part of the breaker that is visible from outside your panel.

Some circuit breakers offer additional types of protection. One incorporates GFCI protection, described in Chapter 5, Safety. Another incorporates surge protection. A third protects against arcing faults. The latter normally are high-resistance faults that, in consequence, may draw low current and thus will not trip normal breakers.

Arcing Fault Circuit Interrupting breakers are available today for most modern circuit breaker panels. AFCIs potentially offer good protection from the fire hazard caused by a bad connection anywhere along a circuit. The fire hazard from electric arcs might be especially likely to develop in older houses where electrical insulation is worn, or wiring was done carelessly, and electrical connections make inadequate contact. The breakers, unfortunately, are on the expensive side; the receptacles probably also will be. Most such breakers eventually will incorporate GFCI protection along with the AFCI protection.

Note that "combination" AFCI breakers protect far more closely than "feeder-branch circuit" AFCI breakers.

Why Do People Choose Circuit Breakers?

 **CAUTION!**

If you flip a circuit breaker off, you can as simply flip it back on. However, *tripped* circuit breakers whose handles move to a special position when tripped *cannot* be reset simply by pushing the handle towards the ON position, even when it is safe to restore power to their circuits. First, you have to push their handles into the full OFF position, and then you can reset the breakers back on.

The Myth

Customers sure do love the convenience of circuit breakers. (I do too.) However, I think there's an unconscious but dangerous fantasy: if restoring power is as simple as flipping a handle, then if you flip the switch, and it stays on, you don't need to do any troubleshooting or pay for a service call to repair an electrical problem. As mentioned at the beginning of this chapter, this is a delusion.

The Advantages

Every circuit breaker is designed to be "trip-free." This means that you can hold the handle in the "on" position, and if the circuit is not safe the breaker won't feed it any power. Furthermore, in this case, the handle won't stay in the "on" position once you release it.

Having circuit breakers does mean no more hunting for fuses, no more puzzling over whether a circuit is blown (usually), and no more inadvertent overfusing. Conscious or very ignorant overfusing still is possible, but only to someone working inside your panel, or extending or replacing a circuit's wiring with smaller conductors. "Overfusing," incidentally, is used as a generic term, applied to oversizing both fuses and breakers.

The ampacities of your circuit breakers are unchangeable. In almost every type of residential breaker, ampacity is marked on the end of the handle—the part that would face you if the handle were halfway between the ON and OFF position. With one or two old varieties, though, the marking is on the (narrow) sides of the handle, and one odd variety has the marking on its body.

Almost all circuit breakers allow you to turn a circuit off or on by flipping a handle. On a short or overload, the handle moves either to center or to the off position.

Don't count on breakers working just that way all the time, though. While, in most brands, the

One ancient design, the "Pushmatic," (manufactured by Bulldog, then by I.T.E.) uses a square button that you punch to turn it back on when it is off, and punch again to turn it off, instead of flipping a handle. Whether it is ON or OFF—and OFF is indicated the same way whether it has tripped or you turned it off—you have to look alongside the button for the quite-small letters OFF or ON. You look on either side of the button itself for the ampacity.

handle moves into a position midway between ON and OFF, sometimes, especially but not exclusively in certain brands, the handle will just move a hair away from the full ON position when tripped. Only by jiggling it will you learn that it's not latched firmly in the ON position and thus confirm that it has tripped.

To save users the annoyance of not being sure whether a breaker has tripped, three brands have colored indicators that show when it has. In each case, the handle also moves on tripping and the handle is used to turn the breaker off or on. The indicator is just an extra visual cue.

The first brand, Wadsworth, is long obsolete, but this does not mean you need to get rid of your panel just because it's a Wadsworth. Its breakers are a good design, although incompatible with any other brand of panel. In case the panel cover label is missing, you can tell you have a Wadsworth because the breaker body is brown, and there is a square red button near the middle of the body. When the breaker trips, the handle moves away from the ON position and the button pops forward just enough that you can't mistake the fact.

The second brand is Square D. Its "QO" line, which made its breakers' reputation, has a square plastic window on the front, to one side of the handle. When a QO trips, a red color appears behind this window as the handle moves away from the ON position. If you have a QO panel, no matter how old it is, it can accommodate most modern QO breakers.

The third brand is Trip-and-Light. With these breakers there is no question of whether a breaker has tripped or which one it is. When it trips, the handle moves and a light shines from the breaker body. Trip-and-Light is not a brand of panel; it is an aftermarket manufacturer whose breakers were Classified by UL as working in various manufacturers' panels. The Trip-and-Light instructions specify just which ones.

Comparing Fuses and Breakers:

Will Updating to Circuit Breakers Make Life a Lot Easier?

What's better: a fuse box or a breaker panel? Age is one factor. If you have a fuse box, it is far more likely to be old and not built to current Code requirements. It may contain deteriorated components. Furthermore, besides the fact that fuses are less convenient to change, older fuseboxes may have invited overfusing, which means overheated insulation on your branch circuit conductors. Even this, however, does not tell you whether you should change to a circuit breaker panel at this point. Install fustat adapters in all Edison-base fuseholders and you eliminate the risk of future overfusing.

Four different circuit breaker designs; even if these had the same internal characteristics, you couldn't substitute one for another.

For busbar

For mounting

Circuit breakers from the bottom; this shows what they use to clip or bolt on to the busbars.

Fuses

It may not matter that it takes a little more time to unscrew a fuse and screw a new one in as opposed to resetting a breaker. The major difference between restoring power with fuses and circuit breakers is that you have to be sure you have replacement fuses on hand. There's a far less expensive response to both of the fuse characteristics that irritate people: uncertainty as to whether a fuse has blown, and the nuisance of having to replace it rather than just flipping a handle. The answer is threefold: label your circuits accurately, keep known good fuses on hand in the sizes you need, and throw out blown fuses immediately. Replacing a suspect fuse still is a bit of a nuisance, but it won't take so very much longer than flipping a breaker. So long as your fusebox is good, replacement fuses should be quite inexpensive.

Circuit Breakers

As a circuit breaker panel becomes out of date, however, parts become quite expensive indeed—sometimes even unavailable. Still, I wouldn't go back to owning a fusebox, other factors being equal.

There are a number of reasons to prefer breakers. With breakers, overfusing is much less common. Breakers have a further safety advantage. Multiwire circuits and 240 volt circuits are circuits that feed power to a load or loads from

I received an emergency call from a customer who had lost perhaps half the circuits in her house. First, I noted which circuits were involved. Then I removed those circuit breakers necessary to permit a close look at a length of the busbars. By observing their discoloration, I determined that the MAIN circuit breaker probably had developed high internal resistance. (An alternate explanation that I pretty well ruled out was a bad connection between that breaker and the busbar, or the breaker and the incoming hot wire.) The panel design was modern and the brand of breaker certainly was still being manufactured, although the manufacturer was now the subsidiary of a subsidiary. I contacted the manufacturer's representative to find out about ordering a replacement MAIN breaker. He told me that the panel was easily 15–20 years old; no luck. "Replace the service," he advised.

both busbars. When a "two-pole" breaker that feeds one of these circuits trips, this fully kills both parts of the multiwire circuit, or both sides of the 240 volt circuit. If a 240 volt circuit or multiwire circuit is fuse-protected, normally only one of two fuses will blow. The other fuse still feeds 120 volts to the line, ready to zap the unwary troubleshooter. This is a plus on the side of breakers.

On the other hand, circuit breakers are electromechanical devices that are bound to get tired. When they do, they can decalibrate. This means that they can trip earlier or later than they should. In the worst case, they can fail to trip altogether.

You can reduce the risk that a breaker will decalibrate in two ways. First, exercise your breakers. Turn each one off and back on manually at least once a year. Second, avoid shorts. Some lazy or ignorant people identify a circuit by shorting its wires and seeing which breaker trips. Doing so carries a risk of injury, and also a slight risk of decalibrating a breaker. In addition, it can damage any electronic device that was plugged in to the circuit at the time.

Easily-recognizable tripping and quick resetting are matters of convenience. A distinctly functional advantage to circuit breakers is that, within their design capabilities, the higher—and potentially more dangerous—the current, the faster they trip. This is not true of fuses in quite the same way.

One safety advantage to a fuse is that when it blows, it blows. Then you put in a brand new one. Of course, the holder that you screw or clip it into can get old and worn.

Fuses too can decalibrate, but the consequences are far less severe. You will learn that slow-blow fuses utilize a spring dipped in solder. If the circuit is heavily loaded, near to the point of overload, the solder can soften a bit, and the spring start to pull away. The next time the circuit is heavily loaded, or the time after, even though the current flowing does not exceed the circuit's rating, the spring may pull all the way out of the solder, breaking the circuit. Fuse manufacturers justifiably point out that the inconvenience of premature operation is nowhere as dangerous as failure to trip in time. The latter failure only can occur with a decalibrated circuit breaker.

The fact that fuses, as well as all currently-manufactured residential circuit breakers, respond to heat rise, does mean that both will trip slightly faster when it's hot in the vicinity of the panel, and slightly more slowly when it's cold.

Backup Protection

One problem I commonly find in inexpertly-installed or ignorantly-modified loadcenters in older houses, especially in many that were installed before the 1980s, is a lack of coordinated protection. Coordinated protection is a Code-required type of safety backup to rescue you in case something goes very wrong with your wiring and the fuses or breakers that should protect you don't do their job. This issue is overlooked by many electricians; most nonelectricians, including handymen, are not even aware of it.

MAINs and Branch Circuit Breakers

If a 15- or 20-amp branch circuit breaker (single-pole or two-pole) freezes in the "on" position when it ought to trip—this DOES happen, albeit rarely—there's supposed to be another circuit breaker to catch the ball. In modern wiring this is usually the system MAIN, commonly rated 100 to 200 amps. Sometimes it is a MAIN in the service panel, commonly rated 40–100 amps, that protects a feeder leading to a subpanel. Sometimes it's even a supplementary MAIN located in the subpanel for convenience or added safety. Most commonly this also is rated about 40–100 amps; thoughtful installers usually choose a slightly lower ampacity for it than for the MAIN protecting the feeder at its source. Finally, in "split-bus" panels, described in Chapter 6, Examining Your Panel, it may be a "MAIN LIGHTING" breaker, usually a 60 amp one, protecting a subsection of the service panel from which those 15- and 20-amp branch circuits are fed.

Having a larger rating than the branch circuit breaker, normally a MAIN just sits in the background and lets the lower-rated breaker handle any problems. However, it is just waiting to trip in the event that the fault current flowing gets very large—hopefully, before your house starts to burn. Such backup is rarely needed . . . but when it's needed, it's *needed*. A circuit breaker marked "15," which should trip within a short time if overloaded with a 30- or 50-amp draw, may not do so if it has decalibrated or otherwise gone bad. In this case, as the current flow increases to 60, 80, 150 amps, exceeding the ampere rating of the MAIN, the MAIN eventually trips. This shuts off all the branch circuits, which is an inconvenience and potentially presents some hazard. However, this is far better than a fire. You do not have equal need for a safety net with fuses because decalibration cannot prevent a fuse from blowing, or slow the action. (It can happen in a fuse *box*, however, if your protection consists of mini-breakers or if your fuses have been bypassed with solid metal.)

Going Upstream There is one way that both fuses and circuit breakers can be installed that results in loss of backup, and several ways that are unique to circuit breakers. It is unlikely that installers bypass the safety feature of coordinated protection on purpose, because bypassing it is so risky, but it is easy to do so out of ignorance.

The most heedless way people bypass protection is by shoving another wire in with the wires coming into the house from the meter, upstream from the service MAIN or, in the case of split-bus panels (described in Chapter 6, Examining Your Panel), the MAIN section. Somehow they manage to do this without getting shocked so badly that they are unable to complete the operation. Bypassing protection saves them the price of another circuit breaker or fuse, or at worst of installing a subpanel, but they're risking meltdown. They're ignoring the manufacturer's instructions and basic safety rules.

You can check for such bypassing (see Chapter 6) when you open the panel containing your MAIN, and, if they are separate, the enclosure of your system disconnect. I've found it again and again. If you do find such bypassing, do not put off getting it corrected. While it is present, you have a real fire hazard: not only does that circuit lack coordinated protection, in every case I have encountered the misfed wire was very much overfused as well.

All of this applies whether you have fuses or circuit breakers. Before closing this chapter, there is one more large topic that is important for readers who have circuit breakers or who may want to deal with them in the future. Why just circuit breakers? The reason is that while fuses differ somewhat in shape and design, circuit breakers are far more incompatible. Fuseboxes vary in configuration and design from one manufacturer to another; however, the fuses you put in them differ little. Breakers, on the other hand, differ a great deal from one brand to another. Moreover, two breaker panels that superficially look similar may have very significant internal differences.

Schematics

Location

One of the first things to look for inside the panel is the schematic diagram, previously mentioned in Chapter 2, where you tried to determine the amount of power coming into your house. (For our purposes I use "schematic" to refer to the sketch and the accompanying instructions together.) On occasion, you will notice the schematic located with the circuit directory pasted on the inside of the panel door. Normally, though, it will be further inside, on a side wall of the wiring compartment accessible only after the panel cover is removed. Depending on how crowded with wires the panel is, the schematic should be at least somewhat visible after you remove the cover.

Access

You will discover that panels tend to fill up, not only with fuses or breakers, but with wires. One very unfortunate consequence is that they get most assertively in the way of schematics. Do your best to read them, but don't take risks that frighten you, for instance, by taking the chance of pushing against something live in the process of forcing wires out of your line of sight.

Directory and schematic.

Initially, any electrician with pride "dresses" the wires. To "dress" them is to install them tidily, bending them in square turns that follow the shape of the box (which tucks them out of the way along the edges), and sometimes even using plastic ties to bundle them. The result is that there is clear space around the periphery and easy access to terminals.

That's the ideal. Now imagine a wiring space filled 100% with wires and splices, so that not only can't you see past them, but you might not even be able to pour water through. Take that image and subtract some of the wires and splices, leaving 75% of them in place. Impossibly, insufferably crowded? It's legal!

What Schematics Tell You

There are six reasons to check your panel's schematic and instructions to determine the suitability of the circuit breakers' installation or find out which circuit breakers to install. Leaving more extensive explanations for the next few pages, you'll look for the following: breakers in the wrong places, breakers of the wrong ampere ratings, damaged or dead breakers, breakers of the wrong brand, the need to add circuits, and inadequate AIR ratings (I addressed AIR on page 80). Any of these could require selection of circuit breakers; you won't find it difficult, at least in concept, to install a circuit breaker. It may not be at all obvious that a breaker is in the wrong place; however, it certainly can be dangerous, and it is common because so many overlook the issue.

> Some years ago, I was consulted in connection with a product liability lawsuit. A circuit breaker panel (not a residential one) had blown up in someone's face. Could it have been defective? "Yes," I replied. Could the installer have been ignoring the manufacturer's instructions? "Yes," I had to agree again. What's more, it could have blown up in the poor fellow's face because some *previous* worker had ignored the instructions. Read them not only for your safety but also for others.

Configuration

The schematic tells you several important things. The first is the busbar layout. One issue arises with "split busbars," which are discussed in Chapter 6, Examining Your Panel. In some older panels the busbar connected to each hot leg has more than one section. The schematic indicates which busbar sections are main, or unprotected, and which are for branch circuits. While this issue may not be familiar, it is highly important in the context of coordinated protection, or backup.

What Breakers Go Where Whether in the context of split-bus panels or others, the busbar layout determines what kind of breakers, and how many, go where. Even within one brand and style of panel, there are different varieties of breaker, and different locations where even breakers approved for use in a particular panel may safely be installed.

Here's an example. "Two-pole" breakers occupy two spaces in the panel and two openings in its cover. Connecting to each of the two busbars, two-pole breakers make 240 volts available. This is useful for both 240 volt circuits and multiwire circuits. Two-pole breakers may have either a single handle or two, permanently-connected, handles.

In contrast, one "tandem" or "piggyback" breaker, or, in some brands, two "half-size" breakers, put two breakers in the space normally occupied by one. They feed from the same busbar. There would be no reason for a circuit to have one conductor attached to each. It will not make 240 volts available because there is zero voltage between the two breakers incorporated in a tandem.

A further complication is the use of handle ties. These are manufacturer-supplied components that connect the handles of two adjacent 120 volt breakers—not necessarily full-sized ones. (The use of two-pole breakers and handle-tied breakers to control multiwire circuits will be discussed in Chapter 8, The Walk-Through.)

> Ignorant or lazy installers may have substituted nails or other makeshifts for the handle ties. This is NOT okay. The nails should be replaced, and the panel checked for other instances of corner-cutting.

If the instructions say, "Do not install any breakers in positions 1 and 3," and you find any there you

⚡ **CAUTION!**

Sometimes a simple confusion can take on the appearance of a far-more-dangerous error. If the installation shows errors at spaces 1 and 3, or 17 and 19, check for the possibility that the panel cover was installed upside down. This could make the erstwhile spaces 1 and 3 actually spaces 18 and 20, for example, and spaces 17 and 19 actually 2 and 4.

have quite a problem. If the panel says (either in words or symbols), "Put half-size [or tandem] breakers only in slots 12–18," you shouldn't find any in other locations. If you do, you've definitely found something for your "to correct" list.

The panel may restrict the use of tandems to those spaces for reasons of heat dissipation or busbar arrangement. The purpose of using tandems is to squeeze more circuits out of a panel. The manufacturer may have determined that the panel can support and withstand only so many circuits, in terms of heat or wiring room or some even more critical characteristic.

Variations Within Brand

You cannot reliably predict a panel's characteristics or restrictions based on the brand. For instance, in some QO panels not all of the spaces are intended to accommodate tandems, and the style of tandem specified on the schematic actually will not fit in the other spaces. You still may find that such a breaker had been shoehorned in place with the injudicious use of a saw or metal snips, or that a type of tandem not intended for the panel was blithely popped in. If tandem circuit breakers have been installed where the schematic says only singletons belong, some ignorant person has in effect decided that he or she knows more about the panel's capabilities than does the manufacturer or the Listing laboratory.

Should you come across any of these errors, you need to correct it. If a 120 volt branch circuit breaker was located in the MAIN section, you need to blank off that space and move the wire that was attached to it to the branch circuit section. Usually when you encounter this error, there are no empty spaces in the branch circuit section. This may require that you add tandems or half-size breakers in the branch circuit section if legal. Otherwise you may need to pig-tail the wire to another, combining two circuits, or even replace the panel or install a subpanel (jobs beyond the scope of this book). In any of these cases, you will need to update the circuit directory to incorporate your changes.

Overfusing

A breaker of the wrong ampacity does not sufficiently protect the wire leaving it. For example, you might find that a 20 amp breaker feeds #14 wire, in violation of NEC Section 310–15. You correct this overfusing by replacing the breaker with one of the right rating; in this example, a 15 amp breaker.

Malfunctioning Breakers

A dead breaker, one that will not reset and show voltage even when you remove the wire attached to it—so there's no load available to trip it—needs to be replaced. If there are no illegalities involved, simple substitution will do the job. I also have encountered breakers whose handles were locked in position from the inside so I could not budge them from the ON position, breakers with broken handles or bodies, breakers with scorch marks, and breakers that were so encrusted with paint that I could not guess at their ampere ratings. I have even come across breakers whose handles had been painted over to the point that the handles could be moved only sluggishly.

One of the most dangerous breaker designs ever identified was a two- or three-pole version of Federal Pacific (FPE) breakers whose internal tripping mechanism sometimes jammed. There is no visible clue when this happens. The experts have proven the problem to exist with absolute certainty only in FPE's commercial breakers, but its discovery helped destroy FPE's reputation.

One of the most dangerous malfunctions you can encounter is a breaker that will not turn off. While you don't want to test-trip a breaker by shorting, you certainly can turn it off manually and then test for voltage at its load terminal or terminals.

Additional Breakers

There are two standard reasons to add new breakers. One is that you want to add lines to feed newly run circuits, and the other is to separate circuits, usually ones that are overloaded. Running cable is beyond the scope of this book. There are, however, two instances in which you may want to separate circuits right in the panel. First, sometimes you have multiple incoming wires illegally attached to a breaker that permits only one wire. Second, on rare occasions you have wires pigtailed to a single breaker inside the panel, but there happens to be room for one more breaker, perhaps by installing a tandem. Remove one of the wires from the double-wired breaker, and treat this wire as though you had removed it from a branch circuit breaker that had been located in the MAIN section.

 CAUTION!

If you are a beginner to wiring, actually working in your panel, as opposed to simply examining its contents, is beyond the level of task you can count on being able to do safely. If you do have the background to be certain you can handle this work, it still needs to wait until you have reviewed the material right up through the end of Chapter 14, Replacing Receptacles. At the end of that chapter, I explain a procedure for removing and inserting breakers and the conductors attached to them.

Picking the New Breakers

Does it matter what brand of panel you own, and what brand of circuit breaker you use? Yes and no. Different manufacturers have different reputations, but the reputations are as likely to result from the panels' design as from the breakers', so your substituting a compatible breaker from a more reputable brand probably won't improve reliability. Some electricians think poorly of Gould/I.T.E., but while I.T.E. breakers are widely interchangeable, I would not bother replacing them unless I had to.

Interchangeability. If you substitute whatever circuit breakers will physically fit in your panel, you will be engaging in a Code violation, albeit one that may be ignored even by conscientious installers who are familiar with the issue. Here's the problem: even when distributors advertise circuit breakers as "interchangeable," the manufacturers into whose panels the breakers are to be installed often disagree.

Panel manufacturers have threatened not to honor their warranties when others' breakers are used in their panels. As of this writing, such threats by the original equipment manufacturers have not been put to test in court.

Some aftermarket breakers are Classified (certified) by testing laboratories as compatible with various panels, as opposed to merely being sold as interchangeable. "Trip and Light" manufactured circuit breakers whose Listing classified them to be installed, quite legally, in a host of panels, as indicated on the circuit breaker package. Thomas and Betts (T&B) does so nowadays. Donna, a technical representative of T&B, pooh-poohed the idea that tripping curves of physically interchangeable breakers differ. She said that when T&B paid UL to Classify T&B breakers to fit in the panels of all the "interchangeable" brands, they didn't require any modification. One thing that increases her credibility is that many brands of "interchangeable" circuit breakers have changed manufacturers several times. When a company sells breakers with differnt brand markings but with the same configuration, I find it very hard to believe that the manufacturing machinery is modified for identical-appearing breakers with different labels.

There is more to determining physical compatibility than merely looking at the front of a breaker or even its general shape. You need to look at the back as well. The part that clips onto the mounting rail and the busbar must conform to the shape of the space between them. Even the same brand has used different, incompatible designs in different panels. See the photo on page 85.

Even if you prefer to play by the book, you may have to assess compatibility; your brand of panel may be extinct. Almost all extinct panels will accept substitute breakers, but there are exceptions. Wadsworth was a highly-reputable brand that is extinct and also incompatible with other brands' breakers. Two other extinct brands, ones distrusted by many electricians, are Federal-Pacific and Pushmatic/Bulldog, the latter at one point manufactured by I.T.E., which may have gained a bad reputation among some electricians simply because it once manufactured Pushmatics. You have no choice but to find breakers that are "interchangeable" or "Classified" as compatible when you need a new breaker in one of these brands—unless you want to undertake a major search, or replace your panel.

Fuses

Fuses are the older type of protection and, to the surprise of some, they are safer in certain ways that I discuss later in this chapter. The essence of a fuse is metal that literally fuses, or melts, when overheated. However, that is the primary means of operation only of "fast-acting" or "single-element" fuses, such as you find in "in-line" fuseholders (capsules with one wire at each end) used today inside a car or computer. You may have such antiques in your fusebox. If so, I recommend replacing them with "dual-element" (also known as "slow-blow" or "slo-blo") fuses, which offer almost the same protection and are much less likely to blow when it's not really necessary. There are screw-in ("plug") fuses, which are readily replaced, and cylindrical ("cartridge") fuses, which normally protect heavier loads or provide backup protection. These tend to blow less frequently than plug fuses because they do not commonly protect branch circuits.

Types of Fuses

Plug Fuses

A plug fuse has a glass window in front through which you can see a break in the metal "element" inside (the element is the metal piece that melts), when the fuse "blows" (interrupts the circuit) on a short. Often when it blows on a short you also will notice blackening of the window. When a quick-acting fuse blows on an overload, the mechanism is the same as when it blows on a short. When a dual-element fuse is overloaded the process is different; this is why it pays to have a dual-element fuse, unless the need for quick protection is very great and the specific load is known. The dual-element fuse contains an internal spring stretched so its end dips in a hard dot of solder. The solder is part of (in series with) the circuit. When the fuse carries too much current, the solder starts to heat and soften. If the overload is momentary, the solder cools and firms back up. If it persists, the spring pulls out and interrupts the circuit.

Fuse openings

Busbar stab-power connection

Fuseholder from side-front.

Fuseholder module from top.

Cartridge Fuses

Cartridge fuses are opaque cardboard cylinders with metal end caps, and some-times with flat metal extensions on the ends of the caps. Except in the case of some unusual modern designs, there is no visual sign when they blow. In homes, they are found mostly in black pullouts. These are square-front bakelite (a nonconductive material) with metal "pull-tabs" allowing you to withdraw them easily. Pullouts, which also come in a version that does not include fuses, are plugged in to the live panel, serving as disconnects—the equivalent of switches. The fuseholder version has clips to hold two cartridge fuses which are part of the current path through the pullout, from the LINE to the LOAD side of the pullout. This means that power is interrupted either when a pullout is pulled out or when one of the fuses it holds blows.

Clearly, a cartridge fuse is safe to touch once you pull its pullout completely out of and away from the panel. Those that are not in pullouts *should* have been disconnected from power once the switches controlling them, which in modern design are the latches of the doors concealing them, are released—but always test before trying to remove any fuse that remains inside a panel. (Voltage test-ing is covered in Chapter 3.) Cartridge fuses are removed by pulling, which sometimes requires a fair amount of force. Reinserting them is easier because you are pushing. If you do need to purchase a replacement, bring the old one along to make sure that you get one of the right shape, as well as ampacity. If you prefer, you can bring the pullout with you instead.

Cartridge fuse assortment and pullout fuse block.

Cartridge fuses are rarely used on circuits other than those feeding heavy-duty 240 volt equipment. They may be used to protect your panel, a section of your panel, a subpanel, an electric stove or dryer, a water heater, or a large air conditioning or space-heating load. You also will find them in separate fused disconnects such as sometimes are installed at the outside compressors for central air units. You'll sometimes find this last use even at houses with circuit breaker

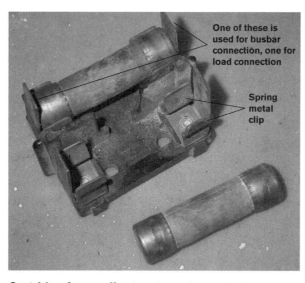

Cartridge fuse pullout with one fuse pulled out.

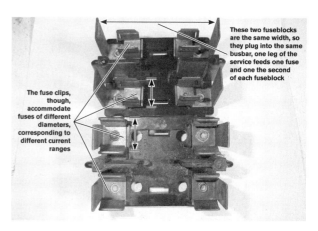

Two fuse blocks, same width for same busbars, but with different clips for different cartridge ampacity ranges.

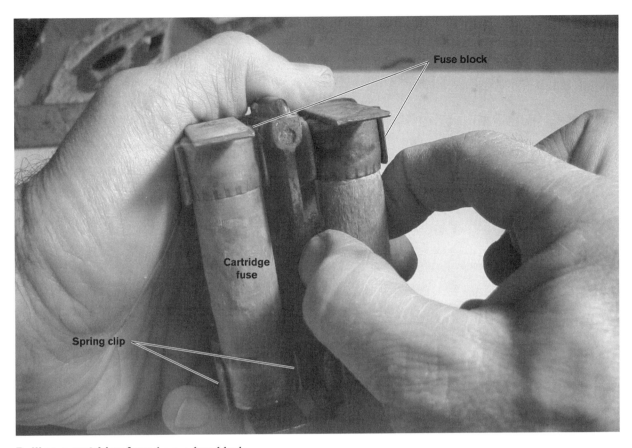

Pulling a cartridge fuse from a fuse block.

 **CAUTION!**

The label on a compressor that specifies, "Protect with maximum 30 a fuse" does *not* mean "protect with maximum 30 a fuse *or circuit breaker.*"

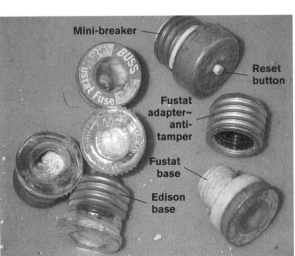

Fuse assortment, including fustat, Edison base, single-element, fustat adapter and mini-breaker.

panels. This can either be because fused switches can be less expensive than circuit breakers, or because the equipment itself specifies fuse protection—some do, and that's the manufacturer's decision. One special advantage of a fused disconnect is that fuses are readily available in a wider range of ampacities than circuit breakers. Alan Nadon, a very experienced electrician and Chief Inspector for Elkhart, Indiana, protects his sump pump with a six amp fuse. The 15 amp breaker that protects the pump circuit would overfuse the pump motor itself, allowing it to burn out if it stalls—without any violation of the NEC.

Protection Against Overfusing

Plug fuses are far more common than cartridge fuses. These normally protect branch circuits: 15 amp, 20 amp, rarely 30 amp for a large air conditioner, or, more likely, a clothes dryer. They come in two diameters. Edison base fuses are as fat and long as a standard light bulb's base. Fustats are skinnier, the diameter of a "candelabra" light bulb's base. The length of a fustat depends on the fuse's ampere rating. This makes fustats noninterchangeable. All Edison-base fuses are interchangeable. This is why Edison-base fuseholders are a thing of the past; they make it easy to overfuse, which means to install a protective device that allows too many amps to pass unimpeded, leaving a home under-protected from overloads.

New installations of plug fuses must have fustat adapters in the fuse-holders. You will find such installations mainly in single-circuit "subpanels" used to feed equipment that specifies fuse protection. Usually these subpanels are right by the equipment. A moment's thought will show you that so long as there is a fuse immediately ahead of the equipment, the manufacturer's instructions have been complied with. It has to be irrelevant whether there is a circuit breaker farther upstream. Incidentally, these fuseholders almost always are either the cartridge type or the Edison-base type. In the latter case, it is worth your while to read the equipment's nameplate and check that it hasn't been overfused if a fustat adapter is not installed.

Here's something especially relevant to the legality of old wiring systems with fused loadcenters. When a circuit has been overfused, anyone interested in obeying the NEC installs the appropriately-sized fustat adapter in the fuseholder to prevent future overfusing. As an electrician, I certainly do so to make sure that the circuit will be adequately protected after I leave.

IMPORTANT NOTE: If you find that one of your fuses has been bypassed, I suggest calling an electrician. Unless the fuse that has been bypassed is in a pullout, you may need to shut off all power in order to remove it safely. Furthermore, it strongly suggests exceedingly ignorant or indifferent behavior on the part of some previous inhabitant or worker. The only exception to this is when an antique panel had fused neutrals. As I mentioned in Chapter 6, bypassing neutral fuses can be a good idea.

While it is true that ignorance of the law is no excuse that will hold up in court, it is evident from my experience that homeowners often are ignorant of this requirement. Here's what's so dangerous about not installing a fustat. Suppose that a fuse blows, and someone installs a 30 amp fuse to protect a 20 amp circuit. This may be a simple oversight, or perhaps the 30 amp fuse is the only one handy. Sometimes the larger one is installed because it "will work better" (i.e., it won't blow). The latter practice approaches that of purely bypassing a fuse—an exceedingly dangerous practice. Most commonly a fuse is bypassed by putting a coin in the fuseholder behind a plug fuse. Less commonly, this gamble is practiced by replacing a cartridge fuse with a conductor such as a section of copper plumbing tubing.

Overfusing and bypassing can cause fire. Substitution of the wrong-size fuse or insertion of a coin or slug behind a fuse is very, very difficult to achieve with a fustat in place. I encourage you to modernize Edison fuseholders, making them safer by inserting fustat adapters.

There are two parts to the task. One is simple: screw the new fuse into the adapter, and then screw the combination into the Edison-base fuseholder. One is more challenging: choosing the adapter. If the fuse feeds receptacles and lights here and there in your house, a 15 amp fustat always will be a safe choice. If, however, you want the fuse to have a higher rating, you will have to make sure that it will be legal and safe. In most cases, the other rating that might be legal is 20 amps. As you'll learn, the two most common wire sizes are #14, for 15 amp circuits, and #12, for 20 amp circuits. To protect a circuit at 20 amps, you must make a reasonable effort to confirm that no #14 wire is connected to that circuit. (What's a reasonable effort? I can't evaluate this for you; you have to weigh the effort it takes against the security you will achieve.) I have worked on older houses where a circuit left the panel in #12 wire, but was spliced into #14 at a junction box. You are very, very unlikely to find this in the exceedingly rare modern house that employs fuses; it is, however, quite common in older houses that have been worked on by many people—many of whom were not electricians—over the years. If in doubt, you can play it safe with 15 amp fuses. (This can result in one technical Code violation concerning "small appliance circuits," but in an old house this is far less important—any inspector will agree—than avoiding overfusing.)

Determining Whether a Fuse Has Blown

When a fuse-protected circuit seems to be dead, how do you know whether its fuse is blown without opening your panel? With a plug fuse, there are three visual clues. You can see a gap melted in a fuse link; you can note blackening of the window that lets you see into the fuse; or, with a dual-element fuse, you may recognize the shortened spring, mentioned above. (Sometimes you also can tell because something rattles inside when you shake it. This, however, is even less reliable a test of fuses than of lamps.) None of these characteristics is present in a blown cartridge fuse.

An ohmmeter or continuity tester will tell you unequivocally whether power can pass through a fuse, or whether it is blown. Follow the procedure I show in Chapter 9, Tools. Note that any electrical test can give misleading results because of poor contact—including that caused by not screwing a fuse in tight enough.

Here is a procedure for dealing with a blown fuse, which I present in the context of the fused switch feeding a heat pump compressor:

1. Kill power to the switch at your panel. You will learn how to identify the circuit in Chapter 8, The Walk-Through.

2. Open the switch, and test everywhere inside with your voltmeter to make sure that you have killed the right circuit, following the procedure I describe in Chapter 3, Testing Voltage.

Minibreakers

Minibreakers are shaped like Edison-base fuses, and turned off by unscrewing like plug fuses, but are not fuses nor, exactly, normal circuit breakers. (Their action is like that of a very primitive circuit breaker, such as will—rarely—be found in car wiring, but never will turn up in houses' protective systems.) Minibreakers do not have inherent time delay like dual-element fuses and circuit breakers; they respond strictly to a magnetic field created by too much current. Resetting a minibreaker merely requires punching a button that pops forward when it trips. This is much more convenient than changing fuses, but minibreakers do not protect the same way as fuses or circuit breakers. Of course, nothing—fuse, minibreaker, or circuit breaker—can usefully be reset or replaced before the fault that caused it to trip has been corrected.

Fustats and circuit breakers are safer than minibreakers because of minibreakers' Edison-base configuration. This configuration prevents the installation of fustats and thus potentially allows overfusing. On the other hand, there is no *good* reason someone would replace a minibreaker when it trips. There *is* a bad reason: someone who doesn't realize that failure to reset probably means the circuit is faulted may conclude that the minibreaker "doesn't work." Consequently, such as person may replace, for example, a 15 amp minibreaker with a 30 amp fuse. For this reason, the minibreaker really ought to be treated like an Edison-base fuse and not be used to protect a circuit that previously has been overfused.

3. Remove the fuses, and test each one for continuity, following the procedure I will explain in Chapter 9, Tools.

4. If any fuse fails the test, replace it.

5. Close the switch box and then restore power.

6. See whether your equipment works, if necessary adjusting a thermostat control to an extreme temperature setting to make a heating unit, for example, kick on. Be prepared to wait for a few minutes; sometimes these have a built-in delay. Because fuses rarely blow without cause, if your replacements blow, it is time to have the equipment, or the system feeding it, checked out.

THROW OUT blown fuses, even though that's not on the top of your list of priorities at the time they blow. It's good to keep spare fuses handy, and if you leave the dead ones around, you may mistake them for good fuses.

8

The Walk-Through

Blueprinting Your System

It was 9 p.m. and suddenly the TV quit. No fuse looked blown. Pete went back upstairs and tried plugging a light in the outlet that had fed the TV, but he got nothing.

What should he have done?

- Go back down and look harder for a blown fuse?
- Start changing fuses, just in case he'd over-looked the blown one?
- Just go to bed?
- Call an electrician to come out, at double-overtime night rates?
- Holler for the fire department?

None of the above. Pete is unusual in that he knows which circuit feeds each outlet in his house, including his living room receptacles. This means that Pete could have simply replaced the fuse, if he had a spare—and Pete's the sort of person who does keep spares on hand in each size he needs. He didn't do that. Why? He knows that fuses don't blow without cause. Pete unscrewed the fuse a couple of turns, breaking the connection, and went to bed, feeling secure that solving the problem could wait till morning.

Do You Know What Each Circuit Feeds?

You may not be as well prepared as Pete. If you look at the directory pasted to your electrical panel and mostly what it says is "Plugs and lights," you could be in trouble when an electrical problem arises. This chapter will teach you to correct lackadaisical labeling, whether of the "plugs and lights" variety or some other. It is important to label correctly both your service panel and any subpanels, for several reasons.

Why Fuss about Labeling?

Why take the trouble to make sure everything is labeled correctly? If you have one or two overloaded circuits, you *know* which fuses keep blowing. If this is not the case, how often do fuses blow? Even if this is the case, how often do *other* circuits blow fuses? There are many reasons to heed the NEC requirement that all your electrical circuits be accurately labeled at the fusebox or breaker panel.

- Circuit directories enable you to deal quickly with problems and emergencies. If a fuse blows or a breaker trips, you will be able to figure out what to unplug or turn off in order to remove the short or overload so that you can restore power.

- If a more complicated electrical problem arises, you don't want to have to shut the entire system down for safety. You want to kill just the circuit involved, as Pete did, so that you can continue using your house without worry until the problem is resolved. You cannot selectively shut off a circuit or circuits if you have no way to know which might be involved.

- In the worst case, if a fire starts, you may need lights for safe evacuation even if you shut everything else down to remove the electrical source of a fire—or to prevent a fire from becoming electrical. You cannot selectively leave one or two circuits on if you have no way to know which circuit is which.

- You won't have "mystery outlets." A colleague complained to me that he had to move roomfuls of furniture before finally locating a hidden outlet the customer hadn't even known about, an outlet that contained a broken wire that was interrupting power to a receptacle that was in daily use. This scenario is not unusual.

- If part of a circuit goes dead, you won't need to pay emergency rates for the electrician to spend time back at the panel identifying which circuit the problem is on.

⚡ CAUTION!

Look everywhere. I've even found a subpanel (installed decades earlier, in an illegal location—under the sink!) that the homeowner did not know he had.

I have spent many expensive hours opening various outlets at some customers' houses in attempts to determine where a circuit had been interrupted. I never have found a wire broken *inside* a cable or conduit, though this could happen. It always is possible that a wire was broken somewhere inside the walls at a splice that is at least twice illegal; once for not being in a box, and once for being buried in the wall. (See "The Case of the Shocking Stove" in the next chapter.) However, it is most probable that both he and I had overlooked the outlet responsible, thinking it was on a different circuit entirely. Nails and screws going into walls do penetrate cables, albeit rarely; however, they tend to result in blown circuits or, in the worst case, live nails or screws, not just dead outlets. And when a nail penetration does blow a circuit, you do know which circuit your dead outlet is on.

How Is Labeling Useful Even When You Don't Have an Emergency?

You gain a number of additional benefits from labeling. First, if you want to shut down power for some other reason, for instance, to replace a receptacle or to scrub or paint in its vicinity, you now will know which circuit to open. Second, if you want to know whether an existing circuit can support a new appliance safely, you now know what loads it already serves. This makes it easy to calculate available capacity.

Preparing a Punch List

There is a third, major advantage to undertaking this project that does not involve preparedness for emergencies. This is that labeling will require a comprehensive tour of your house's wiring, which offers benefits well beyond the fact that you will be able to find all your outlets more easily in the future.

An electrically-oriented tour of your house will enable you to prepare a "punch list" of projects that you need to get done, much like the projects you may have identified in your surveys of your system from the utility connection through your electrical panel. In the course of your survey you may find prob-

Calculating Spare Capacity on Circuit

If you want to plug in a tool for a few minutes, you don't need to do any calculations unless it draws a very heavy load, which usually means that its purpose is heating (like a hair dryer). Just try it, so long as it is not going to bother other loads on the circuit with electrical noise and so long as the circuit does not serve any loads that will be disturbed by the circuit tripping—for example, some dimmers fail if exposed to strong electrical aberrations.

If you want to attach a fixed appliance, first read the instructions and name plate. If it says that it must be on its own circuit, you don't need to do any calculations; just follow the instructions. Otherwise, there's one thing to watch for before you start calculating. A fixed appliance that may be on for extended periods should not draw more than half the circuit's capacity. This means you need to be careful about adding a large air conditioner, for instance, to a 15 amp circuit, regardless of other loads on the circuit. If it draws 8 amps or more, run another circuit to the location.

If neither of these issues applies, go ahead and make a list such as this one.

Circuit ampacity:	15 amps
Present loads:	
4 ceiling lights, each with two lampholders rated maximum 60 watts:	
120 watts ÷ 120 volts = 1 amp per light.	Total: 4 amps
Receptacles feeding stereo	2 amps
Desk lamp marked 100 watts max.	100 ÷ 120 = less than 1 amp
Halogen torchiere, 300 watts	300 ÷ 120 = well under 3 amps
Answering machine marked 800 mA	Less than 1 amp
(mA = milliamperes; "milli" means thousandth)	
Electric pencil sharpener	Negligible
Total present load:	Less than 11 amps
Available:	15 − 11 = at least 4 amperes

This scenario means it would be fine to add a paddle fan or an ultrasonic humidifier, but not a computer, especially one with a laser printer. Certainly not a dehumidifier or electric heater.

lems such as broken or worn receptacles, switches or lights. This discovery lets you schedule their repair on a non-emergency basis, perhaps saving repair for the next time you need to get your tools out or have an electrician over.

If you are willing to invest a little bit more time and effort in your survey, you can test smoke detectors and ground fault circuit interrupters; this is worth doing regularly—perhaps even monthly. If you are willing to invest a fair amount more time in your initial survey, and are sure that you understand how to use a voltage tester, you can combine the basic labeling walk-through with a safety inspection.

What Preparation Do You Need for Labeling?

Labeling your lines should be about the safest, most innocuous of the important jobs described in this book. Before you start, though, make sure of two prerequisites. First, know how to use an electrical tester. Second, make sure that it's quite safe to gain access to your panel. You will be turning circuits on and off. In many cases, this involves swinging open a door that is part of your panel. If you found in Chapter 6, Examining Your Panel, that opening it gives you direct access to live parts because the inner cover is missing, first take care of that.

The core part of the labeling task is filling out a list of which items each fuse or breaker controls. If you don't find a sheet glued in place for a circuit directory, or if you find one that is a thorough mess, print up your own. If it is hard for you to read the numbers embossed on the panel next to each fuse or circuit breaker, darken them or, if necessary, mark them in a permanent, contrasting color. If you simply can't find any numbers on the panel, create your own numbering scheme, duplicating it on the panel and on your directory.

Most panels contain two columns of circuits, side-by-side. Manufacturers tend to label them using odd numbers down the left side, and even numbers down the right. Very rarely, that may be reversed; sometimes, also, there's a 5a and a 5b circuit, or a 5 Left and a 5 Right.

Far less commonly, I have seen the sequence go down one side in order, and then down the other. In other words, the top breaker on the left is Number One, the one right under it is Number Two, and so forth to the bottom, and then the numbering scheme picks up at the top of the right side. You may find the numbering scheme repeated on the schematic. (Schematics are discussed in Chapter 6, Examining Your Panel.)

Once this sort of problem is handled, so you have a numbering scheme for your directory and lines to fill in, you can proceed with creating a proper listing of what each circuit does.

The Task of Labeling

There are three steps to labeling:

- locating, listing, and pretesting everything in your electrical system;
- identifying the circuit controlling each item; and
- finally, writing up a directory or directories.

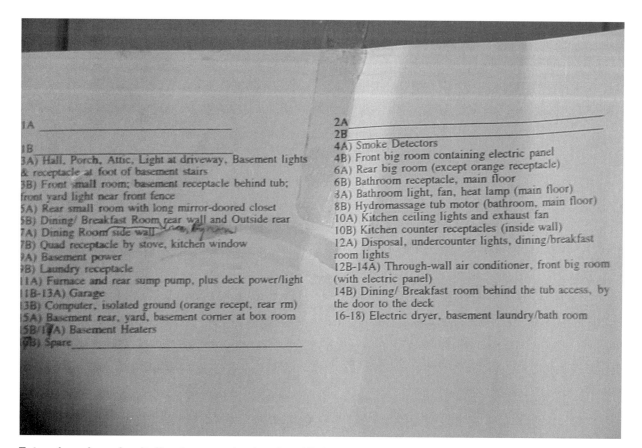

Extensive, clear circuit directory, typed rather than filled in on the form pasted in by the manufacturer.

How Precisely Must You Identify Everything in Your System?

Labeling is a precision task. The more pains you take to do a complete job, the better it will serve you. First, locate all your electrical equipment. Include:

- The thermostat.

- Exhaust fans for the whole house, the attic, bath, and stove.

- The furnace blower.

- Any independently-fed compressor and heating coils.

- Baseboard heaters.

- Any outbuildings such as garages.

- Yard lights and security lights.

- Alarm systems.

- Basements, attics, and crawlspaces.

- Spaces above suspended ceilings—I have found many hidden horrors here.

- The doorbell.

This list is *not* complete. Murphy's Law warns that whichever items you skip may be the next to give you trouble.

Next, make sure that each item works. If a piece of equipment is not operating, how can you tell which fuse controls it? At some point, you may need to interrupt power to any item; therefore, include all of it, not only receptacles and lights. Unless there is a clear reason not to do so, start with all your breakers in the ON position or all your fuses screwed and plugged in. Now check:

- Turn on all lights.

- Test every regular, standard receptacle by plugging in a light, a radio, or your tester.

- Try out fans or heating/cooling equipment by setting thermostats to make the equipment kick on.

- Unused, strange-looking receptacles that don't easily take a normal plug may be rated at 120/240 or 240 volts; test each with a voltmeter. (If your stove, dryer, or air-conditioner is plugged in to one, you can use the appliance as your tester.)

Alternately, you can use your tester for everything—even lights, theoretically.

If any switch doesn't obviously operate a light or fan, or if any duplex receptacle tests as dead, or, especially, if only one of its two parts tests as dead, find out whether or not this means you have a switch-controlled receptacle. These are used in rooms where the wall switch is intended to control a floor lamp rather than—or, rarely, in addition to—a ceiling or wall light.

Any toggle switch or toggle dimmer without marked "on" and "off" positions is a "three-way" or "four-way" switch or dimmer, working in conjunction with another switch or switches to control the same light or lights. (Perhaps obviously, standard on/off switches can have their markings obscured by careless painters.)

How Are You to Remember All This?

You need to create and transcribe notes. Unless your wiring, including any rewiring or other modification, was done very intelligently indeed, it will not lend itself to creating a simple directory. Yes, if you're lucky, a simple list will suffice: circuit 2, master bedroom; circuit 4, living room. . . . Try to use permanent names such as "recep @door in small rm 2nd fl" rather than "Jerry's computer" or "TV in blue room." Jerry may switch rooms or move out. You may relocate the TV. You may paint the room a different color.

In older homes, circuits usually are not nicely separated. The main reason for this is that when someone needed to add an outlet, they drew power from the most convenient location. This location was not necessarily an obviously related or even nearby outlet. If your list is complicated, don't try to scribble it all: "front room, left wall, receptacle closest to the door" and so forth.

> Do you want to trust in luck and your memory? This is an option. Perhaps you thoroughly trust that just about everything works, that the circuit layout is very straightforward, and that you will be able to keep a record in your head of anything that does not seem to function. In this case, you can skip recording your equipment's locations for now. However, I have found that customers very frequently regret having proceeded on this assumption.

A far more sensible approach when you *don't* know that your circuit layout is very logical, is to draw. Sketch your building or buildings, basement to attic, inside and out, wherever electricity is used. Label rooms. Show doors, windows, built-in fireplaces, counters, radiators, sinks and tubs—whatever is fixed, or permanently left in place. Then add everything electrical, from the list above. Yes, you need to move furniture unless you can see the entire space above, behind and under it.

Either look up the standard blueprint symbols in a text, or use symbols that are clear to you. At least in the latter case, put a legend identifying them at the bottom of the page. Someone else may need to use this directory in the future; you may need to refer back to it far enough in the future that you will have forgotten what something meant.

Draw some indication of the switching logic (the arrangement of what switch or switches control what outlets) on your plans, such as wavy lines from each switch to the items that it controls.

How Do You Move On to Making a Directory?

Now you're ready to start identifying circuits. Think about how you are going to record the information you gather. If the layout is simple, you will be able to record it briefly and transfer your records to the circuit directory: Circuit 1, front room receptacles and lights; Circuit 3, smoke detectors; and so forth. A complicated design is one thing; chaos is another. If your house's circuiting is totally confused, a diagram will be of considerable assistance. On the sketch of your house, put the circuit number next to each outlet. It will be much more convenient to use this than a messy list.

> If you have not done so already, make very sure before you proceed that any critical loads have been turned off safely or have extensive power backup. These include items such as computer drives and, most especially, *life support equipment.* This warning applies to more than just critical loads. Circuit testing is a very common time to discover that the battery backup in your answering machine is dead.

First Things First: How Do You Find a Circuit?
Identifying circuits is straightforward. Plug a functioning light or radio into one receptacle outlet, or leave one floor lamp turned on. Then turn circuits off until the light goes out, telling you that you have found the one controlling that outlet. Note the circuit number. Check both halves of each receptacle. In rare cases, they will be fed by separate circuits.

Turn off circuits by flipping circuit breaker handles, pulling out cartridge fuse holders, switching off fused circuits that are individually switch-controlled, or backing out plug fuses one or more turns.

All lights and 120-volt receptacles in residences should be protected by fuses or circuit breakers rated at no more than 15 or 20 amperes. These should be backed up—further protected in case of malfunction, or of a couple of possible kinds of stress beyond their capability to withstand—by a main disconnect or disconnects, usually rated 60, 100, 150, or 200 amperes, which also will kill power to the outlets. This "coordinated protection" is an important safety feature, first mentioned in Chapter 6, Examining Your Panel. The one possible exception to the rule of protection at 15 or 20 amps is a light or outlet that is an intrinsic part of a major appliance that requires a heavier circuit. (An appliance's intrinsic light or receptacle can serve as the means of telling you when you have interrupted or restored power to the appliance.)

If you find a receptacle or light that seems to be protected solely by a fuse or breaker marked 25 amps, 30 amps, or higher, it is likely that someone performed illegal and dangerous work on your system. These sizes are for protecting sub-panels, stoves, dryers, heaters, and air conditioners. Put investigating and correcting this, as needed, on your "to do" list. Sometimes you will find a subpanel fed by that 25, 30, 40, or 50 amp MAIN. The subpanel will contain a 15 or 20 amp fuse or breaker that properly protects the outlet you were testing.

Why Can't Simple Logic Make Short Work of This Project? Do you really need to check each individual outlet? Yes! Don't assume that adjacent receptacles or lights, even ones right next to each other, share circuits. It is fine to use this assumption for a starting hypothesis, but you need to check at each individual outlet.

Are There Any Shortcuts? Here are some techniques that will help you identify circuits. Some are straightforward short cuts; others will save you from having to go back over what you thought you were finished with.

 **CAUTION!**

Again and again I have worked on houses where all the receptacles in a room are on one circuit—except for one of them. Again and again I have found that a circuit was restricted to serving one floor—or even one area, such as the outdoor lights and receptacles—and then discovered a couple of exceptions. This meant that I had to go over my labeling, which was a nuisance. If I had not noticed, and had labeled incorrectly, the result could have been not just a nuisance, but a calamity. People don't always double-check—especially when they see a directory that looks precise.

- This first shortcut is one I do *not* recommend. Some people find out which fuse or breaker controls an outlet by blowing the fuse or tripping the breaker. This is unhealthy for two reasons. First, there is no device designed to do this job safely. It is almost certain that you will at the least draw a potentially dangerous arc. Second, at least in the case of properly-functioning circuit breakers, you will be exercising the breaker in a way that reduces its life slightly. If it is not functioning properly, you could do additional damage—including quite great damage to yourself. This approach is not worth the time savings.
- A patent circuit identifier can give you a significant head start. The inexpensive version, colloquially "a tick tracer with a sender," allows you to plug in a special device that sends a signal through the wires that can be picked up by the main unit without contact, back at your panel. This can save you time, at a price, although it will not eliminate your need to double-check. A tick tracer is not a tool whose indications I trust without verifying them with a definite indicator—turning the circuit off and on and checking the effect at the outlet. The rule, *"Don't put total trust in fancy tools"* has served me well.

- Make sure the outlet you are testing has indeed been turned off. Double-check by turning the circuit protector off and on again. If you have a loose receptacle, your tester, whether a voltage tester, a light, a radio, or a patent device, could have stopped indicating because something wiggled. (If you do have a loose receptacle, make a note to replace it.) Errors during labeling are very frustrating, and can result in danger later on.

- When possible, combine your tests. With circuit breakers, and with the rare fusebox that incorporates switches; "Divide and Conquer" saves time.

    Turn off about half the circuits; if your test light or other indicator at the outlet still is on, restore the circuits you have turned off, and for your next test, turn off half of those that you left on. Follow this logic until you're down to one circuit.

- You need not always run back and forth, back and forth. Use a loud radio as your indicator at a receptacle outlet instead of a lamp or voltage tester, or use an extension cord to bring the outlet nearer, so to speak. When you need to check a light in a distant room, turn off intervening lights to help you see it.

- Assistance almost always saves you time. When checking receptacles far from the panel, try to recruit another person to serve as a relay. Find someone with a loud voice (or use cellular phones).

- If you just haven't managed to figure out what a circuit controls, sometimes the only reasonable choice is to leave the circuit off. You'll find out what outlets it controls sooner or later—or maybe it's a spare. You can find out whether it is a spare by removing the panel cover and looking. If there isn't any wire attached to the circuit breaker or fuseholder, it can't be feeding a load. Just make sure you haven't turned off the deep freeze or some other appliance that's important but out of sight by leaving the circuit off.

Cautions

Avoid an unfortunate aftermath to this process:

- Make sure that every appliance is back on when you finish your explorations.

- Reset your clocks and timers.

- Accept the possibility that if electrical equipment is very marginal, it may fail when you test it. I have seen old fuse holders and deteriorated circuit breakers fail to go back on, even though they had been hanging on (by their metaphoric teeth) until I turned them off for this labeling process.

- For this reason, if you have any reason to suspect your equipment's reliability, don't start this process late on a weekend afternoon or at an equally awkward time for purchasing replacement parts or calling in professional help.

What Has All This Amounted To?

When you finish the process of identifying all circuits, you'll have done more than you started out to do. Not only will you know where all your outlets are and which circuit controls each; not only will you have noted any potential problems such as bad switches, broken lampholders, missing cover plates or loose receptacles; but you will have confirmed that your circuit breakers or, if your panel has them, fuse switches operate correctly. As mentioned in Chapter 7, Protective Devices, such "exercise" may in fact improve the reliability and extend the life of circuit breakers.

You probably will want to incorporate some other testing into the walk-through I have described. For some, though, it is easier to focus on one thing at a time. That's fine.

When a Circuit Opens, How Might You Track Down the Source of the Problem?

At the beginning of this chapter, I talked about Pete, whose fuse blew. He was waiting till morning, rather than just replacing the fuse.

If you simply can't find a tripped breaker or blown fuse, and the problem is not a defective appliance or a switch in the off position, you need professional troubleshooting. If you do find a circuit off, though, there are steps you can take, at least preliminarily.

First, before replacing the fuse or resetting the breaker, eliminate possible causes. Thanks to the labeling you completed, you should know exactly what the circuit controls. Did you have heavy loads plugged in? You know how to calculate the amount of current they draw. If the total was near the circuit's capacity, the cause could be as simple as minor voltage fluctuation, perhaps a brownout.

If not, the usual way to attack this problem is to eliminate what you can eliminate. Start by unplugging all appliances you know to be served by the offending circuit, or setting them to their off positions. Flick off all off/on switches. With three-way switching, unscrew light bulbs. Now, if you can restore power, you know that the problem probably is not in the panel, and not in the wiring inside the walls. Proceed to turn equipment back on, one item at a time. When the circuit trips again, you have found the problem.

Does this approach look a lot like the circuit-tracing tip I recommended against? If an appliance or a light is defective and trips a breaker, isn't it shortening the breaker's life, risking a large problem if the breaker is defective, and chancing an arc or other problem even if you have fuses, or if the breaker is not defective? It is. However, there are two major differences between this testing and that shortcut. First, a defective appliance or light tends to be a whole lot safer to connect to a circuit than most of the improvised oddments I have seen used for test-tripping. Second, to the best of your knowledge the equipment has already tripped the breaker or blown the fuse without causing anything overtly dangerous to happen.

If the circuit stays on in the face of this, you have a mystery. You could respond with watchful waiting, or you could call in an electrician. Unfortunately, it is hard to track down the cause of an intermittent, a problem that presently is not present. It can be done, though—usually.

The worst troubleshooting cases, in my experience, involve hidden connections. In the next chapter, I will talk about one of these cases, where someone extended a circuit by guesswork. It's not always that bad. As you walk through, "blueprinting" your system, note any blank outlet covers. If you have trouble with a circuit, these indicate places to check. Splices go bad, and we expect this to happen. That's why pretty much all electrical connections must remain accessible, by law.

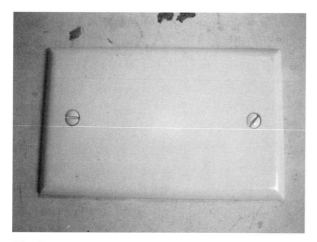

Blank cover, for a flush single-gang switch box.

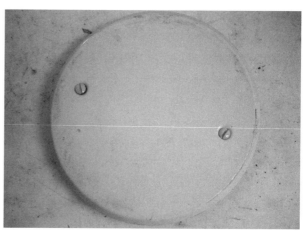

Round blank cover could indicate a former lighting outlet, properly covered.

Any of this will be a lot simpler because, like Pete, at least you know which circuit it must be because your panelboard directory tells you which one feeds the outlet that has stopped functioning.

The Safety Survey

When you walked through your system to label your lines, you probably noticed items that looked questionably safe. You may not have recorded these observations if you were focusing on correcting your panel's directory. Now is the time to go through and make notes about the problems that need correction, based on observations that don't require that you take things apart.

If you notice that a box is missing its cover, or isn't really supported, you've got a problem that definitely needs fixing. If you notice that a light fixture is hanging down on one side, and held up with a wood screw on the other, make a note.

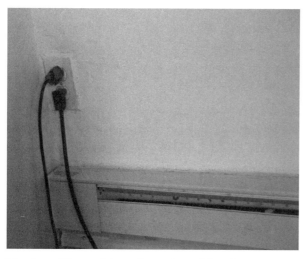

Heater with cord draped over it, cooking, because receptacle is just above it.

Hanging, very old, and quite-illegal installation. There's no box in the ceiling above this light.

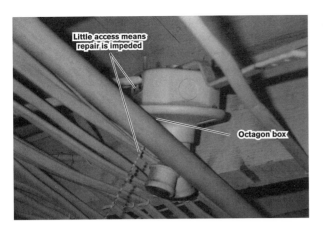

A 4" octagon box supported by a pipe, rather than properly secured; of itself, use of a two-socket adapter is not necessarily bad.

Also, it's worth your while to peek up there and see whether the box that should support it and enclose its wiring is missing. If outlets are installed where cords could be subjected to water or heat, even if you are not going to rework your system in order to correct this, at least make a note not to leave cords plugged in where they will be exposed to these threats unobserved.

Most importantly, you will begin to get some clues as to whether dangerous work has been performed on your system.

I call it, "The Case of the Shocking Stove."

I received a call to rush over on an emergency basis—meaning at emergency rates—because a family got shocked when they touched the stove and the sink at the same time. They didn't want to wait till the next day, leaving the stove and sink alone until I could schedule a visit at merely priority rates; the experience had been just too frightening.

It took me hours to sort it out, and I myself got shocked in the course of it—by something that should have had no connection to the electrical system. First I tested, then I thought, then I checked further. Then we started disassembling the kitchen. We had to remove the kitchen countertop, the undercounter cabinets, and part of the wall before exposing the criminally ignorant "Hey, it works" wiring responsible for the problem. I'll explain what I found shortly.

Insulation Damage

Was Insulation Overstripped? In order to make almost any connection, you have to remove some insulation from a wire. If too much insulation is removed, excess live wire will stick out to shock the unwary. It also may arc. Most commonly it will arc to ground, shooting a spark, or worse, a flare of ionized air and molten metal to the nearest grounded surface. This might injure you badly as you work, and could cause a fire if anything combustible is nearby. Arcing might not occur until later, perhaps much later, when you—or your family, or a future tenant—no longer are on guard.

Look out for bare copper sticking out from any place where insulated wire (other than green) is attached. Keep fingers and tools (such as screwdrivers and flashlights) away from it. Put it on your list of things to repair later. You will learn how to apply insulating tape properly in Chapter 9, Tools and Gear. You will learn how to connect wires, and thus how to redo bad connections, in Chapter 12, Making Connections.

Are There Nicks or Cuts in the Wires' Insulation? Careless stripping and faulty screw installation are also common problems. When you bring cable into a box, you need to remove the sheath of plastic or armor around the conductors. If the installer is not careful enough, the cut through this outer covering can penetrate the wires' insulation as well as the overall cable sheath. One little nick in the plastic (or, in old wiring, rubber) insulation can be deadly, if you chance to touch live wire through the nick. Look very carefully at insulation, from all sides if possible, to find nicks or cuts. Sometimes the cuts are slight-looking, like paper cuts, but you must look out for them. This is especially important if you are going to have to touch the conductor when it might be live. In boxes that use connectors or clamps to secure cables, such nicks may be difficult to spot, lurking right under a clamp or connector. They still can be sources of danger. If one of these hidden insulation breaks that has not blown the fuse on its circuit shifts just the wrong way as you work, it could arc and spit hot metal at you, or expose live wire to your tool or finger.

Has Crowding Caused Insulation Damage? You may encounter a related problem when a box or panelboard is crowded. A screw used for mechanical purposes, to mount something to a box or panel, can nick the insulation of a conductor that is in its way. The screw may be holding on a panel cover, holding a device to a box, or mounting either type of enclosure to the wall. I even have seen cases where screws cut through not only the insulation but also the copper wire itself! Whenever a wire is crowded into a spot where the mounting screw needs clearance, this is very likely to occur.

Sometimes such penetration will trip the circuit. If not, an arc can develop. Worse yet, the insulation can be nicked without quite tripping the circuit, and then the wire with the nicked insulation can be nudged out of the way. If this happens, the unprotected copper may quietly wait for you to remove the enclosure cover to work inside, at which point you will sustain a shock if you touch it. One way installers increase the chance of this occurring is by substituting over-long machine screws for the screws supplied by the manufacturer. Sheet metal screws are even more dangerous because they have sharp points.

Has Your Insulation Deteriorated? On old, old wiring, and on wiring that has been subjected to undue heat, insulation can crack and crumble. When this happens, it no longer can serve as a barrier to keep the electricity in. I just warned you to look for nicks and cuts in your insulation. A crack resulting from age or overheating can be every bit as dangerous as a nick from careless installation.

If you take the cover off a box and see the insulation falling off the wires, you may be amazed by the fact that the circuit continued to function. There are at least three possible reasons that your fuse hasn't blown.

- Perhaps the wires are bare, but not touching anything. Air is an excellent insulator for live parts, so long they are far enough apart.
- The wires may have been protected by insulation—but just barely. The insulation was ready to come apart, but not actually falling off, until you jostled it. (There is no way to avoid the risk that this may happen.)
- The wires may indeed have been shorting, but not quite enough to blow the fuse.

Whatever the reason, when you encounter deteriorated insulation, keep power off until it is repaired or replaced. Make a note to check elsewhere for similar deterioration. (You will learn how to deal with damaged insulation in Chapter 13.)

Ineffective Enclosures

Do the enclosures that should protect you do their job? Chapter 6, Examining Your Panel, goes into considerable detail about examining electrical panels for missing covers and other, sometimes less-obvious, openings. Chapter 13 talks about looking for unclosed openings in outlet and switch boxes. There are, however, incomplete enclosures that neither of these looks at. Moreover, these enclosures are found in places where you could make contact with live parts without opening up electrical boxes or going anywhere near your panel. They pose ongoing, general threats as opposed to risks you take on when you work on your wiring.

Stationary Appliances. Where someone has left a cover plate off an outlet box, it is obvious that you could touch live terminals—or that sparks could fly out, rather than be contained, if something goes wrong. This also is true of fixed appliances, even though at first glance they don't have "outlet boxes." Consider the trim panel at the bottom of an appliance such as a dishwasher. Often this forms the front of a wiring compartment. These enclosures for appliance terminals do not look at all like outlet boxes, and their covers do not look at all like the covers of outlet boxes. Therefore, sometimes people installing or working on appliances don't bother putting back these covers—or put off doing that final part of the job. This leaves parts exposed that could shock you; make sure that replacing this cover gets on your "to do" list.

Fixed Appliances. When a person works on a vent fan such as the one over the stove, he or she is likely to put the cover back on the wiring compartment, because otherwise the unit would be visibly incomplete. I find many other problems resulting from work on these fans, but usually the wiring is covered.

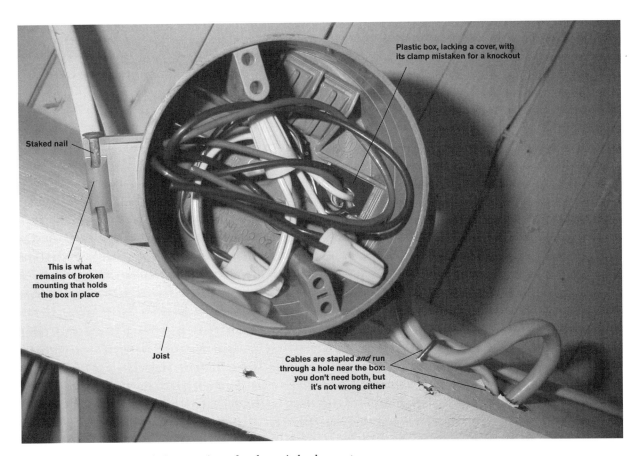

Plastic box, lacking a cover, with its clamp mistaken for a knockout

Staked nail

This is what remains of broken mounting that holds the box in place

Joist

Cables are stapled *and* run through a hole near the box: you don't need both, but it's not wrong either

Plastic box with internal clamp, misused—clamp is broken out.

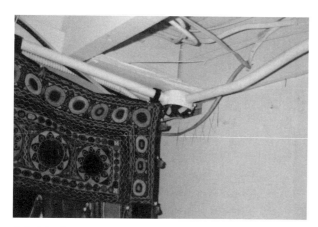

A 3.5 inch octagonal junction box without a cover to protect the combustible embroidery hanging near it.

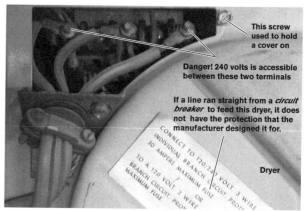

This screw used to hold a cover on

Danger! 240 volts is accessible between these two terminals

If a line ran straight from a *circuit breaker* to feed this dryer, it does not have the protection that the manufacturer designed it for.

CONNECT TO 120/240 VOLT 3 WIRE INDIVIDUAL BRANCH CIRCUIT PROT 30 AMPERE MAXIMUM FUSE OR TO A 120 VOLT 2 WIRE BRANCH CIRCUIT MAXIMUM FUSE PROT

Dryer

Dryer terminal area; 240 volts exposed by missing splice compartment cover.

However, when people replace garbage disposals, it is different. They often leave live parts exposed, for convincing reasons. Working on your back underneath the sink, as sometimes is necessary, is inconvenient and uncomfortable—even painful. Furthermore, there is a very small mounting screw for the flat metal cover to the minuscule wiring compartment on the underside of a disposal. When people lose this screw, or simply have trouble getting the plate back in place, they may leave the cover off. Their logic sometimes runs, "Nobody is going to be crawling under here where they could get shocked," or "Nobody's going to crawl under here and look at what I did."

Unfortunately, sooner or later somebody—very possibly you—*is* going to be put at risk. It is extra-dangerous to touch a live wire under the sink, for two reasons. First, there are many parts close to your torso—or to your arm, if you are just reaching under there for something you store under the sink—that usually are grounded, such as water pipes and the disposal itself. Their proximity makes it very easy for your body to complete the circuit. Second, if you are jammed under the sink where you might be more fully exposed to wires in the open enclosure, it will be difficult to flinch away, in the event that you do feel the tingle of electricity beginning to flow through your body.

Ungrounded Appliances

Can An Appliance Short to Its Enclosure Without Blowing a Fuse? When an appliance shorts to the sheet metal on its outside or a tool shorts to its metal case, you should not get shocked. Instead, if it has a three-prong plug, the short should blow the fuse protecting the circuit into which you have plugged it. If the metal is not grounded, though, you *can* get shocked by touching it if you are in contact with any ground.

In the homes of your parents, grandparents, and great-grandparents, electrical appliances were connected by two-prong plugs, were ungrounded, and could indeed shock you if they shorted to the sheet metal.

In the more recent past, appliances with three-prong, grounding plugs were developed, and for a long time these were the safest. Appliance and tool safety has improved even more over recent years. Some that are not designed with grounding means are nonetheless very safe. An appliance or tool that is marked "double-insulated" is nearly the safest type available; it is designed so that internal parts cannot short to any of the few metal parts in the case. The only type of appliance or tool that is safer yet is the type that has the equivalent of a GFCI as part of its attachment cord. For instance, modern hair dryers have an immersion detection and interruption feature, with a RESET button.

Some appliances are not available in double-insulated versions, so they still use three-prong plugs. This was the case with my customers' stove, the one that shocked them when they touched the sink. It should have been safe, but it wasn't. When a grounding-type appliance is installed without being properly grounded, it presents more danger than does an older appliance that was not designed with a grounding plug.

Why does a modern appliance that should be safer than the old-fashioned type become even more dangerous when this safety feature is defeated? In modern appliances with accessible metal parts, all such parts are bonded together and connected to the third, grounding prong of the three-prong plug. This way, if there's an internal short that impresses any voltage on those accessible parts, the electricity has a reliable path to complete the circuit safely by flowing to ground back at the electrical panel. This blows the fuse. However, this will happen *only* if those metal parts are grounded. Without this final connection to ground, everything that was bonded together to protect you instead becomes live and will shock you when there's an internal fault.

Why Aren't Your Appliances Grounded When You Plug Them In? The first reason your appliance may be ungrounded is a defect in its manufacture. In my experience, this is the rarest reason by far. Usually the cause lies much closer to your home.

Someone's Ignorance Can Put You at Risk

Bad Work Upstream; The Shocking Stove Revisited. Most commonly, the reason grounds are missing is that someone did ignorant work. Sometimes grounding is interrupted by wiring errors elsewhere in the system. In the process of replacing a light or a switch or extending a circuit, an installer may have made a bad connection—or ignored the need to make a connection—in the ground wire or other grounding means. One example of a grounding means that has been interrupted is the spiral cable armor pictured at the bottom of page 113.

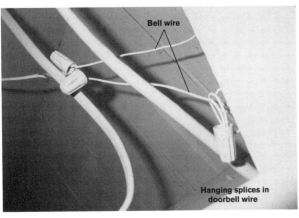

Doorbell circuiting, wired with an exposed wirenut. This open wiring is not a problem because the voltage is so low.

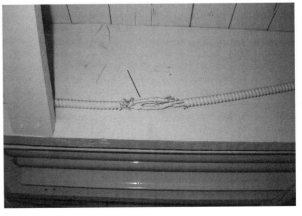

An air-splice in BX; this shows careful work, but it's still dangerous! The air-splice lacks protection and interrupts grounding.

In the case of the shocking stove, someone had extended a circuit from another room in a most dangerous manner. Most of the problems with that person's work do not need to be described here. The relevant aspect of their fumbling was located at a connection that was illegal for two reasons: first, it was an "air splice," one that dispensed with the requirement and need for a box; second, as I discovered after much expensive and messy effort, it was buried in the wall. The shocks were due to the fact that the splices were very, very poor—bad enough that a hot wire had shorted to what should have been a ground wire leading to the stove. This is why the surfaces of the stove became live. What's more, the ground wire going to the stove wasn't even spliced to the one that would normally carry the current from that short back to the source of power; this is why the fuse did not blow when the stove became live.

Receptacles with False Grounds Very often, the reason an appliance is not grounded is that it is plugged into a three-prong receptacle that is not grounded. The hole for a third prong should mean that the receptacle is grounded, but, as I will emphasize repeatedly, the only information you can trust is that which you have confirmed with a tester. The armor of BX is not a reliable ground; the bare wire in romex is not a reliable ground; a metal box is not a reliable ground. Test!

Most of the time, the cause is nothing like the devil-may-care work I uncovered in the case of the stove. Someone may have replaced a two-prong receptacle with a three-prong one, or an entire house's two-prong receptacles with three-prong ones, assuming, in all innocence, that this constituted a very minor repair or even upgrade. They figured that replacing receptacles is a simple task, and one that increases value and safety. There was no need for a tester, presumably, or for knowing anything beyond "this wire looks like it goes here."

> The tool that is most important for electrical work is a voltage tester. This is one of the most important lessons I will hammer on: if you're not using a voltage tester, you're not properly doing electrical work. Once you have learned to use one (see Chapter 3) you become capable of testing reliably to find out whether ground is present, among other things. If ground is missing at a three-prong receptacle, you know you have a problem to investigate.

This is a common, dangerous mistake. There's a whole chapter, Chapter 14, Replacing Receptacles, devoted to some of the *simplest* cases of receptacle replacement—and you need to understand most of what leads up to that chapter in order to truly understand its directions. Replacement is not necessarily simple, and it does not necessarily improve safety. On the contrary, if someone installs a three-prong receptacle without connecting the third prong to a good ground this can mislead users dangerously. Specifically, the appearance of grounding can result in your plugging in an appliance that should be grounded and putting yourself at risk of shock.

Cheaters. You might find a grounding-type appliance, one that came with a three-prong plug, that is ungrounded because someone used a "cheater." Cheaters, means of getting around manufacturers' safety features, have been widely used. When you check your appliances, you may be surprised to find some in place. I have even found cheaters used where grounded outlets *were* available. The installers simply assumed the worst, and subverted what would have been just fine.

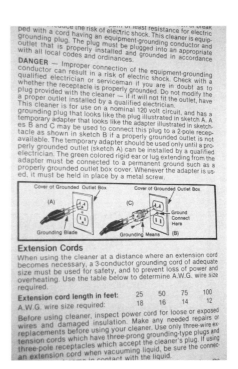

Warning regarding the danger of using even a listed adapter.

Warning regarding the danger of using ground cheaters.

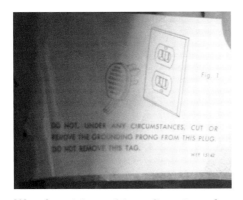

Warning right on this appliance's cordset against damaging the plug by getting rid of the third prong.

Warning regarding the danger of using ground cheaters.

What does it mean to say that you can have a grounded outlet without necessarily having a three-prong receptacle? Isn't it always dangerous to use three-prong equipment at a two-prong outlet? Yes, it often is, but not necessarily, as you will see next.

Three-Prong Adapters. All of these are means of using two-prong receptacles to accommodate three-prong plugs. One type is the outlet adapter. Some appliance manufacturers authorize its use, others forbid it explicitly, insisting that you install a properly-grounded three-prong receptacle—they don't even acknowledge use of a GFCI as an alternative—where you have a two-prong receptacle. The adapter allows you to plug a three-prong plug into a two-prong receptacle. It is intended for use at a grounded outlet, and even there it only does its job when installed right. Use it at an ungrounded outlet, or don't install it right, and it becomes a cheat—fooling users into running three-prong appliances ungrounded.

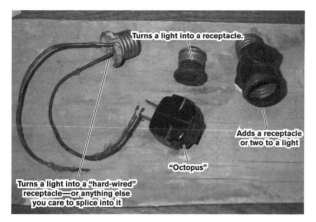

Turns a light into a receptacle.

Adds a receptacle or two to a light

"Octopus"

Turns a light into a "hard-wired" receptacle—or anything else you care to splice into it

Assorted taps, including octopus adapter which allows one receptacle to become three (too heavy a load will cause the octopus adapter to overheat). These allow you to use outlets in ways for which they were not designed. You are far safer adding more receptacles when you need them.

Broken Prongs. Broken prongs are more difficult to deal with. An installer with a three-prong plug sometimes just breaks off the third prong in order to insert it into a two-prong receptacle. In this case, not only will you need to find a grounded receptacle for your appliance, you will have to replace its plug.

Extension Cords. Sometimes the problem of connecting a three-prong plug to a nongrounding receptacle is solved more dangerously. You may find that an installer plugged the appliance into a two-wire extension cord. Plugging the cord into the wall presented no problem now. In terms of safety, though, the installer compounded the problem. Extension cords are intended, and designed, for temporary use. If you use one to hook up a heavy appliance long term, you risk fire, shock, or malfunction: the payment for abusing electrical equipment.

There were two versions of the basic three-prong adapter. Each allowed you to plug a three-prong plug into it, and had two prongs to plug into a receptacle. One version had a pigtail, a short length of wire, attached. It had a hook on the end, which you were to install under the receptacle's cover plate mounting screw. This type, whose body is on the left in the illustration below, has been superseded. The more modern version, which is on the right in the illustration below, has a fixed tab to fit under the mounting screw, rather than a pigtail.

A three-prong adapter cannot be used safely until you have confirmed that the mounting screw *is* grounded. This is something you determine by testing. If there is no ground, you cannot attach a three-prong appliance safely. If and only if there is a ground, you can proceed. Make sure there is no paint between the grounding tab, the mounting screw, and the cover plate, and tighten the cover screw as much as possible without squeezing the tab out from under or causing damage. You also need to test whether the three-prong adapter shows correct polarity once installed as explained in Chapter 14. If it shows reversed polarity, you need to remove the adapter and install it in the other half of the duplex receptacle.

3-prong adapters.

Making What You Have Do More Than It Is Designed to Do

The last thing to put on your "to do" list is to add more receptacle outlets at locations that might be overloaded. One clue to this is the use of temporary adapters, such as the "octopus," named from the many lines leaving it like tentacles, and the cord strip so familiar to computer users. One version of the octopus adapter is shown in the illustration at the top of this page.

The multioutlet cord strip *can* be all right, *if* the circuit supplying it is adequate, the receptacle it is plugged into can take the load, the cord strip itself is heavy enough—#14 conductors usually will be adequate—and the cord is routed where it will not be pinched or otherwise harmed.

Parts—What You Have, How to Choose Them, How to Use Them

9

Tools and Gear

Tools for Doing Basic Electrical Work

This chapter covers two broad subjects: picking good equipment and using it correctly and safely. My goal is to steer you away from radically misusing tools and from making do with inadequate tools, especially those that won't keep you safe.

I advise you to focus on *must*-haves: tools that are essential to doing the basic electrical jobs discussed in this book. It is easier to do good work and to stay safe while doing it when you have good equipment. Reliable, simple basics such as your voltage tester, pliers, extension cord, and screwdriver, are your most important investments. Add up the cost of good-quality basic tools, and invest in them. If you are going to saw or drill, add these power tools to the basic hand tools. Don't skimp: get plenty of replacement blades and bits. Splurging on tools that have additional, nonessential features, such as automatic wire strippers and digital voltage analyzers, comes after you've gotten yourself a good set of the tools you will rely upon most.

I recommend specific tools for one or more of these various reasons. My experience has taught me that they are

- Safer than many other choices.
- Handy for more jobs.
- More widely known or available.
- More reliable.
- Or just better designed and manufactured than some of the other selections I have tried.

I am not advising against all alternative choices; others may prove perfectly suitable. Just evaluate them carefully to make sure that they give you what you need. Any tool can be obtained in a shoddy or defective version.

Some final notes. First, no tool is perfectly safe to use, and you can misuse even the best tool, especially if you misunderstand how it should be used. Second, every manufacturing line misses a few defective units. If at all possible, check tools from even the most reputable brands before you take your purchases home.

Tool Safety

General Rules for Using Tools

Everything you learn in this chapter will improve your on-the-job safety. However, before I examine even basic hand tools, I must focus on the most important tools: those that protect you directly. (This is why Chapter 3, Checking for Voltage, came so much earlier.) I will lead off with a highly condensed summary of the rules for safe tool use.

- Buy and wear eye, ear, lung, and hand protection whenever your work may expose you to dust, splinters, sparks, flashes, noise, caustic substances or rough surfaces.
- Keep your edged and pointed tools sharp.
- Never cut, drill, pull, or swing tools towards your body.
- Never use damaged tools (including cords, cord connectors, and ladders).

If you choose to use damaged or inappropriate tools at some point (just as you may choose to work live, or tired, or rushed), recognize that you're risking your life and health, and employ the extra caution that suggests.

What Safety Equipment Do You Need, and What Are the Basic Rules for When to Use It?

OSHA does not inspect homeowners doing their own work, but then no inspector has *your* vested interest in *your* own skin. Personal Protective Equipment is one of the most important categories of electricians' gear. Using it is even more important for the inexperienced.

Protecting Your Eyes

You need safety glasses or goggles to protect your eyes, however imperfectly, against sparks and flying bits of material. These materials include drops of melted copper from arcing, wood shavings, bits of metal, and especially masonry. Always wear goggles when chiseling masonry or wood, or drilling brick, concrete, or cinder block. Blinking does not work reliably to keep these destroyers out of your eyes. Your gear should have side shields, or wrap around to protect your eyes from matter flying in from above, below, and the sides as well as from the front. Plastic-lens glasses will provide a little of the protection you may need—more than blinking. I find that the best, side-shield safety glasses provide minimal protection when I am exposed to material coming at me from above or at eye level. Goggles are cheap—buy some.

Protecting Your Ears

Loud noises should hurt. The pain they cause you is a danger signal. Once loud noises stop hurting badly, your hearing is beginning to suffer damage. If you're going to use power tools, correctly worn hearing protection is important. Moreover, if a wire in your vicinity shorts, it may do so with a very loud Bang!! causing pain that lasts hours. If you hammer metal, even to remove embossed, ready-to-remove parts, once again it may cause your ears to ache for hours. The noise that you generate may be no problem at all, and usually is not uncomfortably loud; however, this is unpredictable, depending in part on how well the parts are cut out, and how many of them you need to remove.

Hearing protectors come with instructions, and you will hear the difference when you use them correctly. The earmuff type is the hardest to misapply. The in-the-ear compressed foam earplug is the cheapest—yet it tends to be the most effective, when used correctly!

Protecting Your Lungs

It is very likely that you will not encounter any serious respiratory hazards on the job. However, do minimize your exposure to plastic pipe solvent, to paint and lacquer thinners, and to cement dust. The latter is implicated in silicosis, a terrible disease. For years, few construction workers worried about what they were breathing except when it made them cough. In fact, drilling through studs and joists merely raises "nuisance dust," most of the time. Do realize, though, that most respirators, from the disposable variety widely referred to as the "bra cup" to the fancy, multistrap variety with replaceable cottony inserts, are only designed to deal with nuisance dusts. They will make you more comfortable about what you are inhaling, and certainly are worth wearing. However, serious challenges, such as handling old asbestos pipe wrap or removing lead paint, require specialized, expensive equipment—and training in how to use it. Even construction firms stay away from this until specialized hazard abatement contractors have given them the all-clear. If you're not sure what you're dealing with, stop and get help.

Protecting Your Hide

Construction work means chips and sparks and spinning tools. Wear non-flammable clothing that covers your torso and limbs and is not loose enough to get entangled. If you are long-haired or bearded, make sure that your hair is restrained. Working around your house, your body may be more frequently protected by simple denim pants and elastic hair bands than by work gloves, steel-toed boots, and hard hats.

Protecting Yourself from Electrocution

Take an extra step to protect yourself when you work in a moist area, or one that resembles a construction site (perhaps as a result *of* your project). In these cases, GFCI protection for any power tools, extension cords, or drop lights is extremely important. To a large extent, this protects your entire body against both shock and arcing. GFCIs are discussed in Chapter 5, Safety. Permanent ones, whose installation I discuss in Chapter 14, How to Replace a Receptacle, will protect you adequately. Portable versions have an additional internal feature.

General Tips on Using Safety Equipment

If you don't heed the most general rules for equipment use, it may not serve your needs. The first one is to use each type of safety equipment that applies to the kind of work you are doing. For example, if you are testing all your smoke detectors by pushing their buttons to make sure that they screech, you might well wear hearing protection; whereas if you are testing your GFCIs by pushing their buttons to make sure their RESET buttons pop forward, you would not need it. However, any equipment is far less useful, and sometimes even useless, when the manufacturer's instructions for its use are ignored.

Finally, there are two basic rules:

- Keep your safety equipment clean enough so that it won't make you reluctant to use it, and discard it when it becomes a liability. Relying on scratched goggles can mean you work half-blind. Dirty ear plugs can be disgusting, and moreover could cause infection—though this has not been proven, and concerns me far, far less than the risk of further hearing loss. (With 50 pairs of foam plugs—a batch that is dirt cheap, to use an ironic image—you can insert one pair in the morning and a new one after lunch every day, and have them last until you have done all the jobs in this book.)

- If you mislay your protection, stop. Find it or replace it. And put your tools down until you do so.

Essential Hand Tools

On to types of equipment that more traditionally are thought of as tools. These are the tools that you pretty much have to have:

- A voltage tester.
- A pair of lineman's pliers.
- A flashlight.
- An extension cord.
- At least one slot screwdriver.
- Very likely a portable 120-volt light.
- Very likely a continuity tester, multimeter, or ohmmeter.
- Very likely a knife.
- Very likely at least one Phillips screwdriver.
- Probably a pair of channellocks.

I will discuss each of these. Any other tool may be optional, depending on the jobs you take on. These are the ones you are next most likely to need:

- A drill motor and bits.
- A cable ripper.
- A hacksaw or a hacksaw blade holder.
- A wood saw or reciprocating saw.

If you know how to select and use these, you can get by with skimming the rest of this chapter. The only part that I urge you to spend time on is the discussion of lineman's pliers.

Testers

Testers, especially voltmeters, are the tools most distinctly associated with electrical work, both for electricians and for nonelectricians. Checking for voltage is the very most important type of testing, but it is not the only purpose for which even voltmeters are used. Most readers will need to test or at least check for six things:

Note that I say, "functional" equipment rather than "correctly functioning" equipment. Your brakes can be functional, but make your car pull to one side so that it is unsafe to drive if there is a chance you will have to come to a hard stop.

While a device such as a smoke detector or a GFCI may respond correctly when you push its TEST button, this does not mean that it will respond correctly when exposed to the hazard from which it is supposed to protect you. Even if a detector responds when you give it a real-life test, you have no way to know whether its calibration has drifted.

Reminder: If you store a standard multimeter turned to its resistance setting, the battery will drain over time, and you may not be able to use it to measure resistance the next time you pick it up.

- grounding;

- line voltage, meaning 120, 208, or 240 volts, normally between one of your probes and the other;

- continuity, which means at least one unbroken electrical path from one probe to the other;

- functional smoke detectors;

- a functional carbon monoxide detector, if you have a gas furnace; and

- functional GFCI protection, at least at bathroom and outdoor receptacles.

Voltage and Continuity Testers

You need to decide whether you want to buy single-function or multifunction testers. Any of the testers recommended in Chapter 3 will serve your needs for voltage testing. Very rarely will you need to test to determine whether you have 208 or 240 volts, especially for the purposes of this book, so your tester does not need to differentiate between the two. The "dual purpose wiggy" and the multimeter also test for continuity as well as for line voltage, so either tool will be sufficient for the tests I describe.

If you prefer to use a separate continuity tester, one option you might consider is an electrician's flashlight that comes with two flexible leads terminating in alligator clips. When the leads are plugged in to the rear of the flashlight, it only lights when there is continuity between the clips.

Another question you have to consider is whether to purchase analog or digital testers. Some analog versions are very, very inexpensive; digital ones cost more, as do more rugged instruments and fancier ones. Analog devices should be accurate enough for your purposes. In fact, people can be misled by too-accurate equipment. Many know that there should be no voltage between the ground and the neutral. A digital voltmeter could say that there is a 1.5 volt difference between them; this matters considerably in a 5-volt DC circuit inside a computer, but should be dismissed as negligible in branch circuit wiring.

Analog does have its down side. It *can* be harder to read, and usually is less accurate. Moreover, using an analog meter, you can at the very least blow its internal fuse by setting a dial to the "0–30 volt" range or using the jack openings meant for testing resistance, and then touching the leads to 120 volts. Some digital meters are "autoranging," meaning that it is harder to destroy them by choosing the wrong settings.

General Cautions to Keep in Mind When Testing

It is worth recapping five general principles, from Chapter 3, that will prevent injury when you use any electrical tester.

- *Test your tester* and retest, each time before you use it, especially when you're testing for voltage. Testers die; if you rely on a defective tester, *you* could die. Go to a known source of juice and make sure that it registers on your tester.

- *Be very, very conscious of how you hold your tester*—or any tool that might be used around live parts. The first element to this is to make certain your body is not touching live parts—such as the tool's leads or probes.

Alligator clips

Even with the cover on, you know that this is a three-way or four-way switch because there is no ON or OFF mark

Testing a 3-way switch. Alligator clips are convenient because they allow hands-free testing. With the switch not connected to any wiring, you don't have to worry about the risk of shock.

Alligator clips are by far the most dangerous types of tester contact to use, even when they are partly sheathed. It is very easy to touch metal when squeezing the insulated sheathing to open the clips. Admittedly, alligator clips have two significant advantages over leads.

- They are self-holding. This rarely is true of leads, which can tie up both of your hands. If you are absolutely sure that you are applying the clips to dead metal, this also enables you to leave your hands off when you restore power for measurement.

- It is hard to hold leads with unvarying force; alligator clips achieve this automatically. Readings can vary significantly based on variations in pressure. This said, in basic house wiring there are few occasions where you need such precise readings.

Using bare leads, such as those found on neon testers, tends to be a bit safer than attaching or touching alligator clips to surfaces that may be live (even though they should be dead). Make quite sure that you hold a lead by the insulated portion, rather than by the bare steel closer to its end.

Probes with retractable sheaths, such as are found on wiggies, tend to be safest (after cheap three-light receptacle testers, which I would not rely on). When testing a receptacle, the retractable sheaths over a wiggy's probes keep their metal parts from even emerging until they are in the receptacle's slots, out of finger contact.

- *Be very careful to make good contact.* Bad connections, ones that may dangerously mislead you into thinking that power is absent, may be present when wires or terminals are protected from contact by corrosion, paint, or even the residue of old insulation. They also may be found when there is loose contact. Stick a probe into the slot of a tired old receptacle, and you may or may not get a reading. This is why you will retest once you have access to the receptacle's terminals. However, touch the wire at a screw terminal lightly, and it may give you no reading because the screw is not bearing against it tightly. Push against it harder, perhaps with your hand, or move the switch or receptacle to which it is attached, and Zap! Test one bare wire coming out of a splice, and get no reading. Move the splice and suddenly it makes better contact with the other wire it wasn't quite contacting. This problem often can be found when a bad connection, or some other source of overheating, has caused the insulation on the wires going into the splice to melt or burn. Keep your eye out for such situations.

- Be careful that your tester—or, again, any tool used around live parts—doesn't touch what it shouldn't. If a single lead touches both a live part and a grounded part, for example the hot slot of a working receptacle and at the same time its grounded cover plate, it's going to short, possibly with a resulting arc that can hurt you before the circuit opens.

- Both for the tools' sake and for your safety, use them on the correct settings. Use a multimeter on too low a setting, and you may blow its internal fuse. This would seem to be a minor nuisance. After this happens, though, it may stop giving you any readings. THIS could be dangerous, especially if you don't recognize that there's a problem—perhaps it's not very long since you tested your tester. Also, if you use much too high a setting, the reading may barely register.

Hand Tools

General Rules

In general, all your hand tools should have nominally insulated handles. I say "nominally" because manufacturers will not warrant the coverings of standard tool handles as effective insulation. Nevertheless, electricians consistently use standard tools that have thick coverings on their handles for live work, and do so without the least tingle of electricity. The tools on which they rely have handles that are covered either with solid fiberglass or else rubber or plastic, with no steel showing either at the ends or through nicks and cuts. As I mention repeatedly, you may find yourself working live without any intention of doing so. You have to decide for yourself whether you want to follow the common practice of relying on these normal handle coverings as insulation, or seek out the more expensive and somewhat less readily available tools that are warranted as insulating.

Check for three additional characteristics. Cutting edges should be sharp and, when intended to meet, should meet squarely. Tools should fit your hand and be of a weight that you can handle comfortably. Test for feel; no tool you will use regularly should hurt to squeeze with moderate force. Tools should be of the right size and shape to fit snugly into or around the work, whichever is necessary.

I am willing to name names; just recognize that there are more brands that I never have tried or heard discussed than there are brands I do know. Klein is one brand that electricians have trusted for many decades. Ideal is another reputable one, as is Channelock. Asplundh is a premium brand that is worth seeking if you can afford it. Unlike many standard brands, Asplundh, like a few other companies, vouches for the insulating effectiveness of the plastic covering its tools' handles when undamaged. Klein, as well as some other standard brands, does, however, certify some of their tools' handles as insulating—but you have to ask for these and check the labels.

Lineman's Pliers

Now for specifics. Lineman's pliers, sometimes colloquially called "Kleins" or "side-cutters," are nearly the most basic tool used in wiring. (A voltage tester unquestionably is the most basic.) The term "side-cutters" also is used by some to refer to diagonal cutters, which I mention later. Lineman's pliers are used for holding, pulling, bending, cutting, wire stripping, and sometimes crimping. Electricians use them for at least one additional purpose that is well beyond your need for the jobs addressed in this book: reaming (removing burrs from cut edges).

Criteria to Use in Choosing a Pair. Ease of use is an important factor to consider in selecting any tool. I mentioned hand feel, but there's much more to selecting pliers that you may use a lot. Lineman's pliers come in various lengths from around 7″ to 9½″. Buy the longest you find that is not too heavy for your wrist. Other factors being equal, a longer pair will give you more leverage, and thus be less tiring to use. Another thing to look for is smooth pivoting, so you don't expend unnecessary effort each time you open and close them. (The pivot, between the cutting edges and the handles, generally can be lubricated, but in a good pair will not start out stiff.) Next, make sure that the cutting edges mate well. To measure this, use the light leakage test. Hold the pliers up, jaws held closed but not squeezed forcefully, with a light behind the blades. No light should show through the line where the blades meet.

The ends of the jaws of lineman's pliers, the last ½″ or so beyond the cutting blades, do not quite meet. This part of the jaws is cross-hatched to provide a better grip when you use the pliers for holding and twisting. The cross-hatching provided by cheaper manufacturers is relatively smooth. Cheaper ones also use softer steel, with the result that their cutting blades will distort, or at least grow dull prematurely.

Using the Pliers. You will utilize your lineman's pliers primarily for cutting and holding, using the terms in the most general sense. (I think of stripping as a subset of cutting, and I lump shaping and pinching under holding.)

The rules for using these pliers are there primarily to help you work more effectively, but they also keep you safer. To hold, use the ends of the jaws—with two exceptions. One is when you need to hold or pinch something you can't reach with the ends of the jaws; the other is when you are shaping a conductor. When you cut wire, always hold the pliers so that you can see or feel the blade side, where the wire will be cut. Always cut using the part of the blades near their base, close to the pivot. Most important, recognize that whenever you cut or even touch wires that are part of your house wiring, they *could* surprise you by being live.

The Cutting Function

Shearing. These pliers are the tool I choose for cutting off a piece of wire, for trimming a conductor, or several conductors in combination, and for cutting small screws. I even use them for cutting cable armor, when I'm feeling strong. When you are trimming metal, make sure that any bits that fly off do no harm. In particular, this means that they do not fly into your face or otherwise hurt your body and that they do not fall into live parts, or provide a conductive path when power is restored.

Using Pliers to Strip. Most elements of stripping are the same whether you use lineman's pliers or another stripper that follows the same system (traditionally called a "Stakon tool") that I will describe a little later. When you can use either pliers or a Stakon, I will say, "the stripper"; when only lineman's pliers will do the job, I will say, "the pliers."

It's easier to remove insulation when you're not forcing it around bends. Therefore, before you start, if the conductor is appreciably crooked, straighten it with your pliers. Usually you will do this by closing the jaws on the bent part, as though you were squeezing it in a vise.

Here's the standard approach to the actual stripping operation.

1. Decide the length you need, and note the place you will start to strip. If your conductor is long enough, strip off a little more insulation than you expect to need.

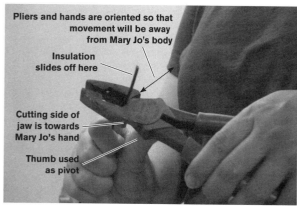

Pliers and hands are oriented so that movement will be away from Mary Jo's body

Insulation slides off here

Cutting side of jaw is towards Mary Jo's hand

Thumb used as pivot

Cutting edge

Stripping with pliers in the reverse of the optimal position. The cutting edge should face the thumb.

Stripping correctly.

2. Gently close your stripper partway into the insulation. Try to feel for when you've cut through the insulation and touched the wire within, and not indented the copper, if it is solid, and not cut strands, if it is stranded. You may need to practice this technique on scrap wire before taking the risk of applying it on the conductors in your house's wiring system.

At this point, you've only cut the insulation at the two points where the blades touch it, if you're using pliers; even a specialized tool doesn't cut through it in a complete circle. Therefore, you might be tempted to cut all the way around by twisting your stripper around the insulation. Doing so should not be necessary, and probably is unwise, as you will see shortly. A straight pull will simply rip away the uncut insulation to either side of where your pliers bit, provided that you have the muscle for a firm enough pull.

If you do decide to twist your stripper to cut the insulation all the way around, be careful not to "ring" (cut into or nick) the wire itself, at least not more than nominally scratching it. With some strippers such as lineman's pliers, if you've closed the stripper very firmly, twisting will severely weaken the conductor by scoring it, somewhat like ringing a tree. If it's a stranded conductor, you may cut off quite a few strands. In other words, you will compound any damage caused by having closed the pliers on it too hard.

3. One hand is holding the stripper. Wrap the other hand around the conductor about an inch beyond the section of insulation you need to strip away.

4. Push the thumb of this hand against the jaws of the stripper.

5. Pivot the hand holding the stripper firmly against the thumb of the hand holding the conductor, and you will slide the insulation off the wire.

Whenever pulling a tool, make sure you are not pulling it towards your body; otherwise, you could strike yourself, and even lose teeth or worse.

This description applies to stripping insulation off a conductor that offers a considerable length for you to hold. This often is true of conductors in large enclosures such as panelboards—but you may very well want to stay out of them because of all the live parts—and certainly is the case with conductors that you are adding in to a system. In most cases, though, the conductor you are trying to strip will be attached to something such as a cable that is secured immovably in a box by a connector or clamp. You probably will not be able to reach the hand that is not holding your tool into the box far enough to grab the conductor. In this case, the little finger side of the hand that is not holding the pliers will bear against the side of the box for leverage, while the thumb pushes against the hand holding the pliers. This principle of using the box for resistance against your pull applies to other approaches to stripping as well.

Here are some alternative approaches that keep your hands off the metal:

- For maximum safety, if you are using pliers you can treat the conductor as though it just might be live. (Most specialized strippers do not have insulated handles, so this only applies to lineman's pliers.) Rather than pushing your hand against the jaw, push it against the handle insulation. This approach applies pressure back of the jaws, and therefore is slightly more difficult than pivoting closer, but it still applies leverage.

- Another safe stripping approach also treats the conductor as though it were live. Don't rely on a pivoting motion. Instead, just pull straight without bracing with your thumb. This is more difficult than pivoting, but it can be done. With this approach, please be extra-careful not to pull toward your body.

- Another option, useful on rubber-insulated conductors—which are just the ones you may find in older wiring systems—is to use the ends of the jaws to pinch the insulation. Once it is pinched off one side of the wire, rip it away. Then go to the other side, and repeat. Usually, two pinch-and-rips will do the job. This option can be handy when you need to strip the insulation off very short conductors inside an enclosure, because it's hard to get your pliers all the way in at the correct angle to use the pliers' cutting blades.

There are a number of specialized stripping tools. Because their use in stripping involves many of the same elements as does using lineman's pliers, I will describe them before teaching you the other ways you may use lineman's pliers.

Specialized Tools for Stripping

Lineman's pliers are not the easiest tool for the inexperienced to use for stripping. There are a number of specialized tools. The simplest, normally just called a "wire stripper," commonly has been known among electricians as a "Stakon" tool. It is a multipurpose tool used for stripping wire, for crimping, and for restoring fouled threads on screws in the sizes most commonly used in electrical work. It has separate paired notches for each size of conductor in the standard branch circuit sizes and for a few smaller sizes such as you may find in light fixtures. So long as you use the right set of notches, you can't make too deep or too shallow a cut through the insulation. While not particularly efficient for the purpose, this tool also can be used for cutting right through a wire. However, it uses a separate part of the blade for this—not the wire-stripping notches.

Far fancier strippers are available. Usually these simply require you to insert the conductor and squeeze the handles. I cannot recommend them, although they are intended to be easy to use. The fancier a tool is, the easier it is for it to suffer damage, and these offer no very great advantage over the simpler tools.

Hitting

Lineman's pliers are jokingly called "the electrician's hammer" because they sometimes are used for hitting. Theoretically, you should use a suitable hammer for any job that involves striking; electricians frequently use the flat, cutting-edge side of their pliers, the area around the pivot, or some other solid metal, for close-up hitting. Because this use directly goes against manufacturers' intentions for the item, according to a representative of Klein Tools, I will say no more about this use.

Nonrecommended wire strippers.

The Holding Function

Ripping. The last technique I described for stripping insulation off conductors with lineman's pliers, pinching, uses them not for their cutting blades, but for the gripping part of their jaws forward of the cutting part. I occasionally use a similarly nonstandard technique to remove cables' sheaths.

Sometimes it is quite difficult to remove the sheaths from tough nonmetallic cables using the basic tool designed for the purpose, the cable ripper. This is especially true when the cables are cold, because cold hardens thermoplastic. This also is especially true when the cables are old, because I find old plastic sheathing (conductor insulation as well) often is tough.

Although this is not a standard technique, I am able to separate the conductors from their sheaths by using two pairs of linesman's pliers. To use the following approach, at least a little bit of two conductors needs to be sticking out beyond the sheath.

Here is what I do, when I'm fresh enough that I'm sure that I have the strength and control to proceed safely. I grab one conductor with each pair of pliers and pull them away from each other to rip the conductors out of the sheath. Of necessity, this is a large motion, so I have to be especially careful that I'm not pulling in the direction of anything fragile, such as—but not limited to—my body. It also requires care not to pull apart more of the sheath than needs stripping. The action of my arms is similar to that when I use the insulation-stripping technique I described just before that of pinching: pulling without pivoting.

Even with test, lockout, and tagout, I'd still hate to stick a finger in this mess. I feel much safer using pliers as I sort it out even though it *should* be dead.

Grabbing. I use lineman's pliers with insulated handles when I need to hold something that may be live, or to pull something apart. For instance, I may need a ladder to reach my work, and not have a good way to aim a battery-operated light. In this case, even though I am not certain I have managed to kill the circuit controlling the outlet I need to check, I may choose to use a plug-in light instead of killing power. In this case, I will be very cautious not to touch any of the wiring until I have tested it and made sure it is dead. This caution means that I will not want to touch the wiring to pull it out to where I can reach it with my tester. My answer is to use my pliers, very carefully, to pull the wiring out to where I can work safely.

I also rely on my pliers many times for their ability to grip. Sometimes I see a cable that is not properly secured in place as it leaves an electrical panel mounted on plywood. I probably want to drive some straps or staples (special ones designed for the purpose, not plumbers' straps, not office staples or carpenters' staples) into the wood to support the cable. It may be difficult to keep the cable from shifting because there is some weight on it—perhaps just the weight of unsupported cable. In these cases, I use

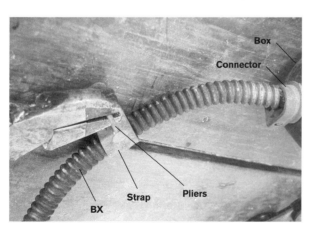

Pliers hold a nail for hammering cable support into wood near box.

my pliers to hold the cable in place as I hammer. It is much easier to take the strain with my pliers than directly with my fingers. Other times, I may simply want to keep my fingers away from where I'm hammering. This is often the case when I am nailing into masonry; masonry nails tend to require more intense hammering.

Shaping. I strongly recommend that you use lineman's pliers when you need to twist or bend wire. Almost any time you strip and cut stranded wire, the strands will spread from their smooth, concentric lay (their lay is the way the strands spiral around each other in the bundle). You will want to twist the

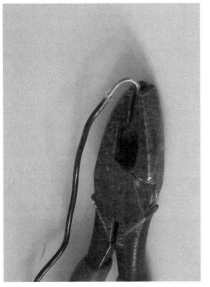

Shaping wire with the end of your pliers. It's not the most efficient position for shaping, but it works.

Squeezing a hook tighter with the side of the pliers' jaws.

Squeezing a hook tighter with the end of your pliers' jaws.

Shaping wire with the side of your pliers' jaws. In this case, the pliers are 90° from the most effective position for initial shaping.

strands tight again before you splice or terminate them. Otherwise, strands will peep out from a connection such as the head of a terminal screw. This could pose a shock hazard when you reach in near the connection, if power is on, as well as reduce the amount of contact the bundle of wires makes with the termination.

These pliers also are very effective for making a properly-shaped loop on a solid wire. Perhaps counterintuitively, I have more control over the loop's shape by bending sideways, holding a wire with the sides of the jaws, than by holding the wire with the end of the jaws and forming the loop over the end of the pliers.

Channels

Let's return to the very general category called pliers. "Groove joint" or "water pump" pliers are colloquially known as "channels," after the Channelock (R) Company. ("Channels" is a "singular plural," like "trousers" or, for that matter, "pliers.") Channels are adjustable and take the place of an assortment of wrenches. The adjustment employs the channels or grooves; this points to the characteristics to use in picking a pair. The grooves must be cleanly machined and fit well so the pliers easily adjust from having one size opening to another, and yet not slip their adjustment while in use.

A representative of Klein Tools suggested that channels are the most appropriate general-purpose tool for use in tightening locknuts. This implies that you have access to both the outside and the inside of the enclosure. Conduit usually is held to an enclosure by screwing a locknut over its end, sticking the end into a knockout till the locknut stops it, and screwing a bushing, or another locknut and a bushing, over the end on the inside. When the conduit is a little loose in a box screwed to a basement joist, I do indeed tighten the connection by twisting the locknut—on the outside. Warning: it's easy for the channels to slip off; don't pinch a finger.

When I have to tighten not conduit but a connector—in which case the sole locknut is on the inside—I often do not find channels to be an effective tool to work with. Therefore, using channels to tighten a connector in that same box, accessible in that same ceiling, often is problematic. I have to work from the outside, which means that the connector had better not contain a cable, or the cable will suffer nasty twisting. The vast majority of the times I have tightened locknuts inside boxes, I have relied on striking a screwdriver against the locknut.

I'll look ahead to work that's beyond the scope of this book. If you are going to take on any tasks involving conduit, tubing, or weatherproof connectors, you will benefit from having two pairs of roughly comparable size, say Channelock 440s, so you can turn one part of a multipiece fitting with respect to another part. Also, on some occasions you may find yourself in need of big bad pliers, such as Channelock's 480 model.

Other Pliers

Other types of pliers are less important. Needle nose pliers can let you fish in tight spaces, and can be used to form tight hooks on wires. However, you rarely—that's not the same as "never"—will need them if you're fairly expert at using lineman's pliers.

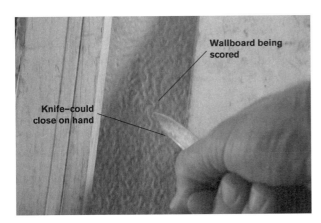

Knife held dangerously in a hand—where it could close on the hand.

There is even less a need for diagonal cutters, also referred to as "side cutters," "side cutting pliers," or colloquially, "dykes," (from DIagonal CutterS). Unlike lineman's pliers, they are designed only for cutting, not grabbing. In some parts of the country, dykes are called "side cutters," confusing though that can be to people who use that term for lineman's pliers.

Finally, locking pliers can be handy as temporary, small vices.

Knives

A knife is not absolutely necessary for the electrical work covered in this book, but it can prove helpful in some cases where no other tool you have on hand will do the job. Knives are dangerous, though. There is nothing unique to electrical work about knife use or knife safety, except for keeping your knife away from live wires. I use a knife for scoring drywall. This is handy when, for instance, I want access to part of an electrical enclosure that is covered.

Electricians use their knives for two main purposes, the most necessary one being "skinning wires." (The colloquial term for removing insulation with a knife is skinning.) You rarely will need to strip insulation off conductors larger than can be accommodated by strippers or pliers when doing the tasks explained in this book. Given the prices of more suitable tools, I certainly do not recommend using a knife to strip wires in the smaller, branch circuit sizes, unless you're going to do very, very little such work. Pliers work fine, with a little practice. If you have not yet developed the skill of using lineman's pliers for stripping, a Stakon is cheap, safe, and easier to use than pliers, so I will not explain the riskier approach of skinning with a knife. For one thing, a knife can cut strands very easily when used to strip small-gage wire.

I also do not recommend using your knife to strip the sheath from nonmetallic cable; it's too easy to nick the conductor insulation underneath, perhaps unawares, as you slice into the sheath. This could result in future shock hazard.

Rippers

Function

There are a few situations covered in this book where you might want to shorten romex. Garbage disposals, many track lights, and most fluorescents are fed by a single two-conductor cable. When replacing the equipment, you have an excellent opportunity to shorten the cable to get to an undamaged portion, should you notice insulation damage (presuming that there is sufficient slack). On other occasions, you will want to strip back cable sheath inside an electrical box, either because the installer left excess sheath or because you have managed to pull in some slack from inside the wall, so as to gain access to sections of conductor with good insulation. In the latter cases, where access is limited, it is especially useful to have a tool designed for stripping rather than have to rely on scoring the sheath with a knife or on pulling it apart with pliers.

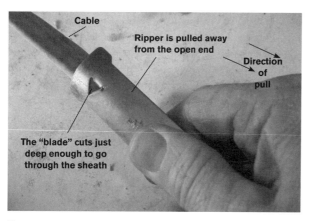

Cable

Ripper is pulled away
from the open end

Direction
of
pull

The "blade" cuts just
deep enough to go
through the sheath

Romex ripper.

Design and Use

If you want to shorten romex, and in the future, install it, you're going to want a ripper. The ripper is cheap, stamped sheet metal, costing a dollar or two. Ripper use is simple: slide the end of the cable through the nonfolded end of the ripper till it lines up so that the short, triangular blade inside the ripper is at the point where you want to start removing the sheath. Holding the cable with one hand and the ripper with the other, squeeze and pull; then pull the slit section of sheath away and cut it off with your knife or pliers. A slightly more convenient version of the ripper, and one that costs little more, does not have a punched opening through which to insert the cable. Instead, the folded end of the sheet metal is bent sideways, which allows you to place the ripper over the cable rather than starting out by sliding the cable through it. What makes this particularly handy is that when the end of the cable has fairly long conductors emerging from it, you do not have to go to the trouble of straightening them in order to get them through an opening.

The ripper is designed for easy use with two-wire, flat romex, but you can take it further, with due care. Three-conductor cable is more or less round, and the conductors within are somewhat twisted around each other, not parallel to the longitudinal axis of the sheath. If you pull straight with the ripper, you may well damage conductor insulation. Instead, squeeze lightly, and follow the twist of the wires inside (which you can make out through the sheath) as you pull. Whenever you've stripped a sheath, especially from three-wire romex, check for damage to the insulation on the wires you've freed.

There are a couple of more elegant versions of the ripper. Some have sharper, replaceable blades. These can be valuable for use in unheated areas during cold weather when the sheath is tougher; in ripping the sheath of older romex, which has toughened up over the decades; and in stripping the sheath of UF, a similar cable described in the next chapter. You probably can do without these rippers, unless you will be installing a fair amount of UF. Other fancy versions simply use different techniques for stripping sheaths. I have found the simplest version, the one I just described, most convenient.

Hacksaws

You will need a hacksaw if you will cut metal such as BX, conduit, bolts and heavy cables. Some, but not all, of these purposes can be served by the use of power tools. The most likely use of a hacksaw in electrical work is cutting back the armor of BX in order to bring sections of the conductors with good insulation into a box or wiring compartment.

Choosing a Hacksaw

As with hammers, channels, and pliers, I must assume that you have a basic understanding of this tool, so my general description will be brief. You should choose a blade that allows three teeth simultaneously on what you're cutting. Most commonly this means 18–24 teeth per inch in residential wiring; use a coarser blade, such as one with 14 teeth per inch, if you're going to cut large-diameter conduit.

Hacksaw aimed incorrectly, between turns of the armor. You can't remove sheath from BX angling it this way without cutting through conductors.

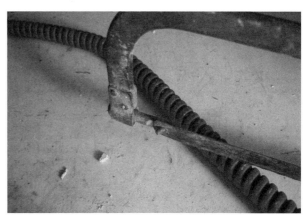

Hacksaw aimed correctly, across two turns of the armor. The blade is perpendicular to the armor's turns and cuts into two of them simultaneously.

It is much harder to cut cable than to use a hacksaw to cut the more rigid items for which it is used outside electrical work. BX, in particular, tries to wiggle in the direction in which you are moving the blade. In cutting the armor of BX, do *not* make your cut parallel to the direction of the twist but perpendicular to it. Make sure that your blade is resting on and cutting through two turns of the armor, not one. When you have cut through *just about* to the bottoms of the armor, past the convexly rounded parts that face out, there will be just a thin remainder of armor protecting the conductors from your blade. At this point, you're done cutting. Bend the cable back and forth once or twice in order to fatigue that little bit of remaining steel into breaking. Then check the insulation on the conductors, tear away the kraft paper if the cable is modern enough to contain that paper barrier, and insert an antishort bushing, a "red hat" (this is described later in the book).

Screwdrivers

The next major tools are screwdrivers. Each screwdriver should reach the bottom of the slot or indentation in the screw head for which it is used, without wiggling or rocking. It is conceivable, but very unlikely, that you never will need a Phillips screwdriver. Older devices tend to rely on slot-head screws. Recently-manufactured ones tend to have combination-head screws, allowing you to use either type of screwdriver.

You will need at least one ¼″ and one ³⁄₁₆″ wide flat blade screwdriver. There's a very good chance you'll need a ⅛″ one as well, if for no other purpose than for inserting into slots on the rears of "backwired" switches and receptacles to release the internal springs holding wires captive. Buy at least a #3 Phillips.

There are five or six additional types of screwdriver to consider purchasing:

- A stubby (very short screwdriver) allows you to turn screws in spaces where you don't have room for a full-length screwdriver.

- An offset screwdriver serves the same purpose. I carry one combination Phillips–flat blade version, and a separate larger (¼″) flat-only version—plus a small ratcheting Phillips.

- A "persuader," an extra-large screwdriver, can be used for extra-large screws and also (with all due caution) for prying.

- A screw-holding screwdriver can come in very handy when you don't want to—or can't—hold on to the screw with your other hand. I have found screw-holding screwdrivers that use claws around the outside of the screw head to be less reliable than the split-blade type, which has a slide that wedges the blade tightly into the screw's slot. A screwdriver with the blade tip magnetized can be helpful, but won't hold a screw nearly as securely as a good split-blade.

- There also is a specialty slot blade screwdriver made by Ideal you might consider buying if you will do a lot of splicing. It has an inset in its handle for driving wirenuts.

I have two more pieces of advice regarding screwdrivers for electrical work. First, a screwdriver that *has* been used for prying to the point that its blade no longer is straight no longer will serve you well as a screwdriver. Second, a screwdriver whose blade tip is hardened will work well much longer than a screwdriver whose blade consists of the same mild steel for its entire length.

Other Tools You May Want on Hand

Your ladder, hammer, and other basics need no particular special qualities for electrical tasks. Clearly, metal ladders conduct. This issue aside, the most important characteristics of ladders are stability and height. A standard locking tape measure, ¾–1" wide, will work fine. Just don't use it near anything that might be electrically live; it will conduct dangerously. Lufkin is perhaps the most-reputable brand name in rules and tapes. You may want to use a level to ensure that the cover plates on your switches are plumb. Plumb installation is essential for the proper operation of the simplest thermostats and of "mercury" switches, described in Chapter 15, Replacing Switches.

Inadvisable Hand Tools

I've named the basic wiring tools, and described the characteristics to seek and to avoid in hand tools. Before moving forward, I'll mention four tools that I advise you NOT to use in electrical work. First is slip-joint pliers, which are useful in a few plumbing applications. They have curved jaws, and a pivot that has two positions, allowing the jaws to have two different opening ranges. The jaws have a nasty tendency to slip between the two positions. Also, they just don't come with insulated handles. The jobs you'd use them for should be performed with either lineman's pliers or channels.

Second are poorly-cutting cutters. Some cheap channels, for instance, have a cutting blade just beyond the pivot point. They're not really designed for cutting wire, and barely hack it. Use the right tool, and you'll do a better job with less likelihood of getting hurt.

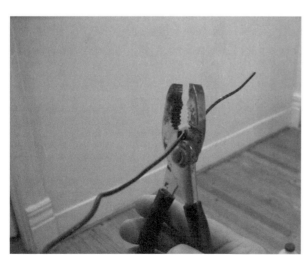

Slip-joint pliers; these could cut adequately—but use lineman's pliers. The latter are right for the job, easier to use, and safer.

Third are "multitools" that function as pliers, screwdrivers, knives, and toothpicks, but can't begin to complete with stand-alone versions of these.

Fourth are, essentially, guns. I can't recommend using explosive fasteners. They're fast, and if you have masonry walls or cement floors and a lot of anchoring to do, their use is tempting. However, they use 22 caliber or larger cartridges, and they're dangerous. I gave mine away, although I had always employed hearing protection when firing, and never had a case where the hardened steel nail ricocheted or shot right through a wall. Others have been less lucky.

Optional Hand Tools

A scratch awl, or "ice pick," can be handy for starting holes, and sometimes for removing knockouts. It can serve by itself for making holes in relatively soft materials, such as drywall, for the plastic anchors used to support light, noncritical loads. A pry bar or "handy-bar," or a small crowbar, can be useful for removing not only nails but also other items you are not too worried about damaging. They sometimes also can be used to "persuade" something into alignment so you can get a screw into it to hold it where it belongs.

Additional Useful Tools and Techniques Not Covered in Detail

Some miscellaneous tools could be of value for a homeowner doing a moderate amount of wiring:

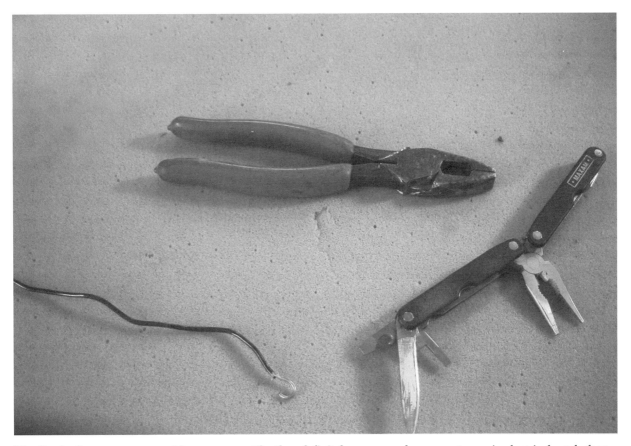

The "swiss" may or may not be more versatile; they definitely are more dangerous to use in electrical work than the pliers.

- A chalk line lets you snap a straight line that's easily wiped away. It's useful for locating the center of track lights and long fluorescent fixtures, and for ensuring that switches and receptacles are the same height.

- Any weight on a string will serve when you need a vertical line, and often will help if you're fishing a line down inside a wall. Of course, in old houses that have settled for many years, anything installed straight may mismatch every structural line visible.

- A "triple-tap" is a set of three (or six) graduated thread taps on one shaft mounted in a screwdriver-type handle. This allows you to create or restore 6-32, 8-32, and 10-32 threads (sometimes also 10-24 and 12-32) in appropriate-size holes. These are the most common threads used in electrical boxes.

Power and Light

Electrical work is so important because it supplies safe, convenient power and light. Naturally, you need power and light where you're working.

Light

You need enough light that you never have to guess—especially when using testers. You have all sorts of options for lighting your work area. Two D-cells are the minimum for a more powerful flashlight. The better the light, the less risk, if only because bad lighting results in eyestrain.

Power

You will need an extension cord for some testing, in addition to serving tools or lights. Make sure your cord is Listed, grounding-type, and as heavy a gage as possible. Most are AWG (American Wire Gage) 16, which is acceptable—barely. AWG 18-gage cords are junk. Longer cords, 50' or more, especially those with multiple outlets on the end, should be 14 gage, or even 12.

If you want the type of extension cord that is combined with a lampholder, be sure to buy one that has a protective cage. Even so, buy a "rough service" lamp for it, which is relatively shock resistant.

There are some useful accessories for extension cords. Consider buying a portable GFCI. These come as plug-in units, or as part of extension cords. If you must use an ungrounded outlet, never, ever use it for three-prong tools without GFCI protection. There are GFCI adapters that intrinsically adapt to two-prong outlets by allowing you to withdraw their grounding prongs.

Reprise

Before I leave the subject of tools, let me repeat what I said at the beginning. Personal Protective Equipment, especially eye protection, is most essential. After that, the tools that you pretty much have to have are a voltage tester, a pair of lineman's pliers, and at least one slotted screwdriver. Any other tools may be optional.

10

Your System's Physical Components

One of the things I loved and still love about being an electrician is working on what normally is unseen: the inside of walls, the guts of buildings and equipment.

I didn't get there instantaneously; there's no "instant electrician—just add water." When I started as an apprentice, one of my first jobs consisted of carrying bundles of steel pipe up many, many floors. Another was splitting rolls and rolls of wide, high-temperature electrical tape lengthwise so the journeymen could wrap it around tightly-curved conductors. This did not feel like an electrician's work. To get my hands into the walls and ceilings as anything other than a closely supervised laborer, I needed to learn how building systems work, starting with the names of the parts. I had to be told, "Carry this up to there" on my first day. With the beginning of understanding, I could be told, "Get a bundle of half-inch rigid, and you and Jeff walk it up to the second deck."

I talked about the concepts underlying your electrical system, the big picture which is the basis of wiring, in Chapter 2, Basics. You were introduced to its layout in Chapters 4 and 6, Evaluating Your Service Entrance and Examining Your Panel. To understand how the concepts translate into a system that you can use, you need to know the hardware. This chapter explains the pieces you will get your hands on.

Cables

This is a formal definition of "cable." Informally, though, sometimes where large-diameter conductors are pulled through raceways, they are referred to as cables even though technically they do not meet this definition. The only application to residential wiring is when people refer to service entrance conductors.

Without wires, you would have no electrical system. Wires, though, can't do it alone. Most houses are wired in cable. To be precise, a "cable assembly" is a more-or-less flexible grouping of two or more insulated conductors, surrounded by a protective sheath. In the sizes used for residential branch circuits, the wires are solid. The sheath holds the wires together and provides some protection from physical injury.

Romex

The cable most commonly used in residential work, especially nowadays, is (colloquially) romex, technically known as type NM, for NonMetallic sheathed cable. The current version, whose conductor insulation safely can withstand the hot temperatures in modern light fixtures, and does not suffer premature aging when embedded in thermal insulation, is NM-B. You'll find the special rules for romex in NEC Article 336. The sheath is plastic, although if you have old enough romex, the sheath will be pitch-impregnated cloth. All the most common versions of romex contain at least a black hot wire and a white. With two exceptions, the white is used as the neutral. If the black is the neutral, someone miswired. (Neither of the exceptions—the "switch leg" or the 240-volt circuit—allows the black to be used as a neutral.) The second most common version adds a third, red wire, also never properly used as a neutral. Any cable manufactured after WWII, or at least from the late 1950s and onward, contains a bare or (rarely) a green-insulated grounding conductor as well. Older romex was ungrounded; that's one way to identify it, and also its biggest problem.

I say the romex of the last half-century *commonly* has been grounded because there are exceptions. I have encountered ungrounded romex that was installed as recently as the mid-1950s. It was manufactured with plastic-insulated conductors but a cloth sheath. As late as 1999, when I reexamined an installation of this during my revision of this book, the insulation and sheath were in perfect condition, and the printing on them was fully legible. You need not presume that ungrounded wiring must be worn out, although you may choose to replace it for other reasons.

Inside the outer plastic sheath of relatively modern romex, you'll find a wrapping of thin but relatively strong brown kraft paper enclosing the wires and their insulation. If you find a significant amount of this sheath exposed inside a box, rip it away if you can do so safely. It is not guaranteed to be fireproof, and it does take up space.

Older romex is different in some ways. Cables manufactured before WWII used rubber-insulated conductors, each rubber coating wrapped with cloth, and no paper. The color coding of those older conductors often is faded by now. For a few years, especially in the early 1970s, a variety was installed that utilized aluminum (plain or copper-clad) rather than copper conductors. There are special rules for making connections to aluminum, so be on the lookout for this variety. Chapter 12, Making Connections, talks about aluminum wire.

BX

For a long period, it was illegal to install romex (and thus other plastic-sheathed cables, discussed below, as well) in houses with more than 3½ habitable stories above average ground level. Once most localities adopted the 1999 NEC, its use in one- and two-family dwellings of any height, and in most apartments, became legal. However, until recently some other cable system was needed. The cable that's been used in taller buildings is BX, technically known as armored cable (generally type AC, with the currently produced version being type ACHH).

Why the change at 3½ stories? Steel armor is safer from penetration and from fire than is plastic, and Code officials had to pick some point, however arbitrary, at which a building was likely to hold too many inhabitants to risk the lesser method. I've seen romex eaten away by rodents in crawlspaces; in these locations I substituted BX, even though the buildings were shorter than three stories. BX has been around since before the 1920s, and its installation in homes used to be very common.

Unlike romex, BX normally does not contain a separate grounding conductor, because the steel armor itself provides a grounding path. There is a thin strip of aluminum (if your house was wired in the 1940s, it might be copper) running just inside the armor to help the armor conduct high current and trip the breaker quickly by overcoming the cable's "inductive reactance." This strip is not itself an equipment ground, although some assume that it is and try to connect it to wires or screws. It doesn't get connected! It has no need to. The bonding wire wasn't present in early BX, which used the same rubber-and-cloth-covered conductors as early romex. Early BX, of which there are various versions, is not as safe as the modern cable. More recently manufactured BX, except for the lead-sheathed version, also has a wrapping of kraft paper around the conductors, providing a slight additional protection.

A version of BX that no longer is manufactured contains a lead sheath under the armor, making it suitable for potentially damp locations. The sheath lasts longer than do the rubber-insulated conductors. Therefore, the fact that you find the sheath in good shape does *not* mean you can stop there with your evaluation. Conversely, if you find perfectly healthy conductors in old BX, or even not

> ### ⚡ CAUTION!
>
> If your BX lacks the thin bonding wire, it dates back to the 1940s or earlier. If it lacks the paper, it dates back to the 1950s or earlier. Neither of these is in itself sufficient reason to replace your cable. But it's not as safe as the modern cable.

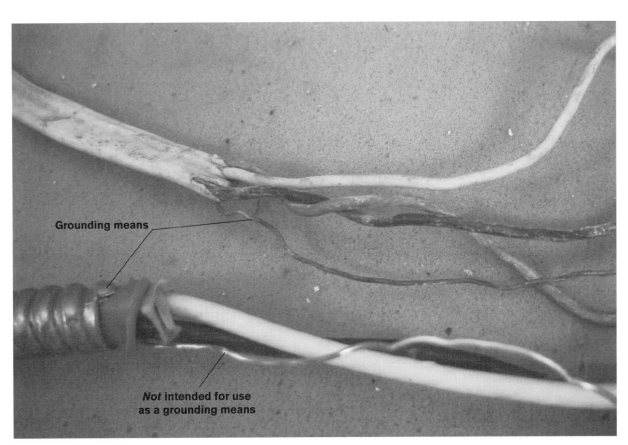

Grounding means

Not intended for use as a grounding means

BX has only insulated conductors; plastic sheaths such as romex and UF have ground wires. The thin aluminum strip inside modern BX sometimes is mistaken for, and misused as, a ground.

particularly old BX, this may not suffice; the armor may have rusted away in spots if it was run in a wet location. This would mean you have to replace the cable. Check for this. Damp locations are bad enough; BX is not legal for use in wet locations, outdoors, or underground.

UF

UF probably is the cable most commonly used for residential branch circuits after romex and BX. *U*nderground *F*eeder cable is a close variant on romex; it has solid, parallel conductors and is commonly used in branch circuit sizes. It lacks romex's paper interior wrapping, and the conductors, with their insulation, are imbedded in solid plastic. UF is legal for use in locations that are wet, underground, or sun-exposed.

Romex is not legal for these uses, though I have found it used this way far too often. The more people who have worked on a house, the more likely someone ignorant has done some of the work, and this is a common example of what such people do. Because romex and UF look similar, you need to make sure that cable used outdoors or underground is *not* romex. Read the legend on the sheath. If you cannot, look at the points where the wires leave the cable, so you can see whether they emerge from solid plastic. You also may be able to distinguish romex from UF, tentatively, by the fact that the solid plastic construction makes UF stiffer than romex.

SE

You saw how Service Entrance Cable (SEC), Types SE and USE or SEU, is used for electrical service entrance conductors, back in Chapter 4, Evaluating Your Service Entrance, but this is not its only use. Inside, used for branch circuits or feeders, it follows exactly the same rules as romex. A service cable's conductors are stranded, and its bare neutral is wrapped concentrically around the insulated conductors, just inside the sheath. As you will learn in Chapter 12, this makes it a little more difficult to attach the conductors in service cable than it usually is to attach solid conductors.

SE cable, unlike romex, is legal for use in the sun or rain. U̲SE or SE̲U cable is a version that is also legal for use underground. This can make it useful for feeding subpanels located in outbuildings that have heavy loads.

SE's use normally is limited to feeding loads such as stoves and central air conditioners. If it is used to feed a subpanel, it has to be type SE̲R, which contains three insulated conductors plus a ground. SER generally is only available in aluminum. Fortunately, in the sizes of conductor used in SER, the use of alu-

SER is required where service cable feeds an electric stove or dryer because the neutral needs to be insulated, unlike the ground. Between World War II and the adoption of the 1996 NEC, though, SE cable, which has only two insulated conductors and a bare combination neutral/ground inside its sheath, was permitted for these circuits, so long as they were fed—any distance—from your service panel rather than from a subpanel. If this is what you find, don't worry about it; there are no records of fatalities associated with this exception. It still has been ended, though; from now on, these installations follow the same grounding rules as any other wiring.

minum is not a serious concern, so long as the cable is installed properly and the conductors are kept quite dry.

Raceways

These were the common cable types. Some old wiring and some very classy new work uses "raceways" instead. A raceway is something you can pull wires through (or, in some cases, lay them in). Conduit and tubing are round raceways. There are various common raceways, some found more commonly in old wiring, others not at all.

Conduit

"Rigid metal conduit" (RMC) or "galvanized" (GRMC), is electricians' pipe, though it is not usually referred to as "pipe" in the trade, but as "rigid." Galvanization has not been around as long as conduit. In older homes, you are likely to encounter black conduit, which is enameled (painted) rather than galvanized. There is nothing at all wrong with this, though it has not been manufactured in many years. It probably did not hold up nearly as well as the galvanized version in wet locations such as outdoor and underground wiring, but I am assuming that your main concern as you work your way through this book is with the inside of your home. The pipe both protected wires and provided a reliable ground.

Black conduit looks very much like plumbing pipe, but conduit systems are very different from water and gas systems. Bends in electrical conduit are made with slow, gradual sweeps of several inches' radius, so that wires can be pulled through without getting caught or scraped as it is pulled in the bends. Plumbers use and have used pipe far more widely than electricians, but their pipe should not be, and never should have been, used as part of your wiring system. While plumbers install and bend copper tubing, plumbing pipe is not bent. Bends in plumbing pipe are made with short, threaded female fittings that make sharp turns. This is why it's not right for wiring; it can create the same types of problem as improperly-bent EMT (which is discussed below). Many amateurs have used plumbing pipe and fittings for wiring. Some electricians ran wires through gas pipe in the very early days of electric lighting as they replaced gas light fixtures with electric ones. Should you find any pipe other than electrical conduit serving as part of your electrical system, call in a pro. Something is very wrong.

Gas pipe poses an additional danger. It is exceedingly unlikely that you will find wiring run through gas pipe; I have found only one or two electricians who encountered this. However, it is far, far more likely that in an older home a gas fixture was wired for electricity, or at least a gas fixture was replaced by an electric fixture in the same location. If this is the case in your home, it is quite possible that this gas pipe was used to support the outlet box and fixture. Beware—it may be connected to your gas system! Do not unscrew the cap from anything that looks as though it may be a gas pipe unless you are prepared to deal with gas; doing so is well beyond the scope of this book.

PVC

You will almost never find "nonmetallic conduit," (formally, Rigid NonMetallic Conduit, RNMC, and commonly called electrical PVC, PVC conduit, rigid PVC, or plastic conduit) inside older homes. It was not in use until the last few decades of the twentieth century, and normally it only is installed outdoors, underground, and in other wet locations. Clever electricians recognize that it can be much more convenient and much less disruptive to run wiring on the outsides of old houses, where doing so is aesthetically acceptable, than to tear up walls and ceilings or to spend long hours fishing cables through those walls and ceilings blindly. PVC is flexible, and can be bent—permanently—by hand, if it is heated enough to soften. If any PVC was mounted on the outside of your house, check to make sure that it does not have any straight sections that are so long that they have pulled apart at their joints as temperature changes expanded and contracted the plastic.

EMT

"Electrical metallic tubing" (EMT), also called "thinwall conduit," has been in use longer than PVC. All the problems I find with EMT installations involve misapplications.

- EMT is legal for use outdoors, but it does rust—it is not galvanized like pipe. However, installing it in such a way that you can expect gradual rusting is not a Code violation, even though painting would make it last much longer. Install it underground, though, and you *are* required to protect it, usually with some substance like underbody coating or roofing tar. If you have EMT outdoors—above ground or underground—make sure it still is intact. It serves as protection for wires, as a means for keeping water out of your electrical system, and as a grounding means. People have died because buried EMT rusted away. Loss of grounding meant that a short to the case of equipment it served did not trip the breaker feeding it, but waited to electrocute someone touching the equipment.

- The connectors that join EMT to electrical enclosures, and the couplings that join lengths of EMT, normally use setscrews. However, standard setscrew hardware is not intended for use with EMT outdoors; that takes other fittings.

- EMT can be bent by electricians, using tools designed for the purpose, after they develop the needed skills. These are two critical elements to bending it properly. Some people bend EMT by putting a big foot on it and pulling up, by levering it against a railing, or by similar creative but dangerous means. Kinks and tight bends can result, and these tear the insulation on conductors pulled through. Like any insulation damage, this is not necessarily evident at the time, but can result in an unpleasant surprise later.

Wiremold

In finished areas where additional wiring was needed but owners were unwilling to cut open walls and ceilings, sometimes a compromise was deemed acceptable, using flattened variations on conduit. All of them serve basically the same function, and all are colloquially referred to as "wiremold." Used to extend existing wiring with minimal damage to walls and ceilings, and considered more attractive than exposed standard round raceways, their presence tells you nothing about the age of your house's in-the-walls wiring. Note: if you see wire, pieces are missing.

Greenfield

Flexible conduit, ("Greenfield" or "flex") which looks like largish BX, has been around for almost 100 years. Its conductors are pulled in by the installer, and it follows different rules than does BX. You are very unlikely to observe any, though, except perhaps as part of your furnace's or air conditioning compressor's wiring. If you look up in your ceiling at the permanently-installed part of a recessed light fixture, you may see some leading into its lamp socket.

Open Wiring

Open wiring and knob-and-tube wiring (these are essentially the same) consist of insulated conductors without any sheath or raceway. This is a very early wiring system whose remnants are disappearing. However, some of it still may be found—and in use! The wires were run at least an inch apart, and were supported by being clamped to, tie-wired to (tied to with wire—probably a short piece of the same insulated wire), or run through pieces of white porcelain. The porcelain hardware is known respectively as cleats, knobs, and tubes. This is such a very old system that it almost always lacks grounding.

There's good and bad. One advantage to this system is that the conductors don't tend to short out, even when the insulation falls apart, because the hot and neutral are separated. What's more, the insulation can last very long indeed. Being open to the air, the conductors dissipate heat very, very well; and heat due to the power conductors carry is a prime destroyer of insulation over time. Add thermal insulation to the spaces through which they run, though, and this changes very quickly. If your outside walls contain knob-and-tube wiring, and insulation was poured or blown in, perhaps during the 1970s energy crisis, any knob-and-tube wiring inside them probably has suffered deterioration exceeding any it knew during the previous 50 years. The Consumer Product Safety Commission has recommended that where 20 amp circuits were run in knob-and-tube that has had insulation added around it, the circuits be derated to 15 amps. Otherwise, functionally, they will have become overfused.

Enclosures

Cables and raceways connect enclosures. An electrical system may function after a fashion without enclosures, but they are essential to its safety. Conductors may be spliced or otherwise joined in these, and enclosures may contain *devices:* switches and receptacles, primarily; sometimes other gadgets for the control or distribution of power.

There are quite a few types, not limited to the following.

- Your electrical panel is one type of enclosure, a cabinet, containing "protective devices."

- The splice compartment of a recessed light, or of a garbage disposal, is an enclosure.

- Your meter base is yet another type of enclosure.

Except for knob-and-tube installations, basically every connection and termination of branch circuit or feeder wiring should be made in an enclosure. Just about every enclosure you install, or otherwise deal with, will be a box. Enclosures that don't support devices such as receptacles are called *junction boxes.*

There are four exceptions to the rule that every connection and device must be enclosed. Splices in low-voltage wiring, such as wiring that operates doorbells, intercoms, telephones and alarm systems, need not be enclosed. However, the 120-volt side, the branch circuit side, of the transformer feeding this wiring must be safely protected in an approved enclosure.

Low-voltage wiring tends to be quite safe without any enclosures. This is less true of the other three exceptions. One is the system of knob-and-tube wiring, which I mentioned earlier. If you have any surviving knob-and-tube, not only may you not find cable or raceway as part of it, you may find splices that were installed, quite legally at the time, without enclosures.

There were lighting outlets and devices, too, that did not require enclosures, but it is highly unlikely that you still will find any of these associated with surviving knob-and-tube wiring. For instance, you are unlikely to come across a two-part bakelite switch that clamps over romex any more than a porcelain switch that clamps over two open conductors. In the late nineteenth century and the first half of the twentieth, installers of farm wiring—that is to say barn wiring but not farmhouse wiring—legitimately used such devices. Unless you are living in an incompletely-converted barn, this should be quite inapplicable. Anyway, these last two systems do not represent the best modern standards of safety. Finally, the NEC permits the use of special splicing devices—even concealed—to splice or tap romex in old work. I think it's unwise to use them instead of accesible junction boxes, because you could be in trouble if you need to find the splices later. I emphasize, moreover, that the devices that are legal to use in this way are not just wirenuts but specialty items.

In Chapter 13, I will talk about evaluating, and to some degree repairing, boxes that are in place. This is one reason I will spend some time now talking about boxes.

Box Types

Quite an assortment of boxes are in common use, and even more can be found in the walls of older houses. Even a particular type of box, for example a "pan-

> ### ⚡ CAUTION!
> Actually, you may well *not* find the splices because they will be hidden in the walls. Fortunately, if they have survived this long they probably were well-enough made that you can rely on them for as long as the system remains in use. The same does *not* go for hidden splices created in more recent times by people wiring ignorantly.

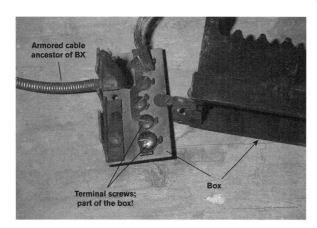

This ancient two-part box with 1-conductor cable was used with direct current on farms. This box had a base and a top part, conceptually somewhat like wiremold boxes.

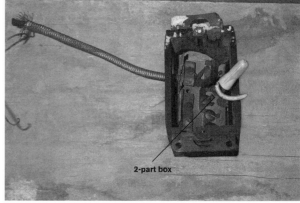

An ancient 2-part box adapted for modern A.C. It was in use for close to a century. Instead of taking the box apart to replace a conductor, someone pigtailed one of the old conductors using a wirenut.

Watch out for this basic mistake made by some very ignorant installers; it defeats grounding. People who are unfamiliar with any wiring system other than romex sometimes replace an existing metal box, perhaps even for the valid reason that it is undersized, with a plastic one. The existing box, in these problem cases, was fed by a metal raceway or, more commonly, in BX. This didn't faze the installers, even though they were unfamiliar with metal wiring systems. In some cases, they just let the BX hang loose near the new box, and bridged the distance with their splices. In other cases, they stuck the BX in through some opening in the new box, not worrying about the need for a connector. Either way, grounding was defeated because there was no system to properly bridge the gap from the grounded armor to the romex's ground wire.

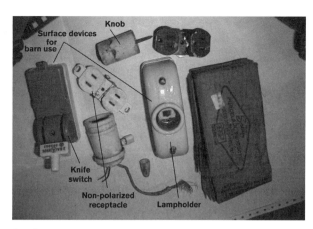

Antique components.

cake box," which is round and only ½″ or ⅝″ deep, may have come in different designs, even with varying dimensions and shapes. There are box extensions that can be screwed onto boxes to add volume. Some boxes contain internal clamps, suitable for attaching specific types of cable—and ONLY those types. Others require the use of connectors designed for securing the particular type of cable being used. The latter types of box, and some of the former, are manufactured with embossed, round *knockouts* that can easily (and on cursed occasion distinctly not so easily) be punched or pried out so you need not drill your own openings. Any that have been knocked out and are not, or are no longer, needed, must be closed, using spring-metal or plastic disks called knockout seals. With

Porcelain knob with wire.

Knockout seal. Spring tabs hold it in the knockout.

Generally speaking, besides wiring terminals and functional mechanisms, a "device" has a yoke, a body, and a face. The face is what you see when you look at the front of a receptacle, with the equivalent being the handle of a switch. The body is the bulk of the device, normally hidden in the box. The yoke is what supports these, reaching from the top of the box to the bottom. A "single-gang" mounting supports one yoke, a "four-gang" mounting supports four yokes, and so forth.

exceptions so rare that you can ignore them, non-metallic boxes are not suitable for use with metal cables or raceways; metal boxes potentially can be used with any wiring system.

In the past, metal boxes were the most common, and came in the widest variety. Modern ones are relatively shiny; older versions were treated with black enamel paint. The box most widely used in residential work, past or present, metal or nonmetallic, is called the switch box. It has various types of systems for mounting to your walls; switch boxes never have been designed for use in ceilings to support lighting fixtures.

Metal sectional boxes or, colloquially, "gem" boxes or just "gems," are far, far more common in older systems than are nonmetallic boxes.

The gem is a versatile servant. It not restricted to use with switches (nor is the equivalent plastic box). It has two tapped 6-32 holes, one at top and one at bottom, whereby it holds one standard switch, receptacle, or similar device. Two or more gems can be combined, "ganged," side by side, as needed, by removing their side sections and using the screws and hooks that had held those sides on to the boxes' bodies to clamp the now open-sided boxes together. This is not true of nonmetallic boxes, which never have been capable of being ganged. Installers always have had to purchase the latter in the exact widths needed: one-gang, two-gang, three-gang. . . .

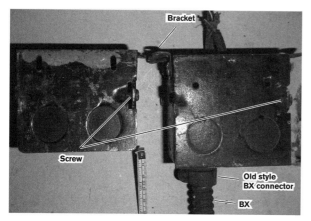

Two gem boxes, different depths, hence nongangable. If the box depths matched, the screws could be backed out, the sides removed, and the 3/4-boxes that remained could be joined by screws to form a single "two-gang" box.

Front flange boxes. Different flanges (brackets) are designed for different set-backs, meaning different wall (or ceiling) thicknesses between the surface and the stud or joist to which the box is nailed.

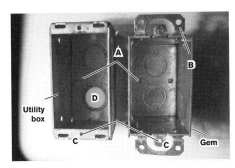

Utility box and gem box; note that the utility box is longer, giving it more volume for the same depth. A: Tapped holes for 10-32 grounding screws. B: Holds against front of wall or, if slid back, used for mounting box to lath. C: Tapped holes for 6-32 screws to hold devices. D: Open knockout.

The octagonal shape is not the result of idiosyncrasy. Round light fixtures, or at least round canopies (the parts that sit against the ceiling) have advantages. Therefore, more-or-less round electrical boxes are the best shape to support and feed them—and be concealed by them. However, a box needs flat surfaces to attach the ends of cable connectors and conduits. The only safe ways to make connections to a round box are by entering through the (flat) back or by including clamps in the box's design. Octagonal walls offer more flexibility.

Note that ground clips are not intended for curved walls.

Perhaps the next most common boxes after gems are round or octagonal boxes. These are required for ceiling lighting outlets. Modern ones all have two threaded holes in ears on their edges, opposite each other, and come with 8-32 screws already installed in the ears. (Gems, incidentally, do not come with screws in their ears because the screws are provided with the devices that will be mounted in them, or with the blank covers used when no devices are installed.) Some old round boxes do not have mounting ears; this can cause you some difficulty when you want to replace a lighting fixture whose mounting presumes the presence of ears. There are means of working around this presumption. Some round boxes are only 3½″ or 3¼″ in diameter, which results in more crowding and more difficulty in finding covers.

Like switch boxes, octagon (they are not called "octagonal" boxes in anybody's usage) and round boxes may come with varying mounting systems, and may contain cable clamps, knockouts for connectors, or both. Like plastic switch boxes, round plastic boxes normally are unsuitable for use with metal cables and raceways. Like switch boxes, round and octagon boxes easily can be overfilled by installers who fail to perform the calculations required by NEC Section 370-16.

Common square boxes are 4″ square or, rarely, 4¹¹⁄₁₆″ square. Rectangular surface boxes, which come in a number of depths, are called "utility boxes," "handy boxes," or "handi-boxes." With very rare exceptions—none of the exceptions have ever been standard—these are 2″ wide by 4″ long. Modern octagons are 4″ in "diameter." Their older versions, some of which are round, use the same fittings.

Extension boxes, or box extensions, are available for utility boxes, square boxes, and octagonal boxes. Octagonal extension boxes really are not designed to extend round boxes, and modern extension boxes for 4″ square boxes have mounting holes that may not match the screw holes in older 4″ square boxes. Therefore, using them for this purpose is of arguable appropriateness.

Extension boxes look almost the same as the boxes they extend. However, the backs are missing, except for what they need in order to be screwed down over the boxes that need to be extended. Extensions for a box may be available in 1½″ and in 2⅛″ depth.

An exposed 4″ square junction box simply takes a solid, flat cover. Not so where a box is used in a wall or ceiling for mounting equipment. A "device ring" with a rectangular opening screws to the square box, and you can mount either one or two switches or receptacles to that "ring," which has threaded holes corresponding to the device-mounting holes in a switch box.

Device rings come in different depths, accommodating different wall finishes. Installers don't have to know how thick your wall will be before they

An ancient, earless pancake box with a fixture-support stud. The built-in stud, or center fitting, is for hanging a light.

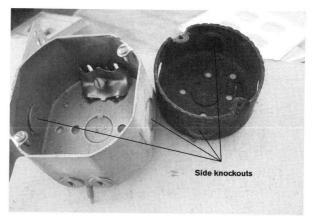

A 4-inch octagon box versus a 3½-inch round box; they enclose considerably different volume. With flat sides, the modern box (a 2½" deep version, here) has more good-contact knockouts, in addition to cable clamps accommodating as many as 4 romex-type cables. Side knockouts, being in curved surfaces, did not make very good contact with locknuts in round boxes; only the bottom knockouts did.

Nonsectional boxes, especially boxes such as square ones and "utility boxes," are made both in rounded-edge and square-edge versions. Given that the outer dimensions of a particular metal box, for example a 2½" deep utility box, which by definition has a 2" by 4" cross-section, are standardized, one with squared edges has a tad more volume. This does not make it a better box—on the contrary. The reason is that boxes with squared edges are not die-formed but instead made of individual pieces of metal tack-welded together. On admittedly rare occasion, a weld will fail and the box will come undone. This disadvantage more than compensates for the extra volume created by the squared edges, which offer little benefit; conductors used in branch circuits are not flexible enough to poke into sharp corners.

"rough in" your wiring, meaning to install the parts that have to be in place well before the walls and ceilings are covered. The rings come in single-gang and two-gang versions, corresponding to whether they are to support one or two devices. This means that if they install a duplex receptacle in such a box, and later you decide that two duplex receptacles allowing four connections would be more useful at that location, the change is relatively painless. Follow the procedure in the next chapter for replacing receptacles in a multigang box. Also, if you add paneling or tile to a wall or ceiling, you don't have to move or replace the box. You can replace the device ring with a slightly deeper one: in electricians' jargon, one with more "throw."

The term, "Plaster ring," or, informally, "mud ring," refers to a ring with a round opening, designed to support a light fixture or smoke detector from a square box. It has threaded holes corresponding to the threaded holes in an octagon box. This ring too comes in different depths to accommodate ceilings of different thicknesses.

A normal utility box versus a narrow old oddball; the latter has almost no safe, legal use. This odd, old handy box may have a switch in it, but you probably won't find a replacement that mounts to its holes.

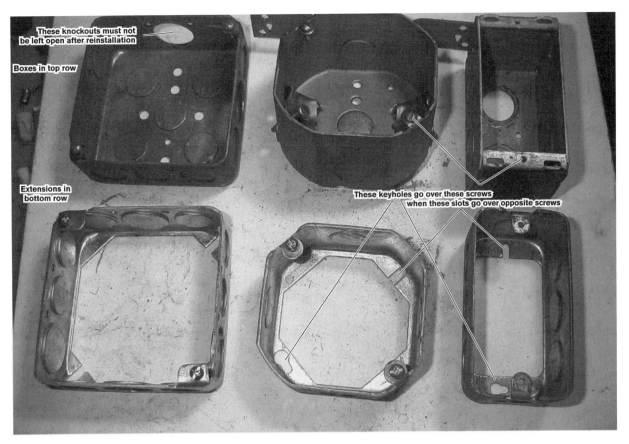

These knockouts must not be left open after reinstallation

Boxes in top row

Extensions in bottom row

These keyholes go over these screws when these slots go over opposite screws

Assorted boxes and box extensions.

When you observe that a box is recessed behind the surface, you have uncovered a problem. You need to bring it forward—flush, if your wall or ceiling surface is wood. Do recognize that replacing a ring requires you to make a 4″ square hole. This requires careful or clever work to avoid an unsightly result, if you are dealing with a single receptacle or switch. One trick is to change a single outlet to one with two duplex receptacles, or, where safe and legal—determining this requires a somewhat advanced understanding—converting from a single switch to a switch with a receptacle partnering it, alongside, sharing its box. A two-gang cover is wide enough that it will cover a carefully-made hole sufficiently large for you to remove the existing ring and replace it.

There is a complication associated even with the task of buying a replacement ring—or cover. (First, be aware that the term, "device ring," is not commonly used. "Mud ring" is a colloquialism used for both "device ring" and "plaster ring.") The complication is that the mounting holes in many older square boxes are configured slightly differently than those in modern ones. Modern mud rings and covers will *not* fit many old boxes. Whether they will fit yours depends on just when your boxes were installed. There was a period when square boxes were manufactured with dual sets of mounting holes to accommodate both ring configurations.

Switch boxes come in varying depths. Square and octagonal boxes come in two standard depths: 1½″ deep and 2⅛″ deep. Old round boxes came in those two depths and also in ⅝″ or ½″ depth. The latter are called "pancake" boxes, and still are available. However, they tend to be very, very crowded.

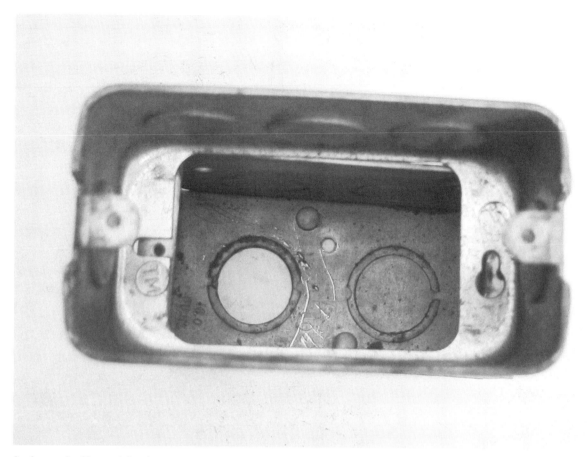

An imperfectly-matched extension ring over a utility box, not screwed together.

Evaluating Boxes Is Easier than Specifying or Installing Them

You don't need to know much about box mounting for the jobs covered in this book. New-work boxes have brackets or staked nails (nails mounted to the boxes so that they are unlikely to fall away in shipping and handling) that let them be securely mounted to studs or joists without wall surfaces blocking their way. This book is not about new work, nor about installation, just about evaluation and replacement. Therefore, I will not explain how to use new-work mounting systems. Ordinary wiring books should not steer you wrong with regard to mounting boxes in the course of new work.

Volume is another matter, and far more problematic. To install or replace a box, you need to understand mounting systems, box types, and—very important—how to calculate how much room you need. The rules on box fill are a little confusing. NEC Table 370-16(a) tells you how much volume you can figure on in standard metal boxes. This number for volume is a bit smaller than the product you would come up with by multiplying the boxes' dimensions, length times width times depth. Like so many con-

A utility box ("handy box") and an extension box, differently contoured but fine to use together. Depth is nearly doubled when you use this extension box. The increased volume helps keep the wiring cool and undamaged but can be a problem when you need to reach in for splicing or taping.

A receptacle on a device ring in the wall, clearly recessed. In this case, the drywall was added over an existing wall and the result is a badly recessed outlet and a gap. Also note that the yoke ears are trapped so the receptacle cannot be pulled forward. This needs to be excavated.

A receptacle on a device ring. Exposing enough of the ring to pull it out and replace it means exposing just about the whole 4", even if you can cock it so it comes out one edge first. This means that all this will need not only patching but *smooth* patching and painting or wallpapering.

struction materials, from "two-by-fours" to "half-inch" conduit, items' names do not give you their actual measurements.

You have to go through Article 370 rather carefully to understand the rules. You cannot rely on the table, or the boxes' markings, alone. The fact that a plastic box, for instance, may have printed on it, "Up to 8 #12 conductors" does not mean that you can legally install two cables, each containing three insulated #12 conductors and one bare. Maybe more are legal, maybe fewer.

Even if you understand the rules, you will face problems if you try to apply them to old boxes. You'll find old gem boxes with standard 2″ by 3″ dimensions

Two receptacles on a two-gang device ring. When you need to replace a single gang device ring, perhaps because it doesn't bring the enclosure far enough forward, using a two-gang ring, and usually, a second device, eliminates most of the patching, and, usually all of the painting or wallpapering. A two-gang cover extends *beyond* this width, which is what you had to cut out to replace it.

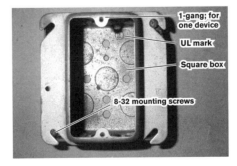

A one-gang ring on a square box— loosen two screws and you can replace the ring. The 8-32 mounting screws are in slots that allow the device rings to be replaced easily with a couple of turns of each screw.

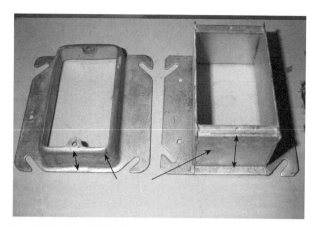

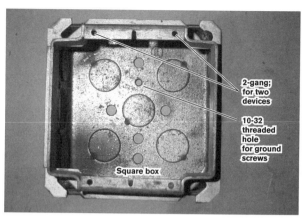

Availability of single-gang rings of different throws (depths) makes it easy to bring an enclosure forward just the right amount. Your receptacle is no longer recessed if you replace the ring on the left with the one on the right. However, you still have a fair amount of cleanup. Hence the advantage of using a two-gang cover.

A two-gang ring on a square box. The main concern in choosing this approach is confirming that there is sufficient volume for another device. You may run into another problem when you try to line them up, as shown below.

in front, where you install devices, but with non-standard shapes inside the wall. Some have strongly-beveled sides, considerably reducing their volume; others have sidewards extensions sticking out alongside them inside the wall. These old boxes no longer even are listed in the table. Therefore, bringing a new cable into an old box can be problematic. If you're just replacing a device—which is as far as this introductory volume takes you—you're covered. Nevertheless, a box that was installed quite legally in the 1940s or the 1980s, and therefore is grandfathered, can cause problems. If it is so crowded that you can't work in it without the likelihood of causing damage, you need to replace the box—a task beyond the scope of this book. Volume calculation is far from the largest of the many reasons that I have excluded this task.

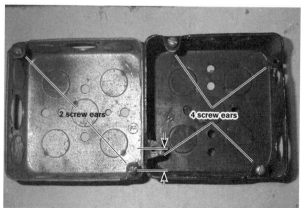

A modern two-gang device ring on an old square box—it doesn't line up. The ring overlaps this box to the left and doesn't cover it on the right, even though the screw going into the 8-32 hole in the box means the sides *should* line up.

Contrasting square boxes, old-style and modern-style. Note how the new box's screw ears and the old box's ears are not in the same places.

An old cover on a modern square box. The old cover is hooked on a box screw in the only way that will allow the one side to align—but it means the other 3 sides don't line up, and none of the other holes can reach the box's other screw. But don't discard the cover!

An old cover on old square box. However ugly this cover, it *does* fit the old-style box. Screws and slots line up for good coverage. Wait till you know you're rewiring and removing all these old boxes before discarding components that might be irreplaceable.

From Cable or Raceway to Enclosure: Clamps and Connectors

In general, an electrical cable must be secured to any enclosure, either with external connectors or with the clamps that come with, and are more-or-less integral to, many boxes. For quite a few years, single-gang plastic switch boxes have been the only ones that have generally been exempt from the requirement that you use a clamp or connector. (See NEC Section 370-7.) Up to some point in the 1970s, though, (the year and month depended on the jurisdiction that inspected houses in your area) nonmetallic cables enjoyed this exemption when entering any type of nonmetallic box.

You don't need to know how to recognize all the different configurations of connectors and clamps that have been manufactured for use with cables, if only because in most cases all you can see will be the part that sticks inside the box. You do need to learn the basic differences between what you should see from inside an enclosure as you look at clamps and connectors designed for use with BX and with romex. These also are the differences you should keep in mind when someone sells you a connector to use when correcting an installation by someone who omitted clamps or connectors.

There are three.

- Because the armor of BX is the means of grounding, it needs to be pressed or squeezed firmly. Most modern BX connectors gouge the armor with setscrews. The sheath of romex, on the other hand, cannot withstand strong pressure or gouging. A setscrew would penetrate it and probably short the conductors it enclosed.

- If only because the armor of BX is hard and the cut end is sharp, it must be prevented from pushing even a little bit into the box where it could jam into conductors and cut their insulation. The situation is different with romex,

Various clamps. Note that the BX clamp folds underneath, to get a better grip on the armor. The openings stop the armor but allow conductors into the box.

which can shift a little as you install it, or later, in response to your building's settling or vibration, without any harm. Therefore, any BX connector or clamp must have a stop, a positive means of restraining the armor. You usually can't tell, when looking from the inside of a box, what kind of connector is on its other side, within the wall, holding the sheath on the inside, though, you certainly can see whether there is anything to keep the armor from simply sliding in.

- For the same reason, the conductors emerging from the BX are especially vulnerable even as you insert the BX into the connector or clamp. Therefore, an anti-short bushing must be inserted between the armor and the conductor insulation after the armor is cut back, before the cable is installed. This "red hat" can be seen from inside the box. (BX with an inner lead sheath is an exception; the lead can be bent back over the conductors to protect their insulation.)

Raceways are not held to boxes by clamps, although EMT uses connectors. Old conduit always should be held in place by locknuts and bushings. With old conduit there always will be a locknut on the outside, and sometimes one on the inside as well. There should be a bushing on the inside, whether or not there is a locknut.

Any BX or conduit not fed exclusively to metal boxes, or not secured to one of the boxes it enters, goes high on your "to correct" list because breaking the continuity of the metal path defeats grounding. I repeat this, despite the fact that I mentioned it a few pages back, because grounding is so important for reducing the risks of shock and fire.

CAUTION!

A locknut often has a concave and a convex side. When this is the case, the concave side always bears against the box. If you see it facing away from the box wall, the installer made a mistake. Put that on your list of problems to correct and be on the lookout for other important mistakes.

You would think that the issue of metal-to-metal continuity would apply only to boxes, or at most to boxes and to integral housings that come as part of light fixtures. There is an exception, and if you are very, very unlucky, your house may fall into this category. Between 1990 and 1996, a well-known manufacturer, Square D, marketed the Trilliant line of electrical panels. These panels employed nonmetallic enclosures, and were intended for use with nonmetallic cable only. Clearly, you won't find a Trilliant as original equipment in a very old house, but it could have been installed as an upgrade even in a 1920s relic like mine. If you find a Trilliant with BX or a metal raceway attached, it is very likely that someone did an illegal panel replacement. To make the transition from a metallic raceway or cable into a Trilliant, installers needed to add metal junction boxes and nonmetallic cable or raceway to make the bridge.

Supporting Cables and Raceways

Wiring needs to be secured, whether it is run exposed on your basement ceiling or in the walls where it will be hidden. All cables should be stapled or strapped to the building, or run through holes, within a foot of the panel (or of any enclosure). All raceways need to be secured within three feet of each enclosure. (The distances are measured along the cable or pipe, not along the "geodesic," the shortest line.) There are staples and straps designed specifically for use with electrical systems. Bent nails are not considered suitable means for securing cables or conduit, nor, for all practical purposes, are wires, nor is conduit a legitimate means of support for cables. Plumber's strapping is not appropriate,

Wiring can be "fished" blindly through walls and ceilings without being secured, but if the spaces are opened up to run it, it must be secured. One problem I have found frequently in older buildings is that ceilings in basements are removed, and cables that had been fished over those ceilings are left dangling.

and the staples to be used certainly should not have been office staples—not even carpenters' staples. Anything not properly secured goes (with relatively low priority) on your "to correct" list, but with higher priority in your "watch" list of circuits that may have been installed unprofessionally.

Wirenuts

Wirenuts are used in U.S. house wiring far, far more commonly than any other type of connector. Wirenuts (formally, "twist-on solderless connectors") were invented by Ideal Corporation in 1927. A standard wirenut consists of a more-or-less conical shell of thermoplastic, nylon, or bakelite enclosing a conical spring wound in a clockwise helix. When wires are inserted and held while the wirenut is twisted clockwise, the wirenut's internal spring expands to grab them, holding them against each other and strongly resisting any pull. Being conductive, the spring also ensures electrical continuity between the wires by slightly cutting into those outside parts of the wires that are in direct contact with it. The part of the wirenut that extends beyond the spring is called the skirt.

There are wirenuts and there are wirenuts. The early wirenut was made of porcelain, and was essentially just an insulating shell. I suggest that you do not reuse these antiques, should you come across any. Some very cheap—and I don't just mean inexpensive—manufacturers provide wirenuts that are plastic but, similarly, contain nothing to take a positive grip on wires. These are marginally usable for small, stranded wires that hold together very well. Paddle fan manufacturers are notorious for supplying these. I recommend that you dispense with such trash. Like the old porcelain connectors, they serve as little more than insulators. For literally pennies more you can create a vastly improved splice using good wirenuts. Finally, black bakelite wirenuts, being inflexible, are prone to crack if overtightened, hit, or crushed. Therefore, I mildly recommend discarding those old wirenuts, rather than reusing them for your own splices—presuming that you have the right replacements handy. Use wirenuts for conductors up to #10 AWG, or at most #8—the larger sizes only if you are very, very sure of your competence at splicing. Even if you find wirenuts Listed for larger sizes, items called bugs are more appropriate for splicing larger cables.

I talk about other ways wires are connected or attached in Chapter 12.

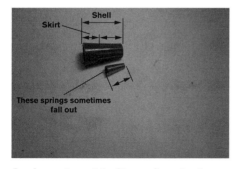

A wirenut next to its spring. Look inside for the spring. If it's not there, then you have a trash "connector." Looking inside, you estimate the length of the spring in order to trim back the wires before insertion.

Choosing Replacement Devices

Just as it is hard to know whether an "extra-capacity" washing machine is bigger, smaller, or the same size as one without the claim, it is hard to figure out which electrical equipment to buy. I will offer a little bit of guidance in the next chapter. There are some product standards that apply, and some tricky Code requirements that make a lot of sense but are not intuitively obvious. For full coverage, you have to read the relevant NEC articles, such as 380 on switches and 410 on lights and receptacles. I will touch on some of the issues.

Switches

Switches come in various ratings. A heavier rating *may* mean longer life. A 15 amp switch is built to handle higher loads than a 5 amp switch, and can be used anywhere in place of the lighter-duty one. Incidentally, at least one of my customers assumed that a switch marked, "15 ampere" could not be connected to a 20-amp circuit. Not so. Unless you're using it consistently to control significant motor loads, the switch's rating indicates how much current it actually can turn on and off—and it rarely is installed where it will control the whole circuit. However, if a switch DOES control a large motor load such as a very powerful garbage disposal, it may indeed need a higher ampere rating than you would expect, depending on its classification. See the NEC. Normally, whether a switch is marked, "AC only," "AC/DC," or "tungsten-rated" is irrelevant.

The next issue is at the extreme outlet limit of what you might evaluate: make sure that 240-volt loads do not have single-pole disconnects. If someone installed 240-volt baseboard heating and wired it through a standard snap switch, it could be a source of danger. In the event that people use the switch to disconnect it for servicing or repair, they may be seriously injured because they thought they had cut power to the unit. That is why if for some reason someone wants to install a switch to control a 240-volt load, the job requires a two-pole switch (properly a "double-pole, single-throw" switch), that interrupts both hot wires. It has four terminals, or five if it has a grounding screw. Even though a single-pole switch would suffice to turn the load on and off, using one for that purpose would be less safe and therefore is illegal. If you do come upon this violation, correcting it definitely goes on your "to do" list.

Normal dimmers are intended only to control permanently-installed 120-volt incandescent lights. Using one to control the motor of a ceiling fan is potentially dangerous and certainly illegal. However, there are motor controllers, suitable for use with some paddle fans, that are installed just like dimmers. There also are options for controlling fluorescents and high-intensity, low-voltage lights. However, no dimmer is intended for use with receptacle outlets; there is no way to know what will be plugged in.

Dimmers

A dimmer is a dimmer, the same to an electrician and a layperson. Like switches, dimmers can be either single pole or three-way. The old term, "rheostat," meaning variable resistor, refers to devices that went out of use as dimmers decades ago. It is most unlikely that you will find a functional one in your house.

With dimmers more than with switches, it is important to make sure how much power you need to control. A common mistake is exceeding the design wattage by controlling too many lights for the capacity of your device. When dimmers are ganged, meaning installed in a box containing more than just one, the wattage each one can handle goes down because their ability to dissipate heat is reduced in part by their proximity to each other. The capacity of the most common dimmer is 600 watt, single gang; 500 watt, 2-gang; and 400 watt, 3-gang. It is perfectly okay to install a heavier-rated dimmer than you need: 1000 watt, 1500 watt, even higher.

The best brand of dimmer I've encountered is Lutron, and next is Ideal. Leviton is quite acceptable, but cheaper brands tend to die young. Price is a fairly reliable guide for comparing one dimmer with another. Cheaper dimmers tend to dim over a narrower range, and to dim less smoothly across their range. One worthwhile feature in a dimmer is radio-frequency-noise filtering. Without that, the dimmer could generate static on your television or radio, or garbage on your modem line. If you hear a hum when your light is dimmed, try changing light bulbs. Some just have filaments that harmonize like tuning forks when dimmer-controlled.

Receptacles

Unlike switches, higher-rated receptacles may *not* replace lower-rated ones on a circuit of lower ampacity. Even though a 20-ampere receptacle is built with at least slightly heavier components than one rated at 15 amperes, other factors being equal, the "other factors" are far more important. Hubbell is one brand whose receptacles have an excellent reputation. Regardless of other reasoning, there is a Code rule regarding what receptacles go on what circuits. A 20-amp circuit may feed 20-ampere receptacles, either single or duplex. It also may serve multiple 15 amp receptacles (a definition that includes a lone duplex receptacle, since it allows two appliances to be connected). However, a 15-amp circuit may serve 15 ampere and only 15-ampere receptacles. Keep this in mind when you buy replacements. Finally, standard receptacles and switches have no business on circuits rated over 20 amps.

11

Materials, Listing, and Quality

This is the chapter to read before you buy materials—certainly before you use them. For example, I explain not only what electrical tape to buy, but also how to apply it.

To begin with, I will fill out your information about what to call things. Beyond that, I will help you develop the knowledge and judgment you need to evaluate what was installed, or to perform or to identify what's needed to do a high-class job. I'll start by talking about legal and functional requirements, but I'll go beyond that in my recommendations; the legal minimum is rarely good enough for optimum functionality and reliability. Considering the cost or trouble of any installation in terms of labor, it doesn't make sense to settle for marginal materials.

Don't expect to find all the materials used in your electrical system mentioned in this chapter. I leave out any significant discussion of:

- systems that are very, very dated, such as open wiring and knob-and-tube wiring, and soldered splices; and

- systems that, whether dated or modern, are beyond the scope of an introductory book.

Listing and Approval

What is the legal minimum? Using approved materials in accordance with manufacturers' instructions and NEC requirements. Approved means acceptable to your area's official electrical inspector. For most electrical materials, by which I mean materials designed specifically for the electrical trade, approval requires that materials be *Listed* (in a directory that mostly is owned by inspectors) and *Labeled*. Labeling, which is your main concern, means that they bear the mark of a *N*ationally *R*ecognized *T*esting *L*aboratory (NRTL). Underwriters' Laboratories, Inc., is the most widely known NRTL.

Foreign countries have their own testing laboratories, which mark equipment they evaluate. Their Listing and Labeling do not indicate that equipment is legal or safe to use in U.S. wiring; for one thing, foreign countries' electrical systems may make different demands on components. An exception is the Canadian Standards Association, which marks some equipment "CSA-NRTL," signifying that it passes U.S. standards as well as Canadian.

It is unfortunate that you need to look for Listing labels, but it is necessary because there is no law against selling non-Listed materials. Perhaps some day the equivalent of dramshop doctrine, which can hold bartenders responsible for the effects of what they sell irresponsibly, will find its way into law with regard to electrical suppliers. An example of the danger is that non-Listed cover plates for your receptacles—bought perhaps because they're less expensive, or because they're more attractive—may forfeit nonflammability.

Here's a nasty twist to the Listing issue. There are Listed products and there are Classified components; both bear testing laboratory labels. This can fool you because Listing and Classification have different meanings. A non-Listed product, say a lighting fixture, can be made with one Classified component, say a switch, that bears a laboratory's mark. However, the fact that the fixture you see in the store incorporates this one labeled component doesn't mean the product as a whole could bear investigation. A good inspector *may* catch such abuses. The best way for you to protect yourself in this regard is to check for a Listing mark on the package or the instruction sheet when buying equipment that has many parts. Small items such as connectors are okay so long as a Listing mark is stamped somewhere. To check a very small part, such as a wirenut that has no room for a legible mark, just look on the box it came in.

Some categories of material do not need Listing. These include items such as common hardware—for example nails, screws, and paints. Staples and straps may be listed, and paints or coatings applied to electrical materials such as cable sheaths, where coating is legal, need to be suitable for the use for which they're intended. Hardware is covered below.

Listed and *Suitable*

As important as it is to check for a Listing mark on the package, it is equally important to make sure what you've bought is right for the use you intend. For example, a barbecue starter may be safe for lighting the coals, but not for lighting your oven—that's not covered in the instructions because it's outside the uses that the manufacturer intends.

Underwriters' Laboratories, Inc.® (UL) and the American National Standards Institute (ANSI) have established standards with which legitimate manufacturers comply. Part of that compliance consists of the manufacturers' providing instructions for product use. Users, in turn, need to comply with these instructions. When there's an accident and the possibility of litigation, violations of such requirements (including the requirement that you GET inspection from your municipality or other jurisdictional authority) could be a significant factor in deciding culpability. Don't make assumptions such as "more is better," or, worse, "this looks right." At least check to see whether the NEC has anything to say about the equipment you are considering. An outlet box suitable for mounting a lighting fixture may look about the same as one designed to support a paddle fan, for example, but they're built differently.

As another example, some cables never should be used as 120-volt wiring in your walls, but nonetheless may have been. These include telephone, intercom, thermostat and bell wire, and flexible cords—all, perhaps, UL-Listed, but none

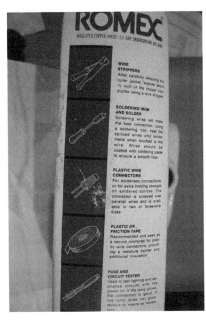

Romex box-instructions on installation; dated, but not so bad.

Listed for this purpose. The fact that they're never installed by people who know what they're doing means that someone quite ignorant worked on the part of your system where you found them. This is an extra reason to put them high on your list for removal and, perhaps, replacement.

I find a great number of Listing violations associated with lighting:

- Fixture conversions. Plug-in, portable light fixtures that customers may want installed on their walls or ceilings were not intended by their manufacturers to be hard-wired (connected directly and permanently to their house's electrical system).

- Moisture vulnerability. In any normal full bathroom, a light fixture that lacks a marking specifying, "Suitable for wet locations" or at least "suitable for damp locations" should not have been in-stalled. This is doubly true for installation outdoors; where rain might strike a fixture lacking a "wet location" marking, the installation is unquestionably illegal.

- Ignoring Ballast Heat. When the installation instructions on a fluorescent fixture say, "Wires within 3 inches of the ballast shall have insulation rated for at least 90° C," and you attach that fixture to 40-year-old wiring, you are asking for trouble. Even if the insulation is in good shape, you have to respect the prescribed clearance—or get a different fixture. Only since the mid- to late-1980s have standard cables contained conductors whose insulation is rated as suitable for an ambient temperature of 90° C.

- Ignoring the Heat From Lamp Sockets. Very many regular incandescent light fixtures have a similar restriction. You'll find it either on the package, in the installation instructions, or right on the fixture itself. Don't despair! A few fixtures still are being manufactured without such restrictions.

- Ignoring the Effects of Specialty Light Bulbs. You'll usually find warnings about temperature restrictions on the fixtures or their packaging. However, I have a light bulb whose packaging carries an explicit warning saying that it only can be installed in fixtures whose lampholders are ceramic; it gets hot enough, far enough back towards its base, that plastic sockets would be damaged.

- Overlamping. This is very, very common, as I mentioned in Chapter 5, Safety. If your fixture is marked, "Maximum 60 watts," and you need a little more light than a standard 60 watt bulb can provide, a halogen bulb may provide a satisfactory and safe solution.

Useful Products

So much for warnings. I talk about useful and nonrecommended products throughout this book, but, before I talk about materials as such, or about how to judge the quality of products such as receptacles, I will mention a few hard-tocategorize items that will help you deal with your wiring. If you are not familiar with basic hardware such as the varieties of screws and nuts, or with wall repair materials, you may want to find a resource to help you with this; perhaps a person, perhaps a book, perhaps a class. One reason is that you may need to patch a wall that was crumbling around a switch box, or reattach a conduit strap whose anchor worked free from a basement wall. The other is that I should not take the space to explain the differences between a flat head screw and a fillister head and a washer (or truss) head. Yet, to evaluate what's been done on your system, you must understand the difference between a machine screw and a wood screw or a sheet metal screw; between a 6-32 and 10-24; between a so-called 2-by-8 and an old-style joist. There are other resources to explain these terms; there are no equivalent resources to fill in with electrical information if I take the space away from old wiring for this. Therefore, I will talk about nonelectrical products more as though I were describing my choices to a fellow tradesman than as though I were teaching a helper.

Mounting Systems

While hardware for mounting your equipment is not "electrical" material, the survival of your installation depends on this hardware. About the only specific NEC requirement regarding support is a prohibition against using a wood plug set into masonry to support electrical apparatus. Its more general requirement is that you use appropriate materials as intended by their manufacturers. This means that if you find loose cables or conduit you don't try to use bent-over nails to secure them in place of the staples or straps designed for the purpose. It also means that you don't use staples or regular nails in masonry—not even in the mortar, which they might penetrate. Regular nails are designed for pounding into wood. If you find this error in your house, check for further misapplications.

How about using masonry nails—either cut nails or, somewhat better, hardened nails designed for that use? These still should not be a first choice, especially for critical uses such as supporting a switch box (as opposed to less critical ones such as securing a cable strap). If you need to resecure equipment that was mounted to masonry, use screws. Get a masonry bit, drill a hole, and set a screw anchor in the wall. Nails are fine for fastening a mounting flange to a wood stud or joist, but sub-optimal in other applications.

For light duties such as securing cable to either hollow surfaces or masonry. I use screws and anchors, though plastic anchors are the least secure of the appropriate, easily installable means of attaching straps on most any equipment. They are borderline, but still okay for mounting boxes that won't be disturbed. Use #8-10 plastic anchors, $3/4''$ long, not #6 by $1/2''$; use a $3/16''$ bit to create snug holes for these anchors, not a $1/4''$.

If you need to secure materials to a hollow surface—whether drywall, plaster, or hollow masonry—the best support system is screwing into a stud or joist. Make sure you use a long enough screw to grab the wood, and one that is large enough in diameter for strength. A largish toggle bolt (whose screws are either $3/16''$ or $1/4''$ diameter) will do for jobs such as holding moderately-sized panelboards and light fixtures. Make sure that the toggle does open fully behind the surface rather than hanging up, either because the bolt to which the toggle is attached is not long enough, or else because the toggle encounters an obstruction.

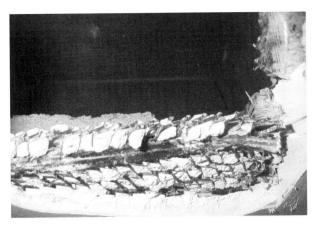

Metal lath is the highest-class plastering system, slowest to lose keys.

This clip is a patent box mounting device. I'd use Madison clips; I know I can rely on them.

A box in a wood lath wall.

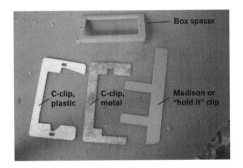

Box support (and depth-adjusting) devices.

Supports that rely on the surface itself intrinsically are less reliable, unless the surface constitutes, or is as solid as, the wall's structure. The larger the area they bear against, the more likely they are to hold. Surfaces vary even in how their surfaces are built up. For instance, plastered walls are supported by a drywall—paper and plaster—undersurface, wood lath (strips of wood that lose their hold), or metal lath, the toughest and most reliable of the three. Unless I can find a stud or joist, or the plaster is over masonry, I'd trust moderately heavy equipment to a metal lath surface. If a wall were wood lath, I'd be more inclined to move the location over a foot or so to find a stud.

Molly bolts simply don't hold as solidly as these fasteners, either to hollow surfaces or to masonry. The same goes for various patent anchors incorporating nails or pins to spread them. There also are a number of special systems for fastening materials, especially to masonry. Explore them if you wish; you can do a perfectly good job merely by using a drill and plastic anchors, and it's easier to do a bad job using the special systems. For a really secure connection to masonry, use a split steel anchor. Make sure that the hole isn't too wide or too shallow for a secure fit to the anchor.

Not all hardware is suitable for use in damp or wet locations. Yes, plain cadmium-plated screws and staples will secure materials used outdoors. And they'll take a long, long time to rust away (though eventually they do—I've seen the results). Meanwhile, unless you paint them, they'll be ugly and probably will leave rust streaks.

Machine screws connecting metal to metal will cause more serious problems in damp or wet locations. Normally, machine screws are used to mount equipment that eventually may be removed and replaced. When they rust, removal becomes near-impossible. Better hardware stores sell stainless steel, nylon, brass, or at least galvanized screws for use in such locations. Think carefully about this one: many a location has suffered rusted screws because it was under a window, or on an outside wall below ground level, or in a bathroom. It would have stayed dry if nothing ever leaked.

"Plastic Tools"

One category of valuable materials sometimes is referred to as "plastic tools." Duxseal®, a duct sealant, is a nonhardening Play-doh®-like compound you'll use to seal the hole if someone put wiring through an outside wall and didn't follow up to restore the integrity of your wall. Roll it between your hands, push it in, pound it into shape; there's no special technique that electricians learn in apprentice class. I do suggest that you check that it sticks to your surfaces. I have had to clear away crumbly mortar before I could seal around a cable run through a brick wall. I have not noticed any differences in quality between different brands of sealant.

Sometimes the holes are too big for duct sealant, which is basically for sealing narrow gaps and bridging uneven contacts. If you have any more substantial repair work outside, you'll need patching cement for masonry. A permanent caulking compound such as silicone sealant can fill in most places you might use duct seal, and some smaller openings where you would have a hard time applying duct seal. Caulk also can be very unobtrusive; I've used it around wall fixtures mounted on brick next to front doors; while it's not invisible, it does a great job of preventing a fixture from swiveling, and keeps moisture and bugs out of the wiring compartment. Occasionally, manufacturers' instructions will tell you to use "RTV compound." The acronym means *Room Temperature Vulcanizing* compound, which is simply a silicone sealant, or butyl, or some other rubbery variation. It does not have to be specifically Listed for electrical use so long as you are not using it as an integral part of the electrical system without explicit instructions to do so. If you did try to use it to repair, say, a broken outlet box, you would be taking a chance.

Electrical Products

Tape

While weatherproof caulk might belong to plumbers, and duct sealant is a heating system installer's product, electricians use both. We don't use duct tape, surgical, masking, or household cloth tape. You'll need PVC (vinyl) electrical tape. Using nonelectrical tapes as electrical insulation is illegal, inappropriate, and quite possibly dangerous.

You may encounter installations using electrical tapes other than vinyl, such as black friction tape or rubber tape, in houses dating back to the 1940s or earlier. Friction tape is pitch-impregnated cloth. Often it was wound over rubber—yes, literally rubber—tape, which by now will present a solid lump. You will not need either of these types of tape, even to repair what's there.

The standard vinyl tape is black, though any color will do for most applications—except for green. Occasionally, you'll want to use tape to color code a wire. This requires a roll each of white and black ELECTRICAL vinyl tape, and perhaps one or two other colors as well. I will talk about this possible need in Chapter 15. I recommend against using white tape to repair any conductor other than a neutral, but it can indeed be useful for clarifying which faded conductors in a box are the neutrals.

Besides getting the right kind of tape, and looking for its Listing mark (on the package or inside the cardboard it's rolled around), you want to check its quality. This means checking two things: resilience, the ability to stretch a little without any holes opening up, and without it ripping; and stickiness. Eight mil thickness is better than 7 or 6 mil, other factors being equal. Quality really does differ. A good brand is 3M.

There's a specific, optimal way to apply electrical tape as insulation; it can compensate for minor voids or less-than-sticky spots in the tape. When you are wrapping tape around a conductor, each turn should overlap the previous turn exactly halfway. This way you get a continuous double thickness, which guarantees that there are no voids or gaps in your taping, as there could be if you were applying each lap next to the other, just barely covering. Also, if there is any defect in the roll of tape you're using, the spot it covers will be protected by the next or previous lap as you apply it. Keep tension on the tape as you wrap it. This helps you control exactly where it lies. Incidentally, vinyl tape is the only type expected to stick to the surface you're taping. Friction and rubber tapes were applied stickier side UP, and stayed in place by sticking to themselves.

Other Repair Systems

Sometimes it is difficult to do the insulating job that is necessary, using tape. There are some alternatives, although they rarely are used. One is heat-shrink tubing, introduced in the 1960s. Buy shrink tubing of a diameter just large enough to get over what you're insulating; cut it longer than what you're insulating because it shrinks lengthwise as well as radially. Then heat it; it shrinks when you apply heat. As you apply the heat, be careful that you don't start a fire, or scorch your way through or embrittle any nearby insulation. Another option is electrical varnish, also called, "liquid tape." It's messy, but effective. However, there are cautions in the instructions.

Finally, there's the question of what to do when the sheath protecting non-metallic-sheathed cable gets damaged. Most inspectors will accept the use of tape to repair the PVC (or, in older cable, cloth) sheath, at least in dry locations, but vinyl repair compounds are available.

CAUTION!

Apply vinyl repair liquids over tape, as otherwise they may dissolve plastic insulation on the wires inside the sheath.

Other Electrical Equipment

Tape was a transition from "plastic tools" to your wiring. Here is some information about the constituents and designs of other electrical components you'll buy, to help you make those choices. Cover plates offer a good lead in.

Metal covers never should be used over devices or on enclosures that are not grounded. I explained how to test for grounding in Chapter 14.

Covers

Cover plates, both those for switches and receptacles and those used to blank off junction boxes, are available in metal. Most also are available in thermoplastic, nylon, and more exotic materials such as ceramics. Metal holds up better than plastic, but plastic is easier to clean and imposes a milder grounding/bonding requirement. Lexan and nylon plates are tough, available at a considerable premium. They never should need replacement. Stainless steel plates are permanent, too.

Older Materials

Nonmetallic is the formal term, used in the NEC. In normal use, everyone calls non-metallic materials "plastic"—except for wood, which you still may find used as part of light fixtures or even as cover plates. However, these cover plates have backs of metal, and the fixtures don't use wood near the splices, the sockets, or the light bulbs. Old systems used a number of other non-metallic materials. For rigid construction such as boxes, manufacturers used porcelain, slate, "ebony-asbestos" (a compressed composite), and bakelite. For flexible designs such as cable, they used cylindrical woven cloth (the form I have seen is called loom), pitch-impregnated cloth, rubber, cloth-covered rubber, asbestos-rubber material, or asbestos fabric covered with cloth. I've run across all of these still in use. I have not yet run into some other materials they have used—masonite and mica—but this is not to say it is impossible that you will find them.

Painted steel plates and even steel screws rust where it's damp; stainless steel or brass plates stand up fine (Brass-*plated* plates do deteriorate.). Standard plastic plates crack easily where there's any physical abuse.

The covers for switches and receptacles come not only in a variety of materials but also in some variety of designs. "Goof plates" are oversize and also deeper than standard. These are intended to cover roughly-repaired surfaces around enclosures, and also to compensate for slight protrusion. There also are intermediate-sized plates that are a little wider and longer than normal, but no deeper. Note that while some suggest using these plates as solutions to the problem of gaps, covering a hole in the wall around an electrical box with a plate does not satisfy the NEC requirement that gaps wider than ⅛" be filled.

Cutting a cover plate to make a combination says, 'unprofessional and ignorant'; beware of this installer's other kludges.

Another issue is using the right cover. I have seen some cover plates chopped so as to more-or-less fit over devices; I have seen them handmade. Neither is acceptable. If you want to install a pretty, handmade, ceramic cover over a switch, by all means set it there, but set it OVER a Listed switch cover.

By using the right cover, I mean using a cover designed for one switch and two receptacles if this is what you have in your three-gang box. They make these. They even make covers for a three-gang box that contains one switch, one duplex receptacle, and one empty space (which needs to be completely covered). You can buy a Listed do-it-yourself product consisting of plate sections that snap together to cover whatever combination of devices you install. Incidentally, a device with a round knob can be covered with a plate that is manufactured with a small round hole in the center or, quite acceptably, with a standard switch plate. Finally, boxes in wet locations take special, gasketed covers.

Devices

As mentioned earlier, switches and receptacles consist of operating and contact mechanisms (the guts); places to terminate wires; bodies and faces; and mounting yokes, the parts that support them by attaching to enclosures. The mounting yoke normally is metal, but some premium devices incorporate the body and face into one rugged unit with the yoke. While I have seen abused receptacles with broken faces, exposing the live innards, these are exceptions. For normal use, installing the type with a nylon unibody probably is overkill; there is nothing wrong with it, though.

(Low) Quality Counts

In some cases, the front of a receptacle will come off with the cover plate. I've had a number of service calls to replace Circle F (trademark: a capital F in a circle) receptacles that had fallen apart even though they weren't terribly old. By "not terribly old" I mean they were wired in thermoplastic romex with a full-size ground. This was 1970s or 1980s cable; I'd hazard 1960s at the earliest. They were falling apart by the late 1990s. Thirty years certainly isn't new, but I've seen far older receptacles that held up better. The spring tension in these Circle Fs was still in the acceptable range. This means that they hadn't suffered heavy use that could have subjected them to undue mechanical stress and heat.

I traced the problem to one weak link; a rivet, as I recall. It held the nonconductive plastic face to the plastic body through the conductive, grounded, yoke. When the rivet loosened, the receptacle's face partly separated from its body. The face, however, didn't even shift, as it was held to the cover plate by layers

of paint. In those cases where it didn't come right off, sometimes I received the service call when the yoke shifted enough that the prongs connected to the hot conductors would arc to it. Crackle...spark.... Sometimes a fuse would blow; sometimes not. Sometimes this intermittent short waited to manifest till a cord connector was plugged in.

I also have been called in because the front of a receptacle separated from its body entirely. From the outside, nothing appeared broken, but appliances plugged in to it commonly acted as though they were making very, very poor contact. It was evident that there was some kind of problem. Every so often, especially when an appliance—even a desk lamp— was plugged in, the circuit would trip. When I removed the cover plate, which was not difficult at all, the yoke stayed very nicely in place, but the body of the receptacle fell away. The only reason it had not fallen back into the outlet box before is that it was wedged into place by the wiring.

Clues to Device Quality

Are there any readily-identifiable differences that make a difference? Yes.

Language The most general indication of manufacture to extra-rigorous standards is the legend, "Hospital grade." (A green dot on a receptacle's face means the same thing.)

"Heavy duty" has a specific legal meaning with regard to receptacles. It means that they will hold onto cord connectors more firmly. This is a good thing, unless someone weak, such as an elderly relative, must struggle to force cord connectors in and out of that tight receptacle. It also has a practical meaning: heavy-duty devices, such as a top of the line Hubbell, may cost 10 times as much as standard ones, such as an anonymous import.

There are meaningful and meaningless terms. "Deluxe," for instance, has no legal nor practical significance. Neither does "specification-" or "spec-" grade, although within a particular brand the latter term may differentiate a manufacturer's bottom-end offering from the better-made part of his line. I said, "may," and here's an example. One venerable, popular brand sells receptacles marked "Commercial Specification Grade," and others marked, to paraphrase, "Professional Use." I was glad I had the chance to examine them side by side. The body and face of one were made of thermoplastic, which is perfectly acceptable. The other used nylon, which is a good bit tougher. One advertised double-wipe contacts, the other triple-wipe (features I describe immediately below). One used side-wiring terminal screws, which are perfectly okay; the other had these screws and capture features (described in Chapter 12). One had captive mounting screws; the other had captive mounting screws plus posiground-type clips, which eliminate the need to bond the receptacle to the box with a separate wire. "Professional use," to them, represented an unmistakably higher standard.

Internal Design

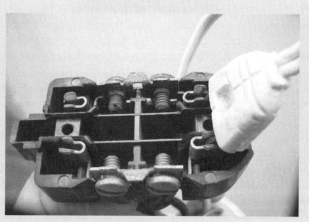

This is what a receptacle, in this case a double-wipe, nongrounding receptacle, looks like with the face and yoke removed.

Plugs make contact inside receptacles three possible ways, called single, double, and triple wipe. Single wipe means that each of the plug's two or three prongs is caught between a leaf of conductive metal and the plastic body of the receptacle: this is just about the cheapest and flimsiest design. Double wipe, by far the most common version, normally means that it's caught between two conductive leaves. Triple wipe introduces a third contact on the side or end of the prong. I have, however, seen an impressively chintzy design, technically double wipe, in which the prong pushed into a slit in a single leaf. This is the one I consider absolutely flimsiest.

The point to keep in mind is that designs with more contact surface produce longer lasting low-resistance connections. Admittedly, most hardware store clerks never will have heard of these distinctions, and this may be true of some electrical distributor personnel as well. If, however, you do find receptacles boasting of triple-wipe design, you now will recognize that they are, at least conceptually, top-class.

Lancaster County, Pennsylvania, inspector Larry Griffith has found a measurable difference in receptacle-to-receptacle resistance based on the size of the terminal screws and connecting links. I talk a great deal more about this in Chapter 12.

Material and Mass What else can you look for? One thing is the size of the terminal screws. The bigger the screw head, the more solid its contact with the wire you connect. Another is the material of the screw. Brass is a much better conductor than steel. And some questionably designed receptacles give you decent-

These three purport to be the same device; design and quality differ.

sized terminal screws for hot and neutral wires, but a tiny one for the ground.

Miscellany

Since this is the one chapter devoted to specifying materials, I'll look ahead to some odds and ends of picking materials to do a good job.

Moisture

I already have talked about damp and wet locations with regard to equipment Listing and Labeling, and to fasteners.

Damp locations are locations that don't get rain or other beating water, but do get moisture. Expose standard sheet metal boxes or armored cable in a damp location, and they will rust. For your work in wet locations, unquestionably, use weatherproof materials made of noncorroding alloy such as aluminum, cast iron, or hermetically-sealed plastic.

Appliances

Save yourself potential harm or loss that could result from buying the wrong heavy appliances. Suppose you bought a new water heater that draws 18 amps. You have a 12/2 cable running to your old 16 amp water heater on a 25 amp circuit breaker. It should work to run the new, 18-amp unit, right? Wait. Read the instructions; requirements are described right on the nameplate. If they specify a minimum #10AWG cable, that's what you needed. If you have a garbage disposal, and it says to install a cord assembly and plug it in, that's what you have to do. It shouldn't be hard-wired unless the instructions explicitly permit that.

Finally, some appliances will say, "protect at 15 amperes," or equivalent language. If so, a 20-ampere circuit won't do. Furthermore, if a washing machine says, "protect with a 15-ampere fuse," even a 15-amp circuit breaker won't do. You're okay if there's a box containing a fuse holder between the circuit breaker and the connection to the washer. This all comes back to obeying Listing instructions.

Before you buy any replacements, I urge you to look at each appliance—open the package, scoot around behind the demonstration model to look at the nameplate, read the installation instructions.

Shopping

Where do you buy tools and materials? Salesman expertise and hutzpah vary greatly. Garrison Keillor, the radio storyteller, invented Sam's Pretty Good Grocery, with the motto, "If we don't carry it, you probably don't need it." In the trades, the complement, "what we carry should be okay for your need" is called "inventory engineering." This term is not complimentary. Unfortunately, you may find it anywhere you shop.

Your local electrical supply house carries most of the materials you'll need, and also the tools, though the latter usually are sold at full list price. You'll find the suppliers in your phone book under "Electrical wholesalers."

Yes, homeowners are perfectly eligible to buy from a wholesaler. Walk in, find the parts counter, and if it has a "cash sales" section, belly up to the bar—or take a ticket, or hail a counterperson if they don't all seem busy. "Cash sales," incidentally, include checks and credit cards, where those are accepted—and they usually are. Cash sales are so called to distinguish them from "open accounts," which serve regular customers who have store credit. You will often, but not always, get parts at a discount. The markup between such discounted prices and what an electrician will charge you for the same materials at your house covers professional judgment in choosing the right materials; time spent picking up the materials; cash investment required to keep common ones in stock—to cut cables to length, and to use a few fittings at a time out of a box of 50; and finally the liability associated with guaranteeing materials sold to you. I find it impossible to predict what will cost less at an electrical wholesaler and what will cost less at a home center. Most of the time by far, though, the sales force at the wholesalers know more about their products than the people at the discount warehouses—even if the former are less forthcoming with advice.

Are Your Wires Safe?

Clues

Insulation

Wires used to be covered with rubber insulation. The rubber was covered with woven, painted cloth. The widely-used colloquial term for these conductors, "rag wiring," refers to this cloth. After the middle of the twentieth century, plastic insulation, with color permanently embedded in it, gradually displaced the rubber. Plastic insulation lasts longer than rubber and keeps its color very well except when charred or nearly charred by serious overheating, overheating at a level that probably would cause rubber insulation to crumble away. Plastic also is a better insulator than rubber of the same thickness—and considerably tougher. Add this to the fact that if your insulation is rubber, it is certainly rather old, and rubber's lower toughness and heat resistance means that if you have rubber insulation, it is more likely than plastic to have suffered damage, damage which can cause you injury. Note that this is a statistical likelihood, but far from a certainty.

Over the years, there have been changes in both rubber and plastic insulations. Newer varieties insulate better and withstand a wider range of conditions. These developments allow the same size of wire to carry more current without damaging its insulation. In the mid-1980s, the chemical formulation of electrical insulation was changed, and it became capable of withstanding far higher temperatures than before without deteriorating. Note, though, that meanwhile studs and joists near the wires may slowly char because electrical insulation is not thermal insulation; all that the improved rating means is that the wires are safe from shorting.

Since the 1990s, most conductor insulation has consisted of colored plastic covered with clear, tough nylon; this is thinner yet than other plastic, and makes an excellent replacement for the old stuff. You will learn to evaluate the condition of insulation in Chapter 13. Testing insulation is one of the most important things this book will teach you to do in dealing with older wiring, especially with truly old wiring.

CAUTION!

The greater temperature resistance of new wiring poses a problem not only for people with old wiring, but for people whose wiring is even slightly old. If your wiring dates to before the late 1980s, most modern lighting fixtures cannot be connected to your wiring without creating a fire hazard. This is discussed again in Chapter 16.

There are other questions you'll want to ask, once you move beyond the electrical panel. For example, is your wiring up to Code, or severely dated? The following features are characteristic of old wiring. This wiring probably has exceeded its design life, and may be considerably deteriorated.

- Rubber-insulated conductors (rag wiring)

- Cloth rather than plastic used for cable sheaths (this too was called "rag wiring"); in some cases, in cable manufactured during a transitional period, the conductors it protects actually will be insulated with an early thermoplastic.

- Armored cable that does not contain an internal bonding wire—or that has the internal lead sheath, discussed earlier.

- Round or octagonal electrical boxes smaller than 4″ in diameter—the boxes probably are overcrowded and therefore the wires' insulation may have suffered overheating from the very first.

- Open wiring on insulators, or concealed knob-and-tube wiring. An exception may exist in parts of the country subject to frequent flooding. There, wiring with the system may be permissible because open conductors dry out much faster than cable or conduit.

- Conduit that is black-enameled rather than galvanized—although if the wiring within has been replaced, who cares (unless the pipe has rusted badly)? Incidentally, on first glance this is easy to mistake for plumber's pipe.

Most of this analysis is secondary. It's of value if you're walking through a home that you're thinking about buying. When you're getting down and dirty, there's only one essential question related to age: Is conductor insulation in good condition or has it deteriorated?

What's the most probable place for you to uncover deterioration? The best places to check are in the boxes serving old kitchen light fixtures, for a number of reasons. First, heat rises. Light fixtures get hot, and kitchens often are hot anyway. Second, kitchen lights frequently are left on for long hours. Third, ceiling light fixtures normally are located in central locations that were convenient for old-time installers to branch out from, so these boxes often are crowded. If under all these conditions the conductors are in good shape, conductors everywhere have a decent chance of being in good shape.

Of course, if the kitchen has been redone and the wiring replaced, that's not where you want to look for deterioration. You'll still look there, though, to see whether the remodeling was done professionally. As you read this book and the NEC, you'll develop standards enabling you to recognize professional electrical installations.

Look also for bootleg grounds or neutrals, for transitions from properly-sized wire to overfused circuit extensions, and for panel covers that are obstructed, even just enough that the screws are blocked from easy removal.

Other Clues to Quality. There are other ratings to consider besides those mentioned in the discussion of receptacles, but they will be covered in the discussion of NEC requirements. Going beyond devices for a moment to general design issues, there are some other things to look for. Any device that comes with junk wirenuts is suspect. Ill-fitting light fixtures, and fixtures with sharp metal edges, show manufacturer indifference. (Yes, it IS appropriate to open the package in the store, if only to make sure all the parts and instructions have been included, and responsible salespersons should give you the okay. Fixtures held together with sheet-metal screws usually are better-designed than those that snap and slip together, though there are exceptions. Fixtures that are bolted together with machine screws almost always are superior yet. Metal generally means better construction than thermoplastic, and porcelain is better than plastic such as bakelite for wirenuts, lampholders, and other heat-exposed items. Finally, when foreign—or domestic—manufacturers don't bother hiring folks fluent in English to create intelligible installation instructions for devices intended for the U.S. market, that's probably not the only place they scrimped.

12

Making Connections

Preparing Conductors for Connection

Ordinarily, when part of a circuit has gone dead, I have to do a good bit of scouting to find the culprit, because in an older home, the wiring often is not laid out in simple, sensible fashion that lets me determine which outlet is the last in the sequence of those that work, and which is the first among those that have stopped working. A line may proceed up to one receptacle, over to a light on the floor below, down to a switch, and then down to a few more receptacles.

Gerry gave me a great clue to help me locate the source of his outage; he told me his problem had started a week or two after he replaced a certain receptacle. After he turned power back on, he had made sure that the circuit worked. Still . . . I opened the outlet he had worked on, and found what had gone wrong. When he had mounted the receptacle back into its box, one of the wires had fatigued just where it was looped around a screw. The outlet, and the ones it fed, worked for a while because the wire stayed in place and made good enough contact to conduct across the break. Then some slight vibration had shifted the wire, and power no longer could get through the gap.

It doesn't matter how much you know about electricity: if you wire without learning to connect wires securely, it's like driving on bald tires. Even if you don't cause a fire, you're courting grief. How? Customers pay for untold hours of service calls spent chasing problems that ultimately are traced to bad connections. These problems include unreliable operation of lights and appliances, tripped circuits, and equipment that doesn't work because the circuit to it is interrupted, even though no fuse has blown. This is why I devote this long chapter to preparing wires, splicing them to each other, and terminating them at screws and other connection points. The following, by the way, is about working on wiring where power is *off*.

Evaluate What You've Got Even Before You Prepare Your Conductors

If your conductors are deteriorated, you may need to call in a pro. I'll define what "deteriorated" means. If the insulation is falling off, crumbling, seriously cracked, it's beyond hope. If it's not as bad as that, kill power to a wire, confirm that it is not live, and then bend it double. If the insulation doesn't crack, it's in at least fairly good shape. If it does crack, it's on its last legs. The same is true if it is heat-blackened or toasted brown. You can either replace it, the safest solution; repair it, which is very uncertain, and undeniably a short-term solution; or live at some hard-to-determine degree of risk. I briefly mentioned some repair options in Chapter 11, Materials.

Prepare the Wires Carefully

Stranded Conductors

There is some difference between the ways in which you need to deal with solid conductors, with stranded conductors, and with combinations of the two. Solid wires are much easier than stranded to attach properly and safely to terminals. It is much harder to splice solid and stranded conductors together properly and safely than to do so with solid and solid or stranded and stranded.

Fortunately, romex and BX in branch circuit sizes do not contain stranded wire, and you will find it in raceways very rarely. You will find stranded conductors in the connections between house wiring and

fixtures and also between house wiring and some specialized control devices such as dimmers and timers. An additional source of difficulty in these applications is the likelihood that the stranded wire will be several sizes smaller than the solid wire to which you are attaching it.

One consequence is that you may have difficulty learning to work with stranded wire. A second consequence is that when you check what others have done you will want to pay extra attention to stranded wire connections.

Strip

Before making nearly any 120-volt connection, you need to prepare the conductors. You are better off preparing their ends freshly rather than reusing the portions that had been stripped previously. Strip the insulation back to expose an appropriate length of bare copper. I'll teach you how to determine proper length in the next two chapters. Normally it is best to strip wires longer than needed and then cut them back to the precise lengths required. It was clear to me that Gerry did not do this.

Look Out for Damage. Make sure the wire is not damaged by the tool you use to cut away its insulation, as discussed in Chapter 9. Damaging wire causes it to have somewhat higher resistance, and can create a connection that over-

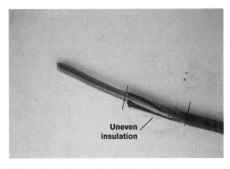

A badly stripped solid conductor.

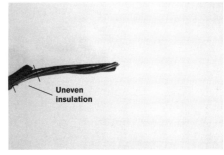

A stranded conductor, insulation stripped too unevenly—this is likely to leave bare copper showing after it is terminated.

heats or gives way—sometimes a considerable time later. Solid wire commonly is damaged by nicking or ringing; improperly-stripped stranded wire commonly loses strands. Look carefully; damage may not be obvious to an offhand glance. I talked about the risk of damaging wires in Chapter 9; however, this is important enough to repeat.

- Nicked wire frequently will break off at a weakened spot as you stuff the wiring back in the box. You may not even know this has happened until you turn the circuit back on. If that happens, you will have one consolation: you are better off discovering this right away than leaving a potential hazard lurking. Unfortunately, you cannot ensure that you do not leave a problem lurking except by being careful as you strip and again as you inspect.

- A stranded wire that has lost strands is less likely to break than a solid wire that has been nicked. Still, where it has become effectively one or two sizes smaller than wire rated for the circuit, it runs hot. Many a fire has resulted from this.

"Free conductor" means the amount of conductor that is accessible and at least theoretically available for use. This is not only the amount you can pull out of the box but includes any within the box that is not covered by a clamp, connector, bushing, or sheath.

If you damage a wire, cut off the damaged section and strip it again, unless the length of free conductor is down to nothing and it is impossible to gain access to more. If you cannot get at any more, the outlet probably will need to be rewired. How do you strip old conductors correctly? To start with, look back at Chapter 9 for a selection of suitable tools and a description of how to use them.

Strip Back Damaged Insulation. When you work with old wiring, examine the insulation and, if necessary and possible, repair it as recommended in Chapter 11 where I discuss suitable materials. You will learn how to evaluate the condition of insulation below.

If a Conductor Is Too Short to Cut Back, Let It Be— If You Can Do So Safely

Sometimes a conductor is just too short for you to cut it back further and strip it again. If you are leaving an old conductor as it is, you need to check three things.

First, decide whether the conductor is long enough for you to make the connection. Sometimes you cannot salvage an existing wire or outlet safely; sometimes even an electrician can't. As you will learn in Chapter 13, Mechanical Condition, on rare occasions you will be able to increase the length of free conductor.

Second, confirm that the wire is not suffering from mechanical injury. This damage is especially likely just where the conductor emerges from the insulation. It could result from careless initial stripping, from metal fatigue due to previous twisting or pinching, from inserting and removing devices, or from screws and nails entering the box. A wire could be damaged even from your bending and straightening and otherwise preparing it so as to be able to make a better connection. You will have to check this again after you are done preparing the wire, and then again after making the new connection.

Third, make sure that the exposed portion of wire is not corroded or otherwise covered. You may have trouble with wire that has been splashed with paint, has gotten sticky from electrical tape, or is even just tarnished. This problem was especially severe in one house that was wired and then left vacant for a year. When I was called in to hook up lights, the previously-stripped wires were covered with so much oxidation that I could not get a voltage reading—even though the conductors were live!—until I burnished them with emery cloth, a sort of sandpaper.

This is worth repeating: it is best to strip the insulation further and cut the exposed wire back, if you have anything approaching a sufficient length of conductor. This is especially true when the wire you can see is in questionable condition.

If you cut back too much, you may have some trouble restripping the shortened conductor. However, this certainly is better than leaving bad insulation that could result in a short, or using wire that you cannot rely on to make a trustworthy connection. The trouble is no different or greater than that of finding insufficient free conductor when you open the box.

Sometimes insulation is unsound from the point where wires enter a box. Worse yet, you can find wires broken off just where they enter—sometimes *inside* their insulation, so that the breaks are not evident.

Getting Stripped Wires Ready to Connect. Before making connections with solid wires, check for mechanical injury and corrosion, whether or not you have restripped them. Then straighten them out as much as possible, unless they are in approximately the configuration you will need for the type of connection you are going to make. Even so, it is best to start with straight wires. The main reason *not* to do so is that the additional straightening and bending could push fatigued wires over the edge; if you don't have access to sufficient length to cut them back further, you might choose to minimize their handling. This is a judgment call, every time.

Before making connections with stranded wires, check for missing strands, whether or not you have restripped them. Then twist the strands together tightly, clockwise. Use pliers to twist them, and when you are done, make sure that you have done no damage with the pliers. If you have, strip back and try again. In a few cases, it might make sense for you to cut the end of the entire conductor off square, to begin; however, it is likely that you will need to do this later anyway, after changing the overall shape of the conductor.

Because knowledge is not enough, you need to practice splicing before you work on your electrical system. Later in this chapter, I will talk about how to evaluate splices; this will enable you to judge when you have the skill to splice safely.

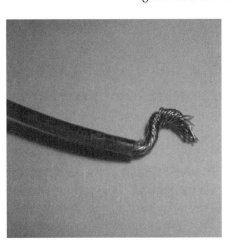

A stranded conductor, twisted tightly and shaped; imperfect, but quite ready for installation under a standard terminal screw.

Splicing Conductors

Some people slice some insulation off a conductor, and figure it's prepared and ready to connect. You know better, having read through an entire chapter on how to prepare the conductor. Similarly, some people stick wires into a connector and twist it, or twist wires together and perhaps wrap some tape around them, and consider the wires spliced. Splicing calls for quite a bit more than that, and this chapter will guide you through the process. Choosing wire connectors and applying them *both* require knowledge and skill.

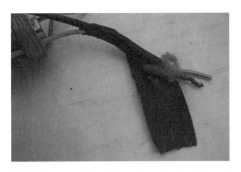

This tape-only splice is quite inadequate. The chandelier these splices fed did not send heat into the ceiling outlet box, but the bad splices could make up for that by starting a fire all by themselves.

The Two Basic Reasons for Splicing

"Connect Wire A to Wire B" is only one reason you will splice. The others go under the name, "pigtailing," which means, loosely, "extending conductors."

- When you need to attach one wire to one screw, and you have two wires to attach, you pigtail. In this case, it means that you splice the original two to a third, and attach that third to the screw.

- When you have one wire to attach to a screw, but the wire is too short, you pigtail. In this case, it means that you splice it to a second wire, and attach this second wire to the screw. This often is necessary when your wiring has deteriorated and must be cut back to prevent it from falling apart or shorting, or when it has been trimmed too short by a previous worker.

- When you need to attach wire to multiple screws, you pigtail. In this case it means that you attach a separate piece of wire, perhaps six inches long, to each screw, and then splice all these wires together with the wires entering the box.

If a Splice Conducts, Why Fuss Further?

A splice can hold together, and conduct, and still be quite dangerous. A high-resistance splice can enable appliances to work, but damage them; worse, it can start a fire. A sloppy splice also threatens anyone working on the box that contains the splice. Consider this customer.

Toby was standing on his six-foot ladder peering up into a very crowded electrical box. He'd just disconnected and taken down a chandelier that had been fed and supported by the box. He was pretty sure that he had killed all power going in to the box, because he'd turned off the chandelier by removing the fuse rather just turning off the light switch.

Some of the connections he could see used modern-looking wirenuts, but other very old wires just came together in what looked, essentially, like black lumps. Just to be sure that power to everything in the box was off, he used his pliers to pull the very old wires away from the box so he could test for voltage.

He almost fell off the ladder. One of the connections had come apart, and the wires sprang out toward him—almost in his face. At least he was wearing glasses. Fortunately, the wires *were* dead. What most surprised him was discovering that what had come apart was a wirenut splice. That's when he called me in. Someone, it appeared, had done some splicing up there without really knowing how, and Toby didn't want to get in over his head.

Styles of Wirenut and Basic Installation

All wirenuts that you are likely to buy are designed to be tightened by twisting clockwise with your thumb and fingers. I'll cover one exception later in this chapter.

Some wirenuts just get tighter and tighter until you no longer can turn them. The old bakelite wirenuts worked this way, and if you tried hard enough to tighten

---

Some Relevant History:
What Are Those Black Lumps That Toby Found?

Before there were such devices as wirenuts, electricians made excellent splices by following a five-step procedure. To begin with, the electrician stripped conductors for a considerable length—usually more than an inch. Next, he twisted the wires together tightly. Twisting connected the wires mechanically, so that they would not easily fall apart. Then he soldered the wires, to enhance their electrical continuity and also to beef up the mechanical strength. After this, he covered the splices with rubber tape, to insulate them. Finally, he covered the rubber with friction tape, which consists of pitch-impregnated cloth, to protect the rubber from abrasion. (Friction tape is a decent electrical insulator in itself.) This is the first splice I learned, and it is the splice my father used for many years, starting in the late 1940s. Some "twist, solder, and tape" splices are still doing fine 75 or more years after they were put together; the insulation on the wires going in to the splices may be in worse shape!

These splices no longer are legal for grounding connections. The heat created by a bad fault might melt the solder, resulting in a high-resistance connection if the wires aren't twisted together really tightly. Grounding connections just are too critically important to take this risk because grounds with poor continuity make fuses take too long to blow.

These splices, or "joints," still are quite acceptable for most uses. They're especially good for use with low-voltage signal wiring because the solder ensures a superior low-resistance connection. The 9-24 volt circuits used in thermostats, telephones, doorbells, and alarms are covered by the NEC (telephone ringing voltage can reach 90 volts!), but I don't address them due to space constraints.

Splicing is a lot simpler now that we have wirenuts. It is not necessarily better, but it certainly is faster and more convenient.

---

them further, the shells cracked. Others tighten to a certain point, and then the internal spring clicks free within the shell as you keep twisting, indicating that the wirenut is correctly torqued, and preventing any further twisting from affecting it.

CAUTION!

If you suspect that the old conductors you want to work on may contain aluminum wire, stop here. At the beginning of this chapter, I urged you to practice splicing before working on any of the wiring in your walls. If you choose to proceed without practicing, this is a place where you could create serious problems. Jump ahead to the discussion of aluminum wiring, at the end of this chapter, before you proceed.

Some have special designs to provide superior leverage for your finger and thumb, while others can be tightened extra-firmly by the use of special tools made for the purpose. Any wirenuts except for very brittle ones (such as those constructed of bakelite) can be tightened more securely by holding them with pliers rather than just using your fingers. If you use this approach, make sure that your pliers do not cut or squeeze through the shell. Otherwise, some of the

spring, or perhaps one of the wires, could be exposed. This could short to a ground wire or, far worse, shock someone touching the wirenut.

The Two Ways to Splice Using Wirenuts

There are two basic ways to bring wires together for wirenut connections:

- pretwisting the wires; or

- inserting them straight and parallel.

Either is fine, when done correctly. Either can fail, as Toby discovered.

The Advantages of Pretwisting

Careful professionals generally take the time to pretwist splices. Here's why:

- When you pretwist properly, you make a connection that is independent of the wirenut. When you insert the wires straight, on the other hand, you rely entirely upon friction with the wirenut's internal spring to hold the wires together. If the wirenut comes off, they may spring apart.

- Even when the wirenut stays on, you can't always rely on this to keep all the wires together. One may slip out, or simply make a relatively high-resistance connection with the others. The reason is that when you have more than three untwisted wires in a splice, one or more necessarily are surrounded by the others, due simply to geometry. Consequently, the spring cuts into only the outer three wires. Any additional wires rely on friction alone to ensure continuity and retention.

- If there is a bit of corrosion on your wires, the spring will cut through some of it. Pretwisting, however, does a better job at minimizing any resistance caused by corrosion that you have not removed when preparing the wires (two chapters previous). The twisting stresses any deposits; and the pliers scrape them.

- Pretwisting lets you check your work visually before you install the wirenut and hide the actual connection from sight.

I consider pretwisting, that is to say *competent* pretwisting, preferable to competent straight-in splicing. However, as you shall see now, straight-in splicing is preferable for most novices.

Some Disadvantages of Pretwisting

That states the case for pretwisted splicing. It also has five or six disadvantages, although only the first four are unequivocal problems.

- Pretwisting takes more skill. A badly done pretwisted splice is far worse than a properly-done straight-in splice—and, unfortunately, more common in work done by untrained installers.

- Pretwisting should be unnecessary. Standardized product tests require that wirenuts do an adequate job without pretwisted wires.

- It is harder for a novice to evaluate a pretwisted splice than a straight-in splice, in the sense of understanding what it should look like before the wirenut goes on.

- The manufacturer's table of wire sizes, which comes with the wirenut package, predicates its numbers on straight-in splicing: a pretwisted splice is fatter, so you need to double-check compatibility. You should do so with either kind of splicing, though. I talk about how to size wirenuts later in this chapter.

- If, at some point, you need to take the splice apart, it is slightly easier to separate the wires if they have not been twisted together. Of course, it's also easier for them to come apart accidentally.

- Because straight-in splices require less skill, novices tend to do better at them, and to choose them over the pretwisted variety. There is one case, though, where there is a certain advantage for these very novices in attempting pretwisted splicing. Getting a wire that has been pretwisted nice and straight can be quite difficult. There will be occasions when you need to change a preexisting pretwisted splice. Sometimes you will take a splice apart to trace the wires. For instance, it may be that a short is tripping breaker 19, and you need to figure out which part of circuit 19, which goes into this splice, includes the culprit. Also, sometimes you will need to remove a wire, add a wire, or replace one that is deteriorated. In these cases, you have three choices:

- Cut back the wires and strip away insulation to access fresh wire for your splice (this is tough if the wires are short).
- Straighten out the wires already in the splice (if you don't do this well, it will be very hard for you to line up the wires so as to get a wirenut started on them, not to mention getting it nice and tight.
- If you cannot cut back the conductors, and, given their condition and shortness, you don't trust yourself to straighten them, you may be best off taking the splice apart as much as necessary, and afterwards retwisting.

Taping is a questionable practice. Because the porcelain wirenuts in use before the 1920s had no internal springs, tape was needed to hold them onto the wires (and pretwisting was absolutely necessary). Some electricians working in the 1930s may have learned this practice as apprentices, and so continued taping even though it was not necessary. However, many amateurs simply tape because they don't trust their splices to hold; or because they strip the insulation back too far, and don't think of cutting the wires back to the right length for the wirenuts they are using. When I find taped wirenuts, it suggests the presence of unreliable work.

Sizing Wirenuts

Whichever type of splice you decide on, in almost every case you will be putting wires together in a wirenut. (Towards the end of this chapter, I will describe a few rarer splicing systems.) In Chapter 10, I referred to wirenuts that are nothing but shells as trash. It can be nearly as dangerous to use a wrong size of wirenut as to use an empty shell. Either choice is likely to result in a loose, high-resistance connection; either is likely to fall off unless you tape it on carefully.

This is not to say that only one size of wirenut is suitable and safe for making a particular splice, even within one product line; however, many sizes will indeed be unsuitable.

Examine two or three sizes of wirenut. Certainly include the wirenut or wirenuts that the manufacturer has specified as suitable for the combination of conductors you are splicing, if you have packaging material with this information. Usually the manufacturer lists two or even three sizes as suitable.

A splice with over-long wires, taped rather than trimmed.

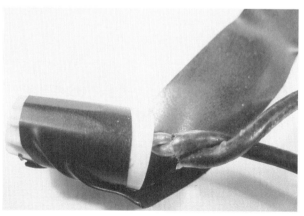

Another splice with over-long wires, taped rather than trimmed.

Take this as your starting point, but confirm which wirenut you need to use for your splice. Does it look as though:

- the internal spring is long enough to be in contact with most of the stripped length of each wire going in?
- the shell is long enough to cover all bare copper?
- the internal spring is wide enough for the wires to reach most of the way into the wirenut, but narrow enough to grab all the wires tightly?

If you are not sure whether a wirenut looks right, don't worry. At the end of the splicing instructions later in this chapter, you will learn how to make sure that the wirenut will not fall off, and that none of the wires will come out. If your splices pass the test, it will mean that you spliced well—and you could not have done so if the wirenut were the wrong size.

Many brands share an unofficial color code:

- small gray and blue wirenuts connect very small wires used for low-voltage signalling and for fluorescent ballast leads;
- orange wirenuts connect small leads from dimmers, GFCIs and light fixtures;

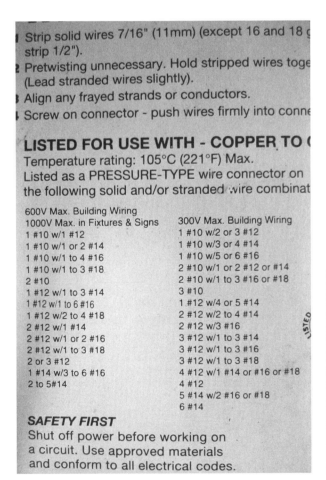

Strip solid wires 7/16" (11mm) (except 16 and 18 g strip 1/2").

Pretwisting unnecessary. Hold stripped wires toge (Lead stranded wires slightly).

Align any frayed strands or conductors.

Screw on connector - push wires firmly into conne

LISTED FOR USE WITH - COPPER TO (

Temperature rating: 105°C (221°F) Max.
Listed as a PRESSURE-TYPE wire connector on the following solid and/or stranded wire combinat

600V Max. Building Wiring
1000V Max. in Fixtures & Signs
1 #10 w/1 #12
1 #10 w/1 or 2 #14
1 #10 w/1 to 4 #16
1 #10 w/1 to 3 #18
2 #10
1 #12 w/1 to 3 #14
1 #12 w/1 to 6 #16
1 #12 w/2 to 4 #18
2 #12 w/1 #14
2 #12 w/1 or 2 #16
2 #12 w/1 to 3 #18
2 or 3 #12
1 #14 w/3 to 6 #16
2 to 5 #14

300V Max. Building Wiring
1 #10 w/2 or 3 #12
1 #10 w/3 or 4 #14
1 #10 w/5 or 6 #16
2 #10 w/1 or 2 #12 or #14
2 #10 w/1 to 3 #16 or #18
3 #10
1 #12 w/4 or 5 #14
2 #12 w/2 to 4 #14
2 #12 w/3 #16
3 #12 w/1 to 3 #14
3 #12 w/1 to 3 #16
3 #12 w/1 to 3 #18
4 #12 w/1 #14 or #16 or #18
4 #12
5 #14 w/2 #16 or #18
6 #14

SAFETY FIRST
Shut off power before working on a circuit. Use approved materials and conform to all electrical codes.

List of wire combinations for one wirenut and manufacturer's instructions for preparing conductors for this one, and installing it.

Red   Yellow   Orange   Small blue

I have never seen the spring fall out of these scotchlok wirenuts

Wirenut assortment. Scotchlok wirenuts are secure—springs don't usually fall out of them.

A wirenut next to its spring.

Question—which wirenut to choose.
Either *may* work okay to join these three wires.

- yellow and the next larger size, red, are the most commonly used sizes for connecting branch circuit wiring.

These sizes are also often numbered using a code originating with Ideal Corporation: 72B, 73B, 74B, and 76B, respectively, for blue, orange, yellow, and red. Beyond these sizes are large blue and grey wirenuts. You probably should not use these latter sizes at this point. They are used to join more conductors or larger conductors than most inexperienced people can connect safely in a single splice.

How to Make a Professional-Quality Straight-in Splice

Here is the background for the general procedure you will follow when splicing; important details follow.

Hold the conductors' insulation with one hand or with pliers—gently in the latter case, so you don't crush through the insulation. You will push the wirenut and the wires together enough to engage the wire in the internal spring, and then start twisting the shell. If the wirenut is sized correctly, you shouldn't have to push hard to get the spring to start grabbing onto the wires.

If you're using your fingers to hold the conductors, get a firm grip on them to keep the portion behind the stripped wire from twisting, while keeping your fingers clear of the wirenut. The general reason you need to hold them firmly is to give the wirenut something to work against. As you turn the wirenut, the more of your effort goes into twisting the conductors the less goes into screwing the wirenut on.

There is a more important reason to hold the conductors firmly when you are working with old wiring. If the insulation on the conductors is at all vulnerable, twisting it greatly increases the chance of a short.

You will hold the conductors not only against their inclination to twist, but also against their habit of shifting forward or back so that they no longer are lined up with each other.

A splice with its twist too far back; conductors were insufficiently restrained, resulting in a less thorough splice.

That was the overview; now for the details. You need to take six steps to create a good splice. You already know the first two from the previous chapter.

1. Prepare any previously connected wires that you have not cut back and restripped.

2. Twist any stranded wires together tightly, whether they originate in one conductor or in several.

3. Line up the insulation on all the wires evenly, and keep it that way as you proceed. Otherwise, the one that is furthest forward may enter the wirenut too far and catch the bottom of the spring, and the one furthest back may cause bare copper to be exposed because the wirenut's skirt cannot cover it.

> With straight-in splicing, your choice of wirenut and the length to which you clip your wires drive each other. First see which wirenut has the best width to accommodate the bundle of wires you are splicing. Second, clip the wires to a length that will allow them to screw into the wirenut's spring as far as possible and not leave any bare copper outside the wirenut's skirt.

4. Clip the ends of the wires square, even. If the ends are not square and even, the longest one may bottom out in the wirenut, keeping the others from seating all the way.

5. Screw the wirenut on tightly. If it is the type whose spring clicks and spins free once it is tight enough, tighten it to that point. If it is the type that just keeps getting tighter, tighten it as much as is comfortable. If a wirenut is loose, it may come off. Even if it does not come off and release the entire splice, it may allow a wire to slide out.

Old and new, copper aligned for straight-in splicing

Insulation too misaligned for a good splice

Wire is lined up, but insulation is not.

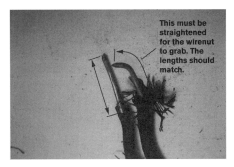

This must be straightened for the wirenut to grab. The lengths should match.

Splicing a bent wire to a straight—better straighten the bent one.

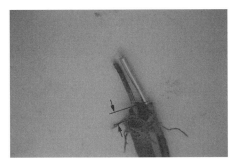

An old conductor and a new, lined up pretty well for splicing.

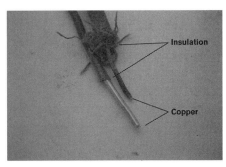

Insulation

Copper

An old conductor and a new, misaligned—both the insulation and the copper.

If a wirenut just won't grab and twist on tight, remove it. Make sure that your wires are square and even. If one wire snuck ahead of the others, it may very well have prevented the wirenut from screwing on. Prepare the wires again and retry. If the wires look right, just go ahead and try another wirenut. If you're curious, look inside the one that didn't work. Sometimes the spring is missing; they do fall out. Sometimes there is a piece of wire broken off inside the end of the spring.

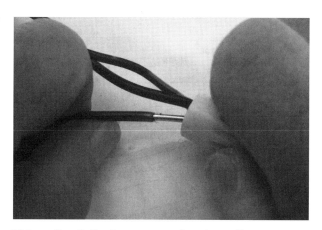

This splice fails the tug test—the wire pulls out.

Sixth, check your splice:

- Tug gently on each conductor to confirm that it is held securely. This always is useful, especially for someone untrained at splicing, and relatively inexperienced.

- Check between the wirenut and the conductors' insulation for bare copper. If just a smidgen shows, you may be able to get it under the skirt with one or two additional twists. If more shows, take off the wirenut, and rethink and redo your splice.

- Check the wirenut itself for damage, if you have used pliers on it.

- Finally, if you are working with old conductors, check for damage to their insulation.

Pretwisting

Pretwisting takes substantially more effort than straight-in splicing. There is only one additional step: cross the wires at a suitable angle and twist them tightly, concentrically, clockwise. This said, there are a number of considerations that cause pretwisting to demand a higher level of skill and dexterity. Even if your wiring is not old, it is more important that you practice pretwisted splicing than straight-in splicing before you apply it on your electrical system. If your wiring is old, experience becomes yet more important.

The General Splicing Procedure, Revisited. Normally, once you know what conductors will enter your splice, you can strip back their insulation and choose the wirenut. This is not the case when you pretwist. Strip back the insulation before you start, as you did for straight-in splicing, but strip it further back than you would for straight-in splicing. You will not be able to choose the wirenut until you reach step 4.

Your choice of wirenuts is more restricted when you pretwist than when you use straight-in splicing. You will follow the procedure for choosing them that you used for straight-in splicing, with two differences. First, the manufacturers' lists of wire combinations that come with packages of wirenuts presume straight-in splicing, so they are of somewhat lower value this time. Second, you will trim your splice not only based on the length of a wirenut that will accommodate the width of the spliced wires, but also based on the need to retain about three full twists in your trimmed splice—more, if possible.

Here is the procedure I gave you above, with some additional differences.

1 and **2.** Prepare the wires as described previously.

3. Line up the insulation on all the wires evenly, and keep it that way as you proceed with pretwisting.

You will not be lining up the insulation with the conductors side by side, though. Hold the conductors so that they cross each other at a moderate angle, say 40–60 degrees. You still need to align the insulation of both, or all of, your conductors. The end of the top one needs to cross the bottom one pointing to the left as seen from the insulated end because you will be twisting them clockwise. This is where the extra step comes in. For the purpose of an overview, though, I will skip past the pretwisting for a moment to complete the remaining steps.

4. After twisting them, clip the ends of the wires even and square. You may find that the pressure of this shearing flattens and thus widens the round mass of wire. This could make it harder to start it into the wirenut. To deal with this problem, you may choose to experiment with squeezing the flattened end of the splice back into its original round shape. If you try this, be careful not to untwist the splice as you wield your pliers on its end.

5. Screw the wirenut on tightly. You may have to push an appropriately-sized wirenut onto a pretwisted splice harder to get the wires caught in the spring than you would if it were a straight-in splice. Fortunately, it is equally hard to simply pull the wires out. Furthermore, with a truly concentric splice of solid wires, you are at very little risk of the wires coming apart regardless of what happens with the wirenut. If stranded wires are involved, you still are at lower risk of wires getting loose than you would be with straight-in splicing.

6. Check your splice by tugging, just as you would check a straight-in splice. The same safety issues are present.

The Twist. Anybody can twist wires. Twisting them *correctly* is the hardest part of this procedure.

Making the wires in a splice concentric, rather than wrapping one around the other, or some around others, is the most difficult part of splicing. It is especially difficult when you splice more than two wires or splice different types of conductors. Here are some tips:

- I mentioned that you need to put two wires together at an angle of 40 to 60 degrees before twisting. When you are splicing more than two conductors, put two alongside each other, and cross them with the other or others. If you try to twist three or more together with one starting out in the middle, that one will remain in the middle instead of becoming part of a concentric spiral.
- To bend something stiff and something soft to equal extents, push harder against the more-rigid one while just nudging the other.
  - The stiffer element going into your splice might be two conductors at an angle with a single conductor in a three-conductor splice;
  - it might be a solid wire being spliced to a stranded one or even a mass of stranded ones;

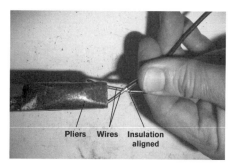

The start of pretwisting. End of pliers' jaw holds wires, crossed at a suitable angle, with insulation aligned, while the other hand holds the insulation close to where the wires cross.

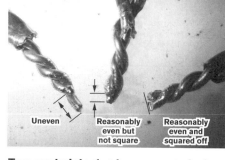

Two pretwists that have come to bad ends, and one that is okay.

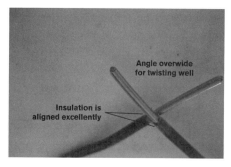

The wires' angle is too wide for easy pretwisting with pliers.

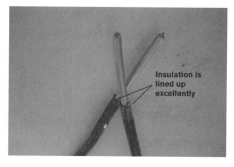

These are at a good angle for twisting.

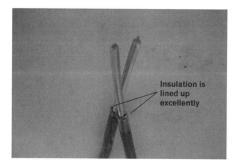

The angle is too narrow for twisting.

Three wires laid out well for twisting together.

A partial three-conductor splice, probably okay when completed, if twisted tightly enough for sufficient turns to hold them together.

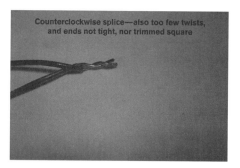

A counterclockwise splice will become undone as you twist on a wirenut.

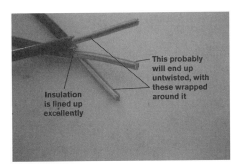

Three wires lined up poorly for pretwisting.

Solid wire spliced with stranded, very carelessly.

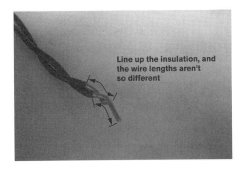

Neither the ends of copper nor of insulation are aligned, and the twist is loose and not long enough; back out the wires, line up insulation, and it may be spliceable, but insulation probably needs to be stripped back.

- it might be a wire of a heavier gage being spliced to a thinner wire.

  To make the splice concentric, therefore, hold your pliers more nearly in line with the less stiff conductors. This will put more bending pressure on the stiffest wires.
- When I am connecting wires of greatly differing stiffness, sometimes I find it helpful to prebend the stiffest one. I shape it into the twisted form that it will hold in the completed splice; this way, I can wrap the other wires around it and still end up with a concentric splice.
- Ensure that your pretwist is tight by finishing your twisting with a pair of pliers, even if you started with your fingers. Sometimes, at this point, rather that twisting my pliers, I squeeze a little, turn my pliers a hair, squeeze again, turn a little more, and so forth. This ends up tightening the twist without putting significant strain on the insulation.

You have basically completed the splice; now you are ensuring that it is a good splice. The wires must be even and the twist must be concentric and tight. If the wires are not twisted concentrically, one wire will be wrapped in the center, and not in contact with the spring. This is your last chance to rectify any lack of concentricity.

You now are ready for Step 4 of the general procedure I described above.

The "Last Shot" Pigtailing Splice

The following is a very, very difficult splice. While it is a straight-in splice, you should skip this one if you found pretwisted splices daunting, and go on ahead to the next essential section, the one on bad splices.

Sometimes when you work on old wires that come from the back of an outlet box, they will be too

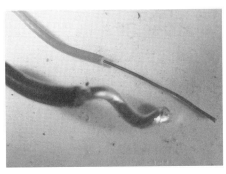

A heavier wire with a skinny one—it is overly pretwisted, given the thinner wire's diameter.

short to splice normally. When, at a customer's request, I am trying to put off replacing old wiring that is in bad shape, this very touchy type of straight-in splice is useful.

So, I have a very short conductor. Sometimes this situation comes up because the metal has fatigued and part of the wire has broken off—perhaps in the course of attempting a simpler splice. Its remaining insulation is in good enough shape to keep the conductor usable, but it is so short that in the ordinary course of events it couldn't be spliced. In other cases, the conductor started out just long enough for a splice, but the insulation was so rotten that rather than bandage it, I chose to pinch it back to a point where I considered it reasonably trustworthy.

I want to pigtail, to add enough length that I end up with a conductor long enough, and in good enough shape, to work with. (Alternately, I want to install a device that comes with leads rather than terminal screws. In this case, substitute the word, "lead" wherever I say "new conductor" or "pigtail.") Here's a method that has worked for me.

1. Choose a wirenut, based on the diameter of the old wire and of the wire you want to use for the pigtail.

2. Cut the wire back to the right length for a wirenut; if necessary, strip back some insulation to expose more bare copper.

3. Strip a piece of good conductor back the same length as the stripped section of the old wire, to use as the pigtail. This stripped section will eventually lie alongside the stripped section of the old wire, facing in the same direction as the old wire so that you can screw a wirenut on it (in other words, it will point *out* of the box).

4. Form a J shape at the stripped end of the new conductor. The straight part of the J, the part that will lie alongside the old wire, needs to include both stripped and unstripped portions of the new conductor, so that it is longer than the wirenut. This will ensure that the curved part of the new conductor doesn't get in the way of the wirenut's skirt.

5. Now insert your pigtail into the box, with the j-hook alongside the old conductor, nestling parallel to it. Align both the ends of the wires and the

Wirenut being put on j-hook for "last chance" splice, inside the box this "j-hook" is next to a very short conductor for a "last-chance" splice.

The j-hook is lined up for a last chance splice in a messy place.

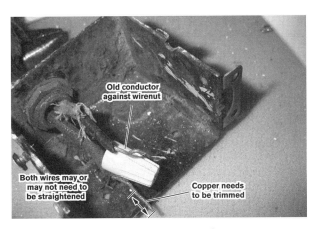

Labels on image: Old conductor against wirenut · Both wires may or may not need to be straightened · Copper needs to be trimmed

Old conductor against wirenut—needs copper trimmed.

ends of their insulation. If you can't do so, you need to redo the new conductor. Take it out and restrip or reshape it.

6. You are ready to begin your splice. Put a wirenut over the pair of conductors.

7. Delicately, delicately, start twisting the wirenut over the pair. Initially, use almost no force, because you don't want to push either wire away, and you have little means of coaxing the old conductor forward into the wirenut. (In most cases, you have no way of getting your fingers into the box to hold the old wire—and its insulation is very, very unlikely to be able to withstand being squeezed by, for instance, needlenose pliers.)

8. Once you feel the spring grab onto both conductors—you will sense an increase in the resistance to your turning—you can twist with *slightly* more force. Be very careful not to turn the wirenut so hard that you will twist the old wire's insulation. It will be subjected to more than enough stress when you stuff your wires and device back into the box.

9. Finally, as with any splice, tug gently to test your splice.

With this type of splice, much more than with those described earlier for use with wires that permit brisker handling, it is important that you make certain that it still is secure after you've done whatever else you need to do in the enclosure—especially anything involving the pigtail.

Does it seem that rewiring the box—a project beyond the scope of this book—would be a lot easier and allow a much more secure connection than this "last shot"? Absolutely. However, rewiring does cost much more and require much more repair.

Bad Splices

You now have a good basic understanding of how to splice correctly. Many don't have this. The most difficult splices to undertake involve mixes of different sizes

J-hook for "last chance" splice, from inside the box.

The "last chance" splice j-hook as viewed from side of box.

of wire or mixes of solid and stranded wires. These are the situations in which marginal splices are most likely to fail. This applies to the splices I described so far, but it applies even more to the many unreliable ways to splice. Each one departs in one way or another from the procedures I described.

Common mistakes include:

- Joining a solid and a stranded wire simply by wrapping the stranded wire around the straight solid wire.

   This defeats one of the two functions of the wirenut—direct spring contact holding onto each of the two conductors. In this case, the wirenut will grab only the outside, stranded wire. This can allow the solid wire to slide out.

- Folding the solid wire over the stranded.

   The installers hope to firmly connect them by tightly squeezing the solid over the stranded wire. Unfortunately, there's no guarantee that the pressure *will* be firm enough, and that it will stay that way. Furthermore, in this case the wirenut will grab only the outside, solid wire. The stranded wire could be pulled out during some unanticipated shifting. Finally, because of the loop at the end of the solid wire, when the wirenut is screwed on, it cannot squeeze the wire into the slightly conical shape of the spring. This makes it more likely that the wirenut will fall off.

Other Splicing Systems

At least three other systems have been used to splice branch circuit wiring, and they're pretty good. As with "twist, solder, and tape" splices, these usually represent professional workmanship. Unlike those, they are not outdated. You may even be able to reuse two of them. Whether you are examining them or reusing, you need to make sure that the conductors in the splice are trimmed to an appropriate length and lined up evenly.

Setscrew Wirenuts

One system is very old: a wirenut that uses an internal setscrew connector, threaded on the outside, rather than a twist-on. After its setscrew is tightened, an insulating shell is screwed on over the splice. As with normal wirenuts, this can be used with solid or stranded conductors, pretwisted or just inserted straight. An untrained installer is more likely to make a reliable splice with this than with a standard wirenut, However, you are far more likely to find this connector in old junction boxes than available for purchase. It is particularly handy for the "last chance" pigtailing splice that I just described. Provided that you can angle the short old conductor enough to get a screwdriver onto the setscrew, it will be easier for you to splice this way than to twist a wirenut on conductors that you have no means of restraining.

Crimp Barrels

The second and least exotic of these systems is the crimp barrel, which is used for splicing grounding conductors. (Theoretically, it can be used for other splices as well, so long as they are then insulated.) It is a copper or brass barrel which

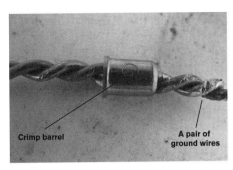

A crimp barrel, on ground wires, not yet crimped.

is crimped—indented—a couple of times, with an appropriate tool. I recommend against your using it, for four reasons:

- Wirenuts will serve you just fine for splicing grounds.

- Crimp barrels cannot be removed without considerable difficulty.

- If you don't indent the barrels correctly, the splices are uncertain.

- They invite installers to trim the ground wires too short—and even when this is not the case, removing the barrels, by cutting the conductors, results in ground wires that are too short.

You are fairly likely to run into crimp barrels. I mention them so that you recognize them, and recognize that they *probably* indicate professional installation. Crimping can be about the most secure splicing system there is. Some types of crimp, or compression, fittings are used by the power company for this reason.

Terminal Blocks

The third system is easier to use, in its way, than wirenuts. It also is more expensive, and it takes up considerably more room in enclosures, which can be a major drawback in crowded old boxes. The basic version is usable only with solid conductors. It employs internal spring clips to secure the conductors that it is Listed for. You simply strip them, trim them to the right length, and stick them in.

There also are setscrew versions of this connector, which are similar, in a way, to lugs. These can be used with a wider range of conductors, but are so much larger, and pricier, that *in general* it would be absurd to use them where wirenuts can serve. However, certain of these do have two very important uses:

- Making connections with larger conductors than you may be able to splice, or may not be able to splice reliably, with wirenuts. Splicing #10 conductors, for instance, can be tough.

- Use with aluminum and copper-clad aluminum conductors. This second use for which *some* of these are Listed is unique and very valuable, as you will see below. To the best of my knowledge, *nothing* else on the market is designed to connect, for instance, two #10 aluminum conductors to a copper #12.

Aluminum Wiring

Everything I have said so far in this chapter presumed that you were working with copper conductors. Aluminum is a whole different story.

Identifying Materials

How can you tell whether conductors are copper or aluminum? Not by color, at least not without experience and probably not even then. Silvery branch circuit conductors sometimes are tinned (solder-coated) copper. You can determine that they are copper in a moment, though, by scraping through the thin coating to expose the coppery metal beneath.

There also are two subtler methods. You can test wires by cutting them and comparing the feel with cutting copper. Aluminum is easier to cut than copper, by a good bit. Bending provides an even more subtle test. For a given diameter of wire, copper is stiffer than aluminum. If you try to splice some #12 wire that looks like solid copper, but feels overly flexible, you might have the bad luck to have a house that was wired, at least partly, with copper-coated aluminum wire, formally "copper-clad." Fortunately, the latter was manufactured during a very brief period. Wiring installed before (or after) the 1970s will not be copper-clad aluminum.

What Are the Problems with Aluminum Branch Circuit Wiring?

Any aluminum branch circuit wiring is bad luck, if only because it hurts resale value. There also are a number of functional problems:

- Dissimilar metals in direct contact can cause corrosion;

- aluminum expands and contracts in response to temperature changes more than copper does, loosening connections;

- aluminum forms a relatively high-resistance oxide coating almost immediately upon exposure to air; and,

- over time, wet aluminum wires melt away, because aluminum oxides are fairly water-soluble.

Aluminum in Perspective

Aluminum wiring may be far from ideal as it's a problem to deal with, even a potential hazard; it may reduce the value of your property and impair your ability to work on your own wiring; but it does not indicate bad wiring practices. If I had aluminum branch circuit wiring, I would replace it, but I might not do so immediately. I would be concerned far more immediately by any evidence of incompetent and careless wiring.

Because wirenuts hold conductors directly in contact with each other, currently-manufactured wirenuts are not Listed for joining copper and aluminum wires together, with one limited exception. Aluminum wiring in small, branch circuit gauges no longer is installed at all. The National Electrical Code still contains rules for its use, but they are largely irrelevant because it no longer is installed; installations are grandfathered. However, if you have aluminum wiring in your home, you have to deal with it.

Some say you can leave well enough alone, if the circuits are not causing problems, but the most certain solution is to replace all your aluminum branch circuit wiring with copper—an immense job. The second safest choice is to bite a slightly smaller bullet and have all wires reterminated in copper. The only marginally safe way to do this is to hire an electrician who is licensed to use the proprietary system, "Copalum," manufactured by the AMP corporation. For further information, call them at 1-800-522-6752. Unfortunately, there may not be anyone in your area equipped to do this; AMP has not been training new installers.

Suppose you need to replace a switch or receptacle? At the very least, the replacement device needs to be marked, "Co/Alr," and any required splice must use a connector Listed for the use. Ideal Corporation makes the one wirenut Listed for some such uses, the pricey purple "Twister Al/Cu."

There are problems involved in its use. Twisters have melted!

- It does not accommodate all common wire sizes, nor all combinations;

- it may accept only one aluminum conductor per wirenut; and

- it could fail if reused

The other option, rather unwieldy and expensive, is to use small terminal blocks, as described above under "Other Splicing Systems."

Connecting Conductors to Terminals

Splicing is only one of the two ways wires are connected. Most device connections rely on terminals. The procedures I explain here may not be as difficult as those in the previous two sections, but the results are equally critical. The skills I describe in this section take some practice, just as did preparing conductors and splicing.

Standard Terminal Screws

The standard terminal screw, the type that has a wire wrapped around it, probably is the most common electrical connector after the wirenut. It probably is the oldest, and it probably is the most reliable. I can find nothing bad to say about it. Some other countries even employ it where we would use a wirenut. It makes a superior connection and it lets you see the quality of the connection better than any other type of terminal.

You will find screw terminals on the sides of a device, or, in the case of some very old devices, on the front, or even on the ends. You will find screw terminals not only on receptacles and switches, but on the backs of many lampholders, especially on plain porcelains; on the ends of most fuse holders and some circuit breakers, and on special devices such as timers.

The rules for attaching wires to screw terminals are unforgiving, but straightforward.

Read through this chapter before assuming that a device that appears to have standard screw terminals actually uses screws in that way. Two types of devices have terminals that involve screws, but the screws are not designed to have wires wrapped around them.

Forming Hooks

After you prepare a conductor, as I described above, start by shaping the stripped wire at its end:

1. Trim the wire back long enough so that it can be formed to go between ⅔ and ¾ of the way around the screw, under the screw head—plus a bit.

2. With your lineman's pliers and the stripped end of the wire pointing in the same direction, grab the end of the stripped wire with the side of the jaws of your pliers. Bend the end of the wire into a hook that is about large enough to wrap most, but not all, of the way around the screw, under its head. The bare wire now should form a J.

3. Hold the hook over the head of the screw and confirm these important features:
   - There is enough bare wire beyond the hook, in the straight part of the J, that the screw's head will not pinch insulation. If insulation gets in the way, it will prevent the screw from making good contact with the copper.
   - The wire is not so long that bare—live—copper will extend beyond the device body after installation.

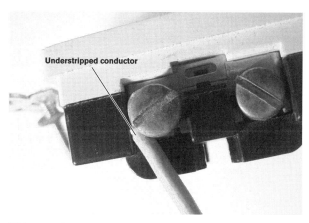

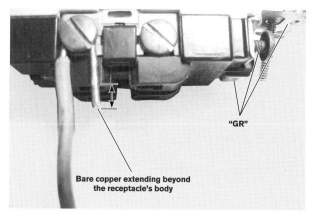

This conductor is understripped—insulation is caught under the screw, keeping it from proper, low-resistance, contact with the wire.

This conductor was overstripped—copper sticks out past the screw and even the device body, threatening shock or short circuit. "GR" signifies that the other screw is the ground. The metal it is screwed into is part of the yoke, which is connected internally to the ground holes.

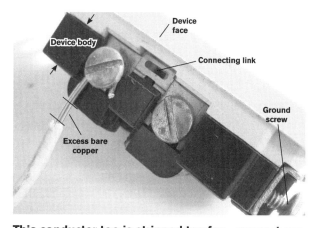

This conducter too is stripped too far—copper shows between the insulation and the device body. The connecting link joins the two receptacles sharing the yoke. For split wiring, which is mentioned later, on pp. 205 and 255, you would insert the tip of a screwdriver into this slot and wiggle or twist it, to break the link.

- The wire will go less than 360 degrees around the screw. If a wire overlaps itself, there is poor contact except where it crosses. At the point of overlap it will be a double thickness, meaning that elsewhere there will be a bit of a gap.

Attaching Wires

You are ready to terminate. Before taking the following three steps, though, look over the hints that follow them. Some involve the way you might form the hook.

1. Slip the hook under the screw head clockwise. After you place the wire, you will tighten the screw clockwise. If you wrapped the wire around the screw counterclockwise (with the open end to the left), you could partly unwrap the wire, or partly squeeze it out from under, as you tightened the screw.

2. Confirm the three important features I mentioned just above, with the hook actually in place.

3. Tighten the screw. One rule of thumb I've heard is to screw it far enough down that it has squeezed the loop flat against the device, and then tighten it another quarter turn (an additional 90 degrees).

4. Confirm that it is tight. If pulling or pushing on the wire budges the loop with respect to the screw, or turns the screw, the screw definitely is too loose. Theoretically, you should use a torque screwdriver, which applies precisely the right amount of force. However, even pros rarely use these except when making highly critical connections such as those at telecom installations.

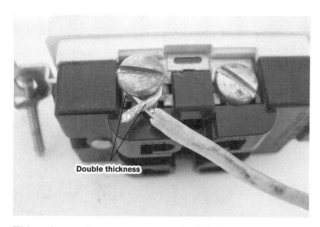

This wire makes poor contact with the screw, because it crosses itself; 360 degrees are too much. The double thickness of wire means that the screw head cannot make good contact with the wire here or elsewhere on its circumference. Using the loop had one positive consequence: it ensured that the wire wouldn't slip off. However, this is far overshadowed by the negative consequence: such a connection has relatively high resistance.

Here are a few hints to make installation easier.

- As you start, form the hook a little more open than might be necessary. It is easy to squeeze a hook that is under a screw head tighter; it is much more difficult to open up a too-tight hook that you have gotten partly under the screw.

- Bend the open end of the hook slightly down (as seen when the hook lies with its opening to the right). Thirty degrees or so works well. In many cases, it is not easy to hook a flat, essentially two-dimensional, loop under the screw's head.

- Back your screw out fully, without forcing. It is staked (locked in place by what are essentially fouled threads at its bottom), so if you work so hard that you remove it entirely, you will find that it is not easy to get back in.

- Begin by working the open end of the hook under the screw head. Once this is in place, you can feed the rest of the loop in place with a little wiggling and pulling.

- IMPORTANT: With stranded wire, make sure that no strands peek out separately from under the screw head. Inadvertent contact with a length of bare copper, even a single vagrant strand that's live, could shock or short.

- Make sure that the wire loops around the screw's body, under the head, are in a tight, smooth arc. Otherwise, it might be squeezed out as you tighten the screw.

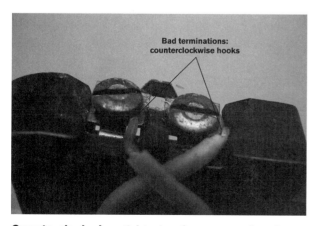

Counterclockwise–tightening the screw pushes the wire away.

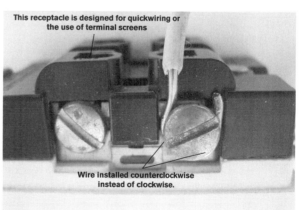

Counterclockwise. In this case, the wire was stripped correctly so the copper does not go past the body. It is properly hooked so that it goes about ¾ of the way around, and is tucked reasonably well under the screw head. However, it was installed counterclockwise rather than clockwise.

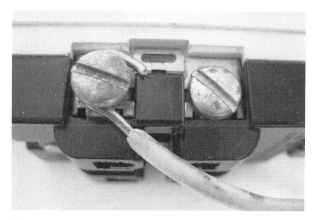

Sloppy wrap–may prevent tight contact.

The lay-in design makes hooking difficult—not impossible, though, nor necessarily bad.

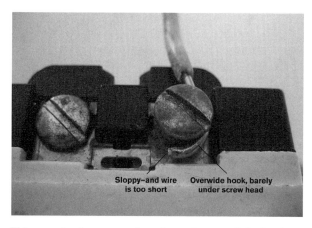

Sloppy–and wire is too short

Overwide hook, barely under screw head

This conductor was stripped too short, and the hook is too open.

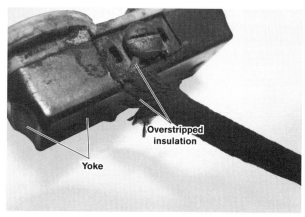

Overstripped insulation

Yoke

An old receptacle with insulation overstripped, or worn and broken off—but not so bad; copper does not extend beyond the device's body. The yoke is more clearly visible here on the back than on some other devices where most of it passes inside.

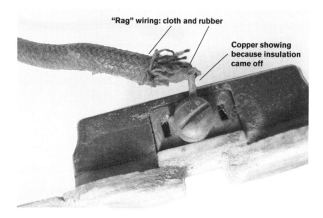

"Rag" wiring: cloth and rubber

Copper showing because insulation came off

An old receptacle with insulation overstripped, or worn and broken off, this time not okay: copper extends beyond the device body—simply because of the angle at which it leaves the screw.

- Put only one wire under each screw because a standard terminal screw is designed to make a reliable, low-resistance connection if and only if it squeezes one wire between the screw head and the base plate underneath.

- Once you have the hook under the screw head, see whether it *is* entirely under the screw head. If not, squeeze it with your pliers to tighten the hook. Yes, you can squeeze it tight enough with lineman's pliers (though other pliers such as needlenose are fine, too). You can do this without removing the loop from its position under the screw head, despite the fact that the ends of the pliers' jaws certainly will not fit under the screw head. Once you have squeezed the loop so that the jaws of the pliers are tight against the sides of the screw's head, the loop must be under the head.

Sloppy loop formation and not pulled as snugly under—in contact with—the screw head as it could be.

- As you tighten the screw, use one hand to pull back on the conductor or the device. Maintaining tension ensures that the hook stays tucked nicely under the screw head.

All of this will take some experimentation. Practice will help you develop the skill you need to get this right the first time.

Extra Touches

Here are two more things you can do for a superior job:

- After you complete all the terminations to a device, wrap electrical tape over them, going all the way around the device. (I tend to go 1½ times around, to make sure the tape grabs onto itself.) This way, if, say, a ground wire pokes forward from elsewhere in the box, it can't short against the hot terminals.
- To do an elegant job, "dress" the wires as I describe in Chapter 14.

Alternate Systems for Terminating Wires

Quickwiring

The next most commonly encountered system for terminating conductors on devices is called "backwiring," "push wiring," "backstabbing," or, most commonly, "quick wiring." (Because the term, "backwiring," also can apply to another system that I describe later in this chapter, I will stick with "quickwiring" here.) This involves sticking a wire—solid only, not stranded—into a hole in the back of a device until it is grabbed by the device's internal spring clip. I don't recommend quickwiring, for reasons that I will mention, but I need to describe it because it is so widely used. You should at least know how to recognize when it has been done correctly and when not.

Prevalence. Quickwiring is very widely used. Terminals for quickwiring are found on the following equipment, as well as other places:
- In most receptacles and switches.
- Occasionally in porcelain lampholders.
- Even a few cheap circuit breakers. The latter, called "water heater breakers," were manufactured in the 1970s.

Some devices—exclusively cheap ones, and I don't just mean inexpensive—allow solely for quickwiring. Many devices offer a choice of wiring methods: both terminal screws and quickwiring holes.

Advantages and Drawbacks

Drawbacks. Quickwiring has four important drawbacks:
- Products designed for quickwiring are very likely to be of lower quality than those designed for both quickwiring and terminal screws. Those designed for both quickwiring and terminal screws are somewhat likely to be of lower quality than products that do not permit quickwiring.

- Quickwiring is not really accepted, even by manufacturers, as a means of making the most important connections—grounding connections— though technically it is permitted for the purpose. This is evidenced by the fact that all halfway modern devices, including quickwired switches and receptacles, have green grounding screws.
- The uses permitted for quickwiring have become increasingly restricted as well. It now only can be used for one size and type of wire: #14 solid copper. Unlike other types of termination, quickwiring may fail when a device originally used with #12 wire is reused with #14.
- Quickwiring also, arguably, is easier to do badly because the mechanism is finicky and the connection is out of sight.

The common view among conscientious electricians is that while quickwiring is fast, even when done well it does not produce as secure a connection as do other methods.

Advantages

- The number one advantage to quickwiring in dealing with old wiring is that it allows you to put off a major project. When the wires are so short that you cannot form hooks for a device's terminal screws, you probably cannot even use the "last resort" splice I described in the previous chapter. Even when you have practiced that splice enough times that you can install it safely, it is difficult and uncertain. When a conductor is so short that you cannot even add a pigtail, you probably have no other way to replace the device to which it is attached, without some recabling. Even when you can reach your pliers in far enough to form hooks, you may very well be unable to hook those hooks around the terminal screws on the sides of your new device and tighten the screws securely. Quickwiring, in contrast, sometimes is possible when you barely have enough good wire to reach the front edge of the box. If you are lucky, you then can quickwire in the process of sliding the device into the box. (You have to test your connections extremely carefully, by tugging gently on the device, when installation has been this difficult.)
- Quickwiring can save time because it eliminates the need to form most of the wires into hooks, although it does not eliminate the need to be careful about wires' shapes.
  - Quickwiring eliminates the need to bring wires under screws' heads and to tighten screws, except for the grounding screw. This too makes quickwiring potentially faster than using terminal screws.
  - Finally, quickwiring provides more available terminals on a receptacle or switch, making it easier to daisy-chain. Pigtailing, though, is far, far more reliable.

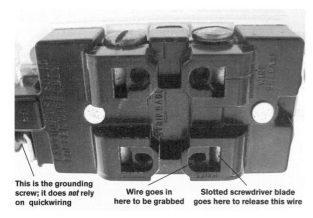

This is the grounding screw; it does *not* rely on quickwiring

Wire goes in here to be grabbed

Slotted screwdriver blade goes here to release this wire

A quickwire receptacle's back showing markings and wire and release holes.

Procedure. Quickwiring is only a four-step procedure, but each of the following steps is critical:

1. Cut each conductor to the right length, as indicated by the strip gauge next to the hole

that it (and only it) will go into. Every quickwired device has a strip gauge. This usually is an embossed furrow showing the exact length to which the copper must be cut.

2. Straighten the wire; it must be quite straight.
3. Push the conductor firmly into its hole until it has gone in as far as it will; essentially all of the bare copper should go in.
4. Tug to make sure the internal spring has caught the wire.

Cautions. Here are some things that can cause problems with quickwired connections. Look for them when installing or examining these terminations:

- **Overstripping.** If this is not trimmed back, excess—and eventually live—copper sticks out the back of the device. Unless power is off and

Carelessness in performing either of these tasks may, and often does, result in poor contact, early failure, intermittent operation, and expensive bills for professional troubleshooting. A master entertainer, Uri Geller, astounded audiences by playing on a common perceptual weakness. He appeared to bend spoons magically while holding them in the air by the handle. His trick is simple. He bent a spoon by surreptitiously pressing it against his body before holding it up for his audience. He then kept the front exactly in line with the audience's angle of vision. Then he slowly rotated the spoon, and the bend began to appear as the spoon's profile was presented. Why the story? I've fooled myself in the same way as the audiences at those magic shows by not being careful enough in examining wires I thought were straight. In this case, the consequence of not checking from all angles is not oohs and ahs but possible malfunction, or even electrical fire.

A three-receptacle round cover from the back.

Wrong—quickwiring with bent wire. Be sure to
tighten unused screws to avoid inadvertant contact.

you are working by flashlight, always assume
that the internal clips, and therefore the wires
touching them, are live.

- **Understripping.** After you strip, straighten,
and trim a conductor, and push it all the way
in, the internal spring clip may not grab it reliably if the stripped wire is too short. Double-
check against the strip gauge after you
straighten the wire.

- **Not straightening well enough.** You need to
straighten a wire far more carefully for quickwiring than you do for straight-in splicing. If it
is crooked, or even curved, the internal spring
clip may only push against it on one side. This
is not good enough to reliably keep it in place
or make a safe, low-resistance connection.

- **Polarity reversal.** Make sure that you use the right holes. The ones in
receptacles for neutral wires will have the word "white" stamped adjacent
to them; the ones in three-way switches for Common conductors (this is
explained in Chapter 15) will have the word Common next to them.

- **Inserting wires into holes not designed for connections.** There will be
other holes, serving a different purpose. Only the round ones are for
wiring.

- **Using quickwiring with unsuitable wire.** To repeat: only solid copper
wires are suitable for quickwiring. Usually only #12 or #14 gage wire
was allowed. More recently manufactured versions are marked to warn
you that #12 conductors no longer may be inserted, only #14.

- **Exposure to live terminals.** Before installing a device manufactured
with both quickwiring holes and screw terminals, screw in and tape any
unused terminal screws; they're live when power is on. Beware of live
screws when you approach devices quickwired by others. Installers who
save time by quickwiring may also save time by untidy work.

- **Bad wire.** Clean coatings such as corrosion or paint spatters. This is far
more critical with quickwiring than with other types of termination.

Disconnecting Quickwired Equipment. The simplest way to disconnect quickwired equipment is to cut the wires. This does, however, leave the quickwiring holes unusable, and usually results in a bit of live copper sticking out
if you reuse the devices—and you lose a little wire.

In the event that you want to or must remove a wire, rather than merely
removing the device, there are other options. First, there are more-or-less rectangular holes marked "Release," "Wire release," or most puzzlingly for holes,
"Push to release." You push a screwdriver into one of these to hold back the
spring clip that grips a wire.

Here is the most common procedure.

1. With one hand, hold the device.
2. With two of the fingers of the same hand (to substitute for your missing
third hand), hold on to the conductor, and push it away from the device.

Removing Quickwired Wires Can Be a Bear

Sometimes you will pick up a small screwdriver and insert it where directed, and the wire will pop out. More often, it seems that you need three or more hands to do the job. The reason for this difficulty is that the spring clip is designed to grab a wire, whereas the release hole is designed to admit, but

not to retain, a screwdriver. Insert a screwdriver at the correct angle, pushing hard enough against the spring clip, and the device will release the wire; release the screwdriver, and the clip will be happy to grab the wire again—immediately.

3. With your other hand, push the screwdriver into the rectangular hole, and, if necessary, work it at different angles until you find the way to release the spring clip.

4. As soon as the wire is released, pull it clear.

You may face additional difficulties. The screwdriver has to be small enough, or else the blade may not fit in the hole at all; perhaps it will enter, but not go in far enough to push back the spring clip; or perhaps it will go in far enough but at the cost of damaging the device. (True, if you are replacing the device anyway, this damage is of no consequence.)

Some older smoke detectors were quickwired directly to the branch circuit, without battery backup. They were designed without "pause" buttons to turn them off in a nonemergency, such as while you took your burning toast outside. Being quickwired, there was no easy way to disconnect them. Hold your ears!

When you are not able to free a wire by working a screwdriver in its release hole, you may need to add the second option. Pull on the device, perhaps twisting it as you pull. This will, in fact, often release a wire even without your using the release hole. Of course, pulling and twisting can damage deteriorated insulation, such as you may find on older wiring. In such cases, you probably are better off snipping the wire where it enters the quickwiring hole, even if the conductor is short.

Other Ways of Dealing with Multiple Conductors

Midpoint Skinning. Quickwiring has not always been an option. It also has not been the only method that saved materials at the cost of quality, or that could cause inconvenience to electricians who came later.

Midpoint skinning is one such shortcut. Here is how it worked. When installing multiple devices in a box, say two or three switches, an old-timer sometimes skinned (removed insulation from) a precise length of a conductor in the middle, rather than at the end. Then he'd wrap the skinned stretch of wire around a terminal screw, and continue on with the same wire to another device or splice. This allowed an installer to daisy-chain using only one screw, the equivalent of looping two wires under the screw. (I address that practice a little later.) If he was careful enough—and the old-timers usually were careful enough—he could

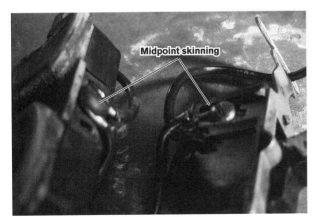

Midpoint skinning—an advanced procedure. This lets the conductors daisy chain from a receptacle with only one termination point per side to a similar old switch.

do this without creating an unreliable connection. The worst consequence of this practice is that it made it quite difficult to pull a single device out of the box for testing or replacement because electricians who used midpoint skinning almost never left a great deal of slack in the lengths of conductor running between the devices.

Don't attempt midpoint skinning. It's too easy for someone new to the technique to skin too much, leaving bare wire that could short or shock. It's also too easy for a novice to remove too little insulation, which means that—unless you bite the bullet, start again and pinch off a little more insulation—either you will wrap the wire far less than ⅔ of the way around the terminal screw or else you will pinch insulation, creating a high-resistance connection. I have used midpoint skinning myself, but far more often I've undone, stripped back, and pigtailed conductors that some old-timer had hooked up this way.

One final point about midpoint skinning: it *almost* always means that a pro did the work.

Doubling up on a Screw—DON'T!

A worse, and quite illegal, practice sometimes used for daisy-chaining was wrapping two wires on top of each other under the head of one screw. If you find this, take it apart and pigtail the conductors.

Daisy-Chaining Devices Equipped for It. Pigtailing is the best choice for connecting multiple conductors or devices, unless your box simply is too crowded. Sometimes, though, pigtailing seems too much of a nuisance, or it seems as though pigtailing would take up too much volume in a tightly-packed box. When this is the case, there are alternate means of connecting multiple devices that almost always are acceptable. The most suitable one to utilize with standard duplex receptacles revolves around the fact that they have at least two screw terminals for hot and two for neutral wires. The "busbar" between the two hot terminals should have negligible resistance. So as long as the receptacle is well made and you connect your wires well to the screws, this will daisy-chain them nicely. In the example where you have two conductors to connect to two receptacles, simply terminate one conductor on one receptacle and the other conductor on the other receptacle. Then cut a length of similar conductor—6–7″ long at minimum—and terminate it on each of the receptacles, as bridge. Now do the same with the neutrals, using a white bridge.

Not all devices are equipped for daisy-chaining. For instance, single, as opposed to duplex, receptacles normally are not, and switches normally are not. In these cases, you need to pigtail.

Capture Features

Each of the next two designs captures wire by screw pressure, rather than with a bit of spring metal. At the same time, neither relies upon the installer's skill at making a good hook. Another positive feature that

While some backwired or insertion lay-in-type terminals involving screws accept more than one conductor, it never is appropriate to wrap two hooks around the same screw.

In certain rare cases in existing wiring, the "busbar" on each side of a duplex receptacle may not be continuous. There is a connecting link, in the middle of each busbar, that is made to be broken out when necessary. Dealing with the wiring configuration requiring this, however, is more or less beyond the scope of this book.

they share is that if you have to replace them, they are least likely to have damaged the wires that you will remove. Therefore, you are least likely to have to cut back the conductors to get to good wire.

Backwiring. The next termination system goes by various names: the pressure plate, pinch terminal, or pinch plate; or, simply—and most commonly—backwiring. These devices have holes in the back, somewhat like quickwiring devices. They have screws at their sides, somewhat like standard receptacles. However, the screws' purpose is to tighten internal plates that squeeze inserted wires between them. The way you can differentiate backwired devices from quickwired is that these have round holes in the back, but no rectangular "release" holes. It is important to note that in a backwired device the screw that tightens a plate probably is not intended for use as a "wire-binding" terminal screw.

You are likely to see these in two places: heavy-duty or deluxe devices, such as better-quality switches and receptacles, and imported lighting fixtures.

Backwired devices are very simple to use:

1. Strip back some of the insulation on your conductors, hold them against the strip gage, and trim them to match the marked length.
2. Make sure that they are straight. While this is important, it is far less critical than with quickwiring.
3. Stick them in the holes to confirm the trimmed length; see whether any bare copper is showing. If so, take them out and cut them back so that you can barely see copper after fully inserting them into the terminal holes.
4. After you reinsert them, tighten the screws that secure the terminations. Backwiring is much like using terminal blocks, which were mentioned in the section on splicing.

Lay-In Terminations. The lay-in, or slide-in, termination uses a terminal screw. However, the wire is not hooked around its head, merely inserted straight under one side of the head, as I describe below. Just as devices with holes in the back

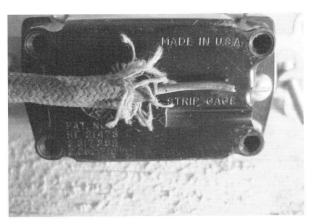

A backwire-type device with old wire against strip gauge—the stripped length is about right.

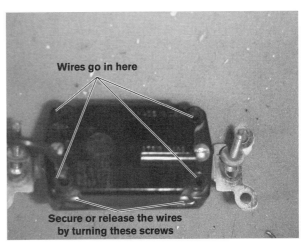

A pressure-plate type backwired receptacle. Note that there aren't any release slots on the back.

CAUTION!

I need to repeat my warning: many imported fixtures are sold without complying with U.S. requirements. If they are not Listed by an American laboratory such as U.L., it may be illegal to install and dangerous to use them.

sometimes are mistaken for the quickwiring type, these can be mistaken for the type with standard, wrap-around terminal screws. However, this one has a strip gage, unlike the type that relies on your forming hooks. You certainly will find this design in fuse holders, and almost always in circuit breakers.

This design does capture less wire under the screw's head than a standard terminal screw. On the other hand, it is much easier to wire correctly because you do not have to form and engage a hook.

Here is what you will find. There is a narrow space or "valley" along one side of a screw, under its head. There is a ward, usually an upturned lug or a ridge, along this side of the screw. It may be continuous with the metal into which the screw is threaded, or it may simply be nonmetallic material, part of the device's body). Its positioning ensures that a wire inserted alongside it cannot slip to the side and escape from under the screw. The difference between this setup and a standard screw terminal is that you will find no easy access to the rest of the space under the screw head into which to fit a hook. Fortunately, with this design it is unnecessary to go around the screw.

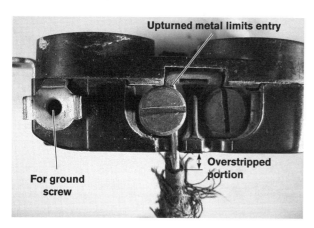

A lay-in terminal; shows a conductor that was stripped too far—bare copper extends beyond the device body.

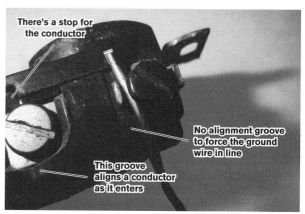

The screw for the ground wire is not a lay-in terminal. The indicators are that it has no stop, and the ward doesn't hold the wire tucked under the screw head. The ground wire here is *supposed* to be hooked around the screw, unlike the hots and neutrals in this receptacle design.

This lampholder has lay-in terminals on both sides; don't confuse the one that is connected to the shell with the one for the center terminal.

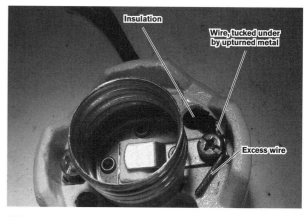

This lampholder shows the lay-in terminal with a captured conductor. Here, excess wire does no harm—it will be sandwiched between two pieces of porcelain.

Not stripped enough, according to the strip gage.

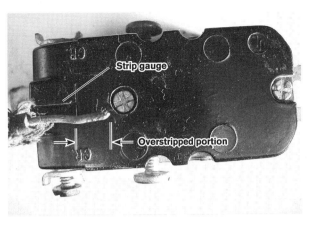

Not trimmed back enough, according to the strip gage.

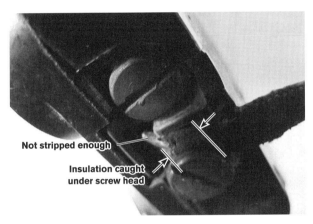

Not stripped enough; the screw's head catches insulation.

The ground wire should—and does—go around this screw; we know it's not a lay-in because there's no stop, and little to hold it under.

This type of terminal is simple to use. Here's the sequence:

1. Trim your wire to match the strip gage.
2. Lay or stick it in the valley alongside the screw, under the head.
3. Check its length:
   - Confirm that it goes back far enough to hit the back of the space.
   - Confirm that it is not so long that bare copper reaches beyond the device's body.
   - Confirm that it is not so short that insulation reaches under the screw's head.
4. Once the length is right, tighten the screw. Tug on the wire to confirm that it is secure.

Grounding Connections

NEC Article 250 covers grounding. Much of the article is irrelevant to the tasks explained in this book. Nevertheless, plow through it and make sure you understand the applications with which you'll be working.

Before making any grounding connections, use a tester and a voltage checker and, if needed, a reliable, independent source, to make sure that a ground is indeed present. I talk about this further in Chapter 14. The fact that your box

It took ten pages, and more than ten hours, to write a report based on one day's consultation for a customer. He had paid many thousands of dollars for crews of professional electricians to add new wiring, every inch of it in armored cable. Because they did not take the time to secure their cables and boxes, some connectors—high-quality connectors—came loose, resulting in very unreliable grounding. This constituted a definite shock hazard, one worthy of documenting in excruciating detail so that he could have the work corrected.

contains a bare grounding wire does not reliably indicate that the other end of the wire ultimately leads to the system ground. The fact that you have a metal box with a metal raceway, or cable armor, attached to it similarly proves nothing.

Normal Grounding Means

Grounding Screws. All modern metal boxes have tapped 10-32 holes at the back. Grounding screws are green 10-32 screws, about ³⁄₈″ long, with washer heads, the better to press evenly and reliably against ground wires. They have either the traditional slotted heads—to allow them to be turned with greater force—or combination slotted-Phillips heads. The heads often are hexagonal, to allow the use of nut drivers for even better and faster tightening.

Some light fixtures include grounding screws that may be smaller, perhaps 8-32. So long as they come with Listed fixtures, they are suitable for grounding (but only at the tapped holes provided for them). Some fixtures have a peculiarity that demands your attention, or you will not be able to make a good grounding connection. The tapped grounding locations are painted. In my experience, this paint always has been nonconductive. It is very important that you scrape or sand away the paint from the area that will be under the grounding wire, to minimize resistance. Besides grounding screws, supply houses sell grounding pigtails. The latter are simply the same 10-32 screws made up with 7-odd inch lengths of green-insulated wire already wound about ¾ of the way around their bodies.

Most inspectors will not complain about your using any 10-32 round-head screw for grounding. In fact, if there is an unneeded cable clamp in the back of the box, held in by a substantial screw, most inspectors will accept your removing the clamp and using the screw that held it—which may be an even-more-substantial 12-32—for grounding. Do be aware that on rare occasions these screws have heads whose undersides are not flat, which means that they will not press the ground wire as evenly as they should, and may not capture it securely. If the ground wire is at any risk of squeezing out from under, the connection is inadequate. Never try to use the same screw to simultaneously press a clamp against a cable and secure a ground wire; that's unreliable (cable has give) and illegal. Finally, don't try to use a wood screw or a sheet metal screw for grounding—or for any other purpose other than for holding electrical equipment in place on a wood or metal surface. Sheet metal screws are very commonly used by well-intentioned but uninformed people in inadequate attempts at grounding.

Having said this, some Listed fixtures, and odd boxes, are supplied with self-tapping ground screws, rather than standard machine screws, for grounding; being supplied with the equipment legitimizes their use.

Treat all screws used for grounding as though they were standard terminal screws used to attach conductors to devices. There is one minor difference; you most frequently will be terminating a bare wire under your grounding screw, so there will be no issue of stripping insulation back too much or too little.

Ground clips. What if there are no tapped holes available for grounding screws, but you need to connect a grounding wire to a metal enclosure? This often is the case in older enclosures, whether or not the enclosures themselves are grounded.

There are two perfectly acceptable choices.

- You may drill and tap (cut threads in) your own 10-32 hole. This rarely is done, but it does permit the best possible device grounding for an enclosure that does not have a tapped hole for a grounding screw.
- You may use a ground clip. At its simplest, this listed gizmo is a folded piece of spring steel with a raised ridge on one side of the fold and a slot in the center. A tunnel is formed on the one side by the ridge.

Its installation is simple to describe, though sometimes not terribly easy to perform well.

1. Bend a length of bare wire into a J.
2. Hold it in the box with the open end flush along the inside wall of the box, pointing out—straight out, rather than at an angle. Its end should emerge from the box a little bit—between ⅛″ and ¼″. So far, this is a little like creating the "last-chance" splice.
3. With your other hand, set the ground clip over the wire and the edge of the box, with the clip's "tunnel" facing in, so that the ground wire is in the tunnel. Use the wire to hold the clip steady so that the clip's opening points straight down along the wall of the box.
4. If you have the finger strength to do so, force the clip onto the edge of the box. If not, hold wire and clip both steady in place with one hand, and take a pair of pliers or, theoretically, a screwdriver, in the other hand and push the clip straight down onto the wall of the box.

Two elements ensure good contact. First, the clip normally has a tooth to bite into the box. Second, the clip presses the ground wire against the wall of the box. All of this depends on the springiness of the ground clip; if you bend it open as you install it, it does not press tightly enough for reliable grounding. Here are some pointers.

- I find that lineman's pliers usually are the best tool for forcing the clip in place, because I can apply even, straight pressure with them.
- Some clips are formed so as to allow you to install them by pushing with a screwdriver, instead. However, I find that this does not work as well (which is not to say that I never use this means). There are two problems. First, a screwdriver may slip; second, it may force the clip open rather than pushing it straight onto the wall of the box. Replace a clip if you force it open for this or any other reason, and try again.

There are limitations on the use of ground clips:

- You cannot use them on round boxes, nor on the rounded portions of octagonal boxes. They only make solid contact with flat surfaces.
- Ground clips are only usable with one or two sizes of wire, usually #14 and #12. Using them with other sizes or with more than one conductor will not provide safe grounding.

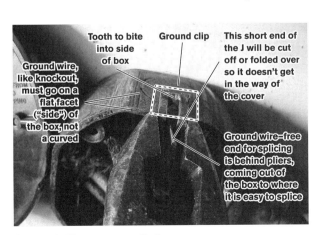

Ground clip being installed.

Labels on image:
Tooth to bite into side of box
Ground clip
This short end of the J will be cut off or folded over so it doesn't get in the way of the cover
Ground wire, like knockout, must go on a flat facet ("side") of the box, not a curved
Ground wire—free end for splicing is behind pliers, coming out of the box to where it is easy to splice

- They cannot accommodate anything thicker than the side of a standard switch box or junction box and maintain their shape, and thus their pressure against wire and box.
- They require a certain amount of thickness; in some cases, I have found them not to be tight enough to grip firmly on sheet metal enclosures.

Here are some problems that you may encounter when trying to install ground clips:

- A corroded or paint-encrusted box needs to be scraped in order for the clip to make good metal-to-metal contact.
- When you try to install a clip on the wall of a flush-mounted enclosure, the wall or ceiling may be too tight around the enclosure—because the installer was careful!—for the clip. If necessary, chisel the ceiling or wall to make room for the clip to be pushed on to the edge of the box freely.
- Clips can get in the way of covers and sometimes even with stuffing tight-fitting devices such as dimmers inside boxes.

If a clip simply will not work, usually you can drill and tap for a ground screw. However, if you are drilling a cast metal box, designed for damp or wet locations, make sure that you can do so without drilling all the way through the box unless the application allows openings in the back. (Some weatherproof boxes were designed to be mounted by screws run through holes punched in their backs.) Modern weatherproof metal boxes ("FS" and "FD" boxes, for example) have holes cast in them for grounding screws; older ones may not. Even if they were installed when grounding was required, the manufacturers may have assumed that installers would use steel conduit; that would ground the boxes by being threaded into them. According to the best authorities, when you use a weatherproof cable and connector with an old box, you cannot ground the box legally by attaching the ground wire to a device that will be attached to the box, and then relying on this contact to ground the box. Thus, if the box is mounted within 7' of a grounded surface, and is fed by UF or (where the cable itself will not be exposed) romex, it needs a ground screw.

Doing Without a Ground

What if no ground is present? Undergrounded switches, and out-of-reach lights, are not so bad. Manufacturers still make receptacles that only will accommodate two-prong plugs, and it is perfectly legal to install them as replacements for worn-out receptacles at most nongrounded outlets. If you do not use three-prong appliances at a location, you might be tempted to go this route.

I strongly recommend against installing these, for several reasons:

- Someone could use a cheater, perhaps making an assumption of grounding based on the fact that the receptacle does not look ancient.

- Later on, you may discover that you yourself want to use a three-prong appliance at the location.

- One of the solutions is quite safe and quite inexpensive, compared to the cost of your effort or of an electrician's labor.

You used to have two choices for installing three-prong receptacles at outlets that lack grounding. One has been eliminated, but today there *still* are two choices, though only one of them is the same as before:

- You used to be permitted to run a ground wire back to the service; most inspectors would accept the work if you ran it back to a subpanel, considering this alternative reasonably safe.

- You used to be permitted to run a grounding wire back to the nearest grounded cold water pipe.

If you come across either of these, you probably are looking at a legal installation. Test, though; the outlet may not be grounded! Very commonly, electricians would run this grounding wire to the nearest cold water pipe, simply assuming that it was a *grounded* cold water pipe. But even if it *was,* there is another reason that you may not find a good ground at your outlet.

> If there is some fault current at the time a plumber works on the grounded metal pipe, but not enough to cause the circuit to open, the plumber could get shocked when making a break in the metal water pipe *upstream* of the fault. Even if there is no fault current at the time the plumber is working, the hazard remains.

The first choice, running a grounding wire back to the service, or to the grounding electrode system near the service, still is acceptable. See NEC Chapter 250 for the details. Using the nearest cold water pipe, however, has a serious problem. First, if a nonmetallic fitting or length of pipe is inserted in the water system, perhaps as a replacement for deteriorated steel or copper, the outlet that was grounded to the water system no longer is grounded—but still incorporates three-prong receptacles or electrical equipment that lacks a safety ground. You no longer can rely on a continuous metal water piping system.

Nowadays, GFCI protection also is an acceptable alternative to grounding three-prong receptacles. The protector can be a GFCI circuit breaker, a GFCI receptacle, a GFCI switch, or a blank-face GFCI. (The latter two are much less commonly installed, and therefore much more expensive.)

Termination at Heavy Appliances

I will not explain how to connect heavy appliances. You can skip this section unless you are very tempted to wire them and want to understand why I do not recommend that beginners do so. I will focus largely on aluminum, but many of my concerns apply to copper as well.

Some loads require much larger sizes of conductor than are used for switches or receptacles. These include compressors for air conditioners and heat pumps; subpanels; and electric stoves and dryers. Because heavy-gage copper is relatively expensive, installers very commonly use aluminum conductors.

I do not explain how to wire with aluminum conductors, for a number of reasons:

- Heavier loads normally mean higher available fault current, thus greater risk of fire and, to some degree, of shock.

- Heavier loads usually are served at 240 volts. You are at twice as much danger should you be shocked line-to-line, at 240 volts, than you would be on a normal branch circuit.

- The rules associated with the neutrals and grounding of electric stoves and dryers are changing.

- While aluminum conductors in the larger sizes can be safe to connect, it would be easier for you to connect them incorrectly than copper conductors.

Many of these reasons apply to rewiring heavy appliances with copper as well.

Never assume that an appliance, or a connector, can safely be wired with aluminum conductors. If such use is authorized, the appliance's terminals will be marked either "Al/Cu" or, very rarely, "Co/Alr," depending on the size of conductor that it accommodates.

P A R T

III

Evaluation, Repair, and Replacement

13

Mechanical Condition

You've learned how to make safe electrical connections. Before you can use this knowledge to replace switches, receptacles, or lights, you need to understand, and perhaps repair, the conductors that feed them and the boxes that protect them from their environment. Boxes guard your system against a variety of threats, ranging from combustible dust to clogging paint to smashing chair legs to gnawing rodents to shorting and mechanism-fouling water. As boxes protect your system from many potential sources of malfunction, they protect your house from fire and they protect you from shock. I briefly mentioned the need for intact electrical insulation and intact boxes in Chapter 5, Safety. In this chapter, I will explain how to gain access to them, how to evaluate them, and, within limits, how to repair them.

What Makes This So Important?

This chapter covers some of the hardest types of problems to repair, so it is understandable that people wish they could skip this material without accepting a major risk. You can't. If the components that feed and protect equipment have deteriorated, or turn out to have been installed unprofessionally, it may be quite dangerous—more so if you are not aware of the problems. In old houses, the insulation protecting conductors from what's outside them, protecting conductors from each other, and protecting your house from them, may be, and often is, damaged. This probably creates the most serious danger, other factors being equal.

The cables or raceways enclosing the conductors, the boxes enclosing connections, and the connectors and clamps holding the cables to the boxes can be in almost any condition, and still your system will work. Everything but the conductors themselves can be severely damaged or even missing, and the system still will work. But it won't be safe. That's why this chapter precedes the chapters concerned with replacing receptacles, switches, and lights.

Overview of the Chapter

How can you find out if the guts of your system are unhealthy? For a safety survey, I use a procedure that I have standardized to a large extent. If you decide to take on any of the repairs I discuss in the final chapters, though, you automatically will take on part of the same exploration. Whenever you replace a receptacle, switch, light, dimmer, indicator, or timer, even in flush installations (those whose guts are inside the walls and ceilings), you have a chance to examine most of the elements that normally are hidden from sight. You will evaluate the condition of boxes, conductors, cables, connectors, and clamps in nearly the same way whether you are replacing devices or simply trying to get a sense of the condition of your wiring.

There is little difference between evaluating flush and surface wiring, but it can be much more difficult to gain access to flush wiring. I could start with either type. In new work in finished areas, flush wiring is used exclusively; even in old work it is the one used most commonly. Most receptacles, switches, and lights are in finished, habitable areas, so I will spend the most time on flush wiring. However, I'll begin with wiring that is mounted on the surface, simply because it is easier to deal with. Surface wiring allows ready access to all the internal elements without damage to walls or ceilings.

I'll start by talking about how to remove covers, and then how to withdraw devices from their enclosures. Once I have talked you through gaining access to the insides of boxes, I will discuss what to look for in examining the boxes themselves, in disconnecting devices, in checking clamps and connectors, and, most importantly, in evaluating and restoring conductors and their insulation, when necessary and possible.

Surface Wiring

In utility areas of the house, such as unfinished basements, there is no need to conceal wiring. This system uses rectangular, square, round, or octagonal galvanized steel boxes mounted on, rather than in, the wall, and on or against joists, rather than flush in a ceiling.

Gaining Access to, and Replacing Devices in, Surface-Mounted Boxes

Even if there are no signs of damage, you will need to get into these boxes to replace worn-out devices and lights. It is not hard to remove or replace devices mounted in utility boxes.

Utility Boxes: Removing and Replacing Covers. Covers are secured to devices in utility boxes by short 6-32 screws. Unless the screws are rusted or paint-clogged, they are easy to remove. If they prove contrary, you can drill them out or literally rip the covers off—they're cheap. To put a cover on a device in a utility box, screw the 6-32 screw, or two, that came out of the cover, or that came with the cover if you are installing a new cover, into the device's matching threaded hole or holes.

Utility Boxes: Detaching Devices. Devices are mounted in the utility boxes by 6-32 screws that go through the yokes into the utility boxes' threaded mounting ears. Most of the time, these screws are easy to remove because the covers have shielded the screws from paint and even from moisture. There is a chance, though, that you will find moisture did get in and rust a screw, or that its threads were fouled, or that it was bent. Try penetrating lubricant; try patience. You also can consider the extreme measures that I suggest later in this chapter for getting into flush boxes that you can't open easily. However, because replacing surface-mounted boxes is relatively painless—distinctly not the case with flush wiring—heroic measures may not be worth your while.

Utility Boxes: Attaching Devices. Here are two extra steps to take when you install a replacement device.

- Device yokes have four ears sticking out to the sides of the oblong holes for the device mounting screws. Break off the ears by bending them back and forth till the metal fatigues, or by cutting them off with your lineman's pliers. Otherwise they could interfere with the cover's fit. Besides the fact that this can make it difficult to get the cover screw in, and besides the fact that it looks a bit sloppy to have the cover plate held away slightly, this makes the enclosure less complete, giving the contents more access to the environment of an area that may not be as clean and dry as are living areas.
- Look inside the box for connectors or even wires near the front that the screws might hit. Unless the utility box is fairly deep, you may need to use your pliers to shorten one or, rarely, both of the 6-32 device mounting screws; alternately, you can just replace the screws with shorter ones. When screws hit metal inside the box as you tighten them, they can foul, making them hard to remove; damaged wires can short or shock.

Square, Round, and Octagonal Boxes: Removing and Replacing Covers. The cover of a square, round, or octagonal box is mounted to the box via 8-32 screws located diagonally across from each other. The mounting systems of most lights use these screw holes. When these boxes enclose devices, the devices are mounted to the covers. This means that removing the covers removes the (still-wired) devices from the boxes.

Like the screws holding the covers on utility boxes, these screws can freeze over the years, whether or not they show clear signs of rust. Unlike the flimsy covers on utility boxes, these are not to be ripped off. Fortunately, a round-head 8-32 is easier to fight free of rust than an oval head 6-32 because its head gives your screwdriver blade more to grip and its slot is not as easy to destroy. I have had some success twisting the heads with the ends of my lineman's pliers. Sometimes you need to remove the screws in order to take off the cover or remove the light fixture. Other times, all you need to do is to loosen them. In one common design, there's a pair of keyhole openings across from each other in the cover plate. After you loosen the screws, turn the plate so as to place the larger part of each keyhole over the screws, and lift the plate off. In another common design, there is a keyhole opening near one corner of the cover plate; at the opposite corner, there is a slot opening in from the edge perhaps half an inch. After you loosen the screws, turn the plate so as to lift the keyhole over one screw, and then slide the slot out from under the other screw. You can simply remove the screws rather than bother with this, but it is not necessarily a good idea; it is a little more work, and, more annoyingly, you may misplace the screws.

Square, Round, and Octagonal Boxes: Removing Devices from Their Covers and Attaching Devices to Their Covers. Devices are mounted to the inside of the covers of these boxes via 6-32 screws going into the threaded holes used to attach cover plates to devices. In the older versions of device-mounting covers, the parts of the devices' yokes with the oval openings (used for the mounting

Three-receptacle round cover.

Three-receptacle round cover from the back.

screws in flush wiring and on utility boxes) are useless, and may have been removed to avoid their interfering with the covers. In the newer versions of device-mounting covers for square boxes, these openings in the yokes are used, with screws and nuts, to provide additional connection between the devices and the covers. (Mounting devices to the covers of octagon boxes is not part of modern design.)

It used to be that one measly screw holding a receptacle to its cover plate was considered an adequate connection for bonding a receptacle. No ground wire was required to bond it to the box. Nowadays this no longer is considered adequate. (By themselves, even the screws and nuts of the new design do not ensure that the device is adequately grounded.)

Eyeballing Your Wiring System

Cables and raceways need to be properly supported, as described in Chapter 8. This is the time to correct any problems in that regard. They should be snug where they enter boxes—you should not see any gaps where conductors show, and cables and raceways should resist twisting, jiggling, or pulling. Be especially

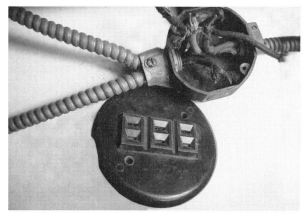

A three-receptacle cover for an octagon box, alongside the crowded box it had covered.

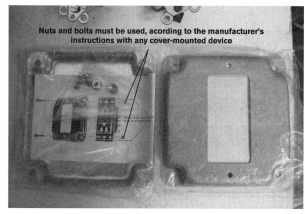

A modern surface cover for mounting a Decora device such as a GFCI on a square box.

KO Seals

Knockouts are round holes in boxes that are standardized to common trade sizes—in residential work, most commonly ½", and much less frequently ¾". Reminder: when an electrician removes a connector or raceway from a ko and doesn't replace it, he or she is required to install the appropriate size of ko seal. The standard version is held in place by spring steel "prongs"; you can install it with them facing either in or out. If your fingers are strong enough, you can push it in; otherwise, tap it in place with a striking tool. (I use closed pliers.) See the photo on p. 221.

careful to check that conduit, tubing and BX are held snugly; bad contact could interfere with safety grounding. If they are not snug, tighten locknuts. With BX, make sure that the armor is pushed all the way into the connector, and then check the tightness of the connector's screw or screws.

Deterioration

Devices always wear out, but surface wiring boxes are tough. You rarely will need to repair one that was professionally installed. It may have been abused, though. The most common repairs I make on these are to restore complete enclosure. The purpose is to keep foreign matter out and, in the event of serious trouble with the connections, to keep sparks in. Usually I need to replace missing covers. Somewhat less commonly, I need to close holes from which cables were removed. If you find that a knockout is missing, you will need to install a knockout seal. In the worst cases, I find that untrained people added cables haphazardly and dangerously, in the process leaving every possible sort of opening.

The mounting of surface wiring boxes usually is tough, too; most commonly, electricians pass screws through holes in their backs and mount them directly to structural members. However, moisture and vermin (the latter usually associated with moisture) cause damage to mountings. What OSHA calls "struck-by injuries" occur next in frequency. When boxes are mounted to masonry, if the masonry is at all damp, the screws can rust away. Similarly, if the boxes are mounted to thin wood against moist masonry, the wood can rot, again resulting in inadequately-secured boxes. In a few cases, I have seen the boxes themselves badly rusted. There is nothing complicated about replacing wood or remounting intact boxes; just make *quite* sure that you have killed power to them before you start. It is far wiser to kill all power to the house because unfinished areas, at least in a basement or ground-level outbuilding, offer increased risk of shock because dirt, moisture, and even cement floors and concrete or cinder block walls ground you better than do the floors and walls in most finished areas.

Back side of surface cover for square box

Receptacle ears interfere with cover—remove them; perhaps even cut off the area

Unneeded mounting screws and yoke

Receptacle

If you don't remove its ears, they block this receptacle's mounting on the surface cover for a square box. Here the mounting screws, and even the parts of the yoke in which they are mounted, are not needed and even can get in the way—or at least take up space. Remove the screws.

Inadequate Boxes

Tough they might be, but older boxes, most especially really old boxes, can be inadequate even if they are undamaged. The boxes often are woefully undersized. A "small octagon," or "17A," is a 3½" or 3¼" diameter round or octagonal box that is 1½" deep. I don't remember seeing a 17A used to support a device, though there have been switches and single receptacles that would fit. Should I encounter one, I will think long and hard about whether to try to find a replacement device if the old one is worn out. This box has not been in NEC tables for many years. When you still could look it up, the relevant table allowed it nothing more than one cable entering and one leaving. When I find one of these fed by a single cable

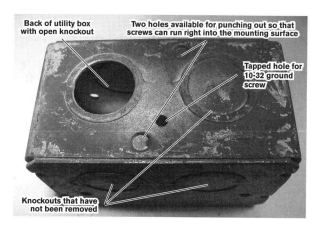

Back of utility box with open knockout; needs a ko seal.

and supporting a suitable light fixture, I am quite comfortable examining it and letting it be. Once it connects two cables (four wires) and supports that same light—a configuration that can be legal under the Code, even considering what the table says—I call it overcrowded. When I don't simply replace it, I add an extension, generally doubling the box's depth. What this gives me, unfortunately, is a box that may have just-adequate volume but that is even harder to look or reach into. This is particularly bad because, given the age of its wiring, access may be important. With 4″ octagons, and with square boxes, an added 1½″ or even 2⅛″ of depth does not hinder access as severely. Overcrowding generally is worst at round and octagonal boxes used to feed lights.

Consider having overcrowded boxes replaced, or reducing their crowding. For instance, I will remove the light that a 17A supports and feed it from some other, nearby source of power. You get one break with regards to the volume constraints that make it difficult for you to replace old *devices:* modern devices generally are more compact than old versions. On the other hand, when you replace an old receptacle with a GFCI—as you may have to by law, or may want to for safety—you reverse this advantage. A GFCI is deeper than most old receptacles.

Wiremold

Wiremold systems, which I discussed under "raceways" in Chapter 9, are transitional in that they are surface wiring used in finished areas. These are not nearly as rugged as other raceway systems. Keep an eye out for loosened or missing pieces, which are especially common where wiremold fittings turn up or down or around corners. The metal molding normally serves as the circuit's grounding means; any missing pieces make it that much harder for a fault to travel a low-resistance path through the raceway to trip the breaker. The raceway must be secured every few feet in accordance with the manufacturer's instructions. For Wiremold V700® raceway, for example, the interval should be about 32″! The clips usually utilized are mounted behind the raceway, which is snapped into them. Unfortunately, while the raceway does stay in them nicely, the clips pull out of the wall very, very easily as all that's holding them usually is plastic anchors.

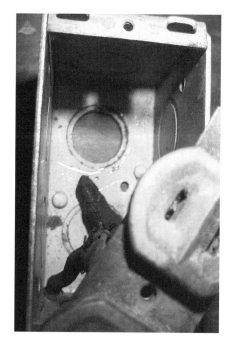

A fender washer or other flat metal would help. Note how this utility box is mounted right on wood, which has two bad consequences. First, this open knockout in the back has a combustible material right there; second, you don't have the depth needed to accommodate a knockout seal.

Wiremold boxes and their contents are far more easy to examine than flush boxes. In finished areas with flush wiring, most devices are mounted directly to outlet boxes (in old wiring, usually gem boxes) in the walls, and the cover plates are secured by attaching them to the devices. Wiremold uses a device-mounting arrangement much the same as used with utility boxes. Thus device replacement is basically the same, except that usually you don't need to break off the ears or shorten screws.

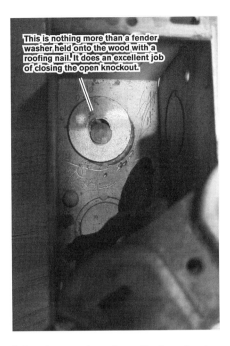

This is nothing more than a fender washer, held onto the wood with a roofing nail. It does an excellent job of closing the open knockout.

A fender washer closes the knockout.

Because wiremold raceway is a slip-fit to wiremold boxes, and this contact is the grounding path, if you have any doubt about the system's continuity it is worth your while to examine this link from the inside. Wiremold boxes come in two parts: a base plate, and a front section that is held to it by screws located in the corners. When you undo these screws and take the front of the box away, you can see where the raceway is hooked into the back. However, try not to take the box apart when your sole intention is to remove the device from the box. A wiremold box uses screws just exactly long enough to reach through the front section and grab the threaded holes in the base plate; it can be quite a nuisance to line up the front section and base plate in order to reassemble the box.

In part because wiremold components are made of thinner metal than many other electrical products, they are more vulnerable than comparable components. When wiremold has been painted once or twice, you may find that the stress created in the process of removing a cover plate from a box loosens the box or the raceway from the wall. The boxes, the wiremold itself, and the elbows (the parts of the wiremold systems used to make turns, since wiremold normally is not bent) seem to be easily kicked apart as people walk by, and even more easily kicked or torn loose from the wall or ceiling.

Flush Wiring

Accessing the Interior

I generally have a harder time getting into flush mounted wiring than into surface. Here's a frequent experience: a switch box containing a receptacle is installed flush with or slightly recessed in the wall. It is covered with a plate that has an opening or openings for access to the device, and one or more screw holes. Several people have painted that cover plate, and it is now one with the wall. (The paint may also have cemented it to the device face, worse luck.) Whether or not you plan to replace the plate, you probably wish to remove it without destroying part of the wall. Removing device covers sometimes is a simple matter of undoing screws. This is the case only when the covers have *not* been painted or wallpapered; and the heads of the screws have *not* been clogged or damaged.

Tapping in a ko seal with lineman's pliers.

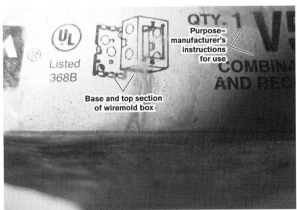

QTY. 1

UL Listed 368B

Purpose—manufacturer's instructions for use

Base and top section of wiremold box

How wiremold boxes are designed.

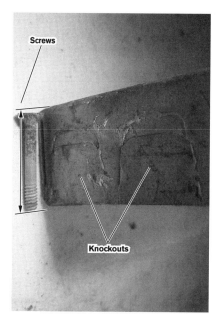

Traditional wiremold boxes are 100% custom in their parts, like Compaq computers but more so. The screws that hold together the boxes are just, just long enough to reach from the top section to the base. Note that the knockouts fit the wiremold raceway (in two sizes, in this sample box) but no other; and there are no konckout seals.

The next steps involve access to the insides of electrical boxes. Before poking around inside an electrical box, you need to make quite sure that there is no power going to it. An important reminder: unless you can turn off all power to the house and work by flashlight, you have to assume that at some point as you disconnect wires you may be working live.

Removing the Cover. Standard cover plates take 6-32 oval head machine screws $\frac{3}{8}''$ long. Anything other than a short 6-32 machine screw should be replaced when you replace the device. If a screw's head is clogged by paint, you probably can chisel the paint out of the slot. In the worst case, where a cover screw is absolutely immovable, it's not a serious problem; you can bend or break the cover plate away from it.

Getting the plate out of the way should, at worst, harm nothing but the plate and, conceivably, the device you were going to replace anyway. In some cases, though, you will encounter a more serious difficulty. It may require some finesse to remove a cover plate without tearing the paint or wallpaper around the plate, or dislodging a chunk of plaster. Except where it is clear to you that the plate is quite independent of the wall, I recommend slicing along the edge of the plate to free possible adhesions. This can minimize the chance of causing such damage to the wall finish. What you use for cutting depends on what cutting tools you are comfortable with. A retractable razor knife works well for me. The corner of a flat screwdriver blade sometimes can do the trick.

Removing the Device

You rarely will find paint clogging the screws that hold a device to a box because they're safely hidden under the cover plate during painting. (This assumes that there *is* a cover plate in place or at least that there was the last time the wall was painted; sometimes, though, painters remove cover plates in a well-intended attempt to avoid gluing them to the wall, and don't even mask.) Nevertheless, you may encounter even worse problems than you had removing painted cover-plate screws when you try to remove the screws that hold devices. If a screw holding a device in place was broken, you'd have noticed. The two most common problems with metal boxes, more common than the problem of broken screws, are screws that are bent or frozen. Often screws have gotten bent because they are too long for the space available behind the box ears they were threaded into. When the installers hit obstructions behind the screws, they kept screwing—and the screws became bent. When I talked about surface boxes, I talked briefly about how to deal with screws that are hard to remove. You may need to call in pros to deal with these because you may end up having to replace the boxes, a task well beyond the scope of this book.

Box Integrity

Once the cover is off and the screws removed, look at the box. First question: is there a box? If you look in the hole and there is nothing but wall and cables or wires behind the device, you have encountered a dangerous situation. If the wires are quite ancient, you may have encountered "knob-and-tube" wiring. Except when this is the case, if there's no enclosure, somebody just did the work wrong—very wrong. The presence of such ignorant work suggests the possibility of "air splices," junctions in the wall without boxes, hidden, unknowable, and inaccessible.

Advanced Ways to Deal with Bent or Frozen Screws

Here I will talk briefly about other methods of attack, simply because you will be dealing with flush wiring far more commonly than surface wiring. The following discussion is only for those whose mechanical competence is far beyond that of a beginner at physical labor.

If you're competent with a torch, you can apply heat to break such screws free from the threaded holes. This may not be an option, though. Fortunately, you usually will encounter either of these problems only with one of the two screws attaching a device to its box. By removing the other screw, you can bend a device out of your way and actually pull it off the screw. One you have the device out of your way you can grab the screw's head with your pliers. This has two advantages. First, it allows you to combine pulling with turning, as opposed to pushing and turning as you do when using a screwdriver. Second, squeezing pliers gives you a much firmer grip than does holding a screwdriver blade centered in a shallow slot.

If this doesn't work, you may be able to stick a pair of needlenose pliers into the wall so you can work behind the box ear. If you can reach behind the ear with narrow cutting pliers, or even with a hacksaw blade, and shorten the stuck screw, that's just so much less you will struggle to unscrew.

If screws are very badly bent, or solidly frozen in place, there's a fair chance that you will twist the screws into two pieces as you try to remove them. What then? Whether a screw was broken before you touched it, or broke as you tried to remove it, you've got to get the stub out. Sometimes you can use pliers to grab the bit of screw that's left sticking out, and continue to undo it. Sometimes you can use that narrow hole in the wall mentioned just above to insert needlenose pliers, grab the stub from behind, and back it out. When backing out a screw, normally you turn it counterclockwise; in this case you really are screwing it further into the wall, so turn the screw *clockwise* so it continues to screw out the back of its threaded hole.

The worst-frozen screws require an electric drill. Be careful not to drill past the screw and damage the wiring. Using a bit small enough that you do not destroy the box ear's threads, drill right into the center of what's left of the screw. If you can accomplish this, the rest of it can usually be removed. Alternately, redrill and tap the hole either for a 6-32 screw again or for a larger, 8-32 screw.

If you don't have a set of taps, or if this sounds too difficult, or if it doesn't work, proceed to the final alternative: having the box replaced. Replacement also is necessary when you find a metal box's ear broken off, or find a plastic box whose screw holes are stripped or cracked or unable to hold a screw firmly for some other reason. Replacing a box in a wall or ceiling is messier and more complicated than replacing a box in surface wiring. Therefore, while it is not significantly more dangerous than other jobs that are described here, it certainly is beyond the scope of this book.

I can't recommend a kludge, or makeshift solution. However, I think it appropriate to differentiate those that are unprofessional, not in keeping with Code, and less than desirable from those that are out-and-out dangerous. When you mount a device in a wall—and I am not speaking of ceilings—your purpose is to secure it in place. So long as the mounting screws are not intended to bond a device to a box, and so long as the device remains fully enclosed, when I see that someone has attached one end of the device directly to the wall where the box ear is missing or stripped, I don't worry about this nearly so much as about other clearly dangerous problems, which I talk about later in this chapter.

Also, beware of installations that squeezed two boxes, whether metal or nonmetallic, under one two-gang type cover, or mounted two right up against each other. The first clue to this may be that the installer trimmed the abutting edges of two cover plates so they would fit close enough. This tells you that the installer did not know what he or she was doing—or was irresponsibly lazy. If a gem was in place, the installer could have added a section—but didn't; if a nonsectional single-gang box was in place, the installer could have replaced it—but may not have had the knowledge and skill. I often find in this type of situation that the person adding the second device also didn't wire it correctly—perhaps didn't even install suitable cable.

Finally, if you find a switch or receptacle half-buried, for instance where molding was installed across the bottom of the box, don't proceed as best you can with-

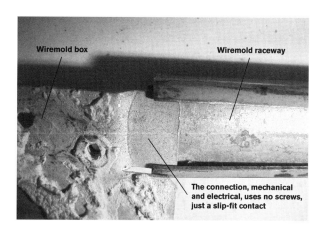

Wiremold box

Wiremold raceway

The connection, mechanical and electrical, uses no screws, just a slip-fit contact

Some wiremold and its box, from the back, showing the slip fit between the two.

out making changes. Whether the bottom of the cover plate was cut off or simply buried, this carpentry needs to be redone or else the box needs to be moved to where access is not blocked. (The latter almost always entails considerable damage to the wall.) In these cases, call in a pro.

Examining the Box's Contents

Once you've actually disconnected and removed your device, you can examine what's hidden behind the surface. If you are merely evaluating your system, and don't feel the need to replace devices, you can simply pull your device out of the way if your wires have enough slack, rather than disconnect them. If you are replacing a device, but want to maximize safety without killing all power to the house, you may be able to keep your examination of the inside of the box eyes-only. If the box and conductors look okay from the outside and from a quick peek inside, in some cases you can replace a device without getting your fingers inside the box. In this case, you will largely limit your examination of the conductors, which is covered a little later in this chapter, to those parts of the conductors that are accessible when you remove them from the device. Actually removing and replacing devices is covered in the next two chapters. The condition of conductors is most critical to your safety.

Wallpaper has no business extending inside an electrical enclosure; cut it back to the edge of the cover plate, or at least to the outside of the box.

Freeing the cover plate from the paint or wallpaper does not get it off the device and out of the way. You may have to follow a similar procedure to separate the plate from the device, whether or not you were able to remove the screw or screws designed to hold the plate on. In addition to cutting paint that glues the plate to the device, you can tap on the device's face or tap the plate around the device's edges. Anything that jolts them could work. Sometimes I've used a drift, a slightly tapered steel cylinder. Once you have done what you can to break adhesions, all that's left is to pry. Sometimes this will destroy plates or devices. Unfortunately, sometimes the paint is what's holding the wall together, so a chunk of plaster will come loose as you work. How much effort will be necessary to separate plate, wall, and device is a matter of skill, experience, and, even more, luck.

Some Details of the Boxes Used for Flush Wiring

Flush wiring uses the same boxes as surface wiring, though with some differences.

• Although round and octagonal boxes commonly are and always have been used to support lights, they rarely were used to contain devices. One reason is that the way the boxes are mounted often results in the devices sitting at an unattractive slant. Another is that the common, surface-style cover plates, to which the devices are attached, are themselves unattractive. When I see a 4″ octagon box cover showing flush in a wall with a device in it, I presume that the owner decided the wall should be finished some time after the wiring was installed, but didn't choose to replace the box with one that is more appropriate. There was nothing illegal about this choice. Still, since the rules for exposed and concealed wiring differ, in these cases I wonder whether there were other cost-cutting shortcuts.

- The rectangular boxes used in walls are gem boxes, not (in ordinary use) utility boxes. Gems are shorter than utility boxes, which means that for the same depth they have less volume. Furthermore, some of them were manufactured with strongly beveled corners, further reducing the space within. On the other hand, some of them have extra sections extending sideways hidden inside the wall for the explicit purpose of adding wiring space.

- Pancake boxes, round boxes ½″ or ⅝″ deep, were designed for use in ceilings and, rarely, walls, to support light fixtures. Being no deeper than the plaster, they were installed where hiding a deeper box from view would have required the installer to notch into the joist or, in the case of a wall sconce, the stud, being used for fixture support. In surface wiring there was no need to use boxes of such shallow depth. Fixtures intended for habitable areas, though, usually require flush boxes.

Some installers did notch joists to install boxes. This is a sub-optimal or even inappropriate practice because of the risk that the joists would be weakened. In old buildings, though, the joists usually were massive enough that they withstood this. Besides, if a joist has survived for 60 years, it probably is handling such insults just fine.

Pancakes are the one type of flush box that is mounted by nailing or, normally, screwing through the back right into wood. In new work, the other boxes are nailed to the sides of joists or studs; in old work, they usually are supported from the wall surface itself.

Missing portions

The first question to ask when you look into any electrical box is whether it is intact. If there are pieces missing, or if sizeable (over, say, ¼″ in diameter) openings have been created and left open, repair or replacement is in order, as discussed. You can close an open ko in a flush metal box the same way as you would in surface wiring.

If a metal box has paired holes at an end, off-round or semirectangular, and one or both have been punched out, most likely a clamp has been removed. There will be a tapped hole in the adjacent wall of the box, near the midpoint between the holes. If you can locate a clamp designed for securing romex to that box, and a screw matching the hole threads, you can tighten that clamp in place to seal the opening. Common plastic boxes with unclosed openings have, alas, no legal fix and should be replaced the next time you have a pro over. The only plastic boxes you are likely to encounter in old wiring that you may be able to repair are those rare exceptions that were manufactured with metal cable clamps. With these, as with the metal boxes with openings where cables were removed, tightening down a clamp may close an opening.

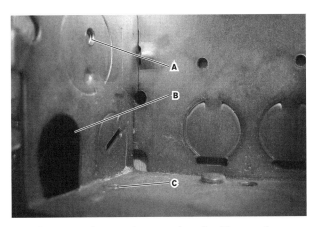

This is a gem box with an unclosed cable opening. A: This knockout has a threaded 10-32 hole for use in grounding. B: This opening represents a missing cable. C: This 12-32 hole is for the screw that held a clamp on the cable that is missing from the hole.

Box pulled forward by chopping away wall in front of mounting flange to insert shims behind the flange.

Recessed Boxes

The second question to ask about an electrical box, or, in the case of a square one, its device ring, is whether it is mounted close enough to the surface. How flush it needs to be varies with the surface, although flush mounting always is preferable, and tells you the installer took more pride and care. In noncombustible finishes, electrical boxes legally can be recessed as much as ¼″ behind the surface. Boxes must be quite flush with, or protruding from, or mounted upon the surface of a significantly combustible finish such as wood paneling, as opposed to such as masonry or tile, or painted or wallpapered plaster or drywall. 1930s ceilings of fiberboard, for example, are quite combustible, even if the last ⅛″ is a plaster coating.

If a square box is recessed too far, you can replace its device ring with a deeper one by removing the plaster that covers the box and ring. If a switch box is recessed too far, sometimes you can remount the box—if it is not fixed in place by conduit—by chopping the plaster in front of the mounting screws, ears, or bracket, or, sometimes, by removing box mounting screws and inserting shims under the box's mounting

> I have dealt with boxes that were recessed more than ¾″ despite having been professionally installed. When installed, they were back about ¼″, not ideal, but perfectly okay. Then the old plaster started to crack, and someone added ⅝″ drywall over it without remounting the boxes.

ears, bracket, or flange. How much plaster you will need to remove depends on how the box is mounted.

When adding boxes after the walls were already in place, installers often mounted metal boxes to wood lath. Unfortunately, this approach is less and less possible when re-mounting such boxes, as is necessary when they are recessed too far. By today, walls that were finished in plaster and wood lath are so old that the lath often is too dried out and brittle to reliably hold screws. Today, you can use an extender that slides partially inside the box to enclose anything from a

This legally recessed receptacle is accessible despite being sunk back from the tile. It is back at least ¼″, but the tile worker has not caused a problem for your electrical system. Once the tiling is evened out and the grouting completed, the electrical box will be complete. The important consideration is that the cover close the box; because the cover is recessed with the box, the depth is not an issue. Be careful, though, that the grouting doesn't get in the way of the cover.

small gap to over 1″ of distance between the surface finish and the box. The extenders are available both in plastic and in metal. There are some difficulties in using these because you have to look through the front bit of the extender to line up your mounting screws with their holes. This is especially problematic when the box is not securely held in the wall. For one thing, if a box tends to fall in, as you push your device-mounting screw towards its hole you tend to push the box away from your screw, which can be maddening. You need one hand to hold the device in place, one to hold and turn the screwdriver, one to hold the screw on the end of the screwdriver, and one to pull the box forward; all while eying the setup—through the hole in the box extender—to keep the screw lined

This side of the box has been removed

Gem box with ear partly slid out from under a screw that holds it to show that it can be adjusted in or out a certain distance

Body of box

Knockouts

Ear for 6-32 device mounting screw

Screw holding the side to the box

Side of box

A gem box with its L-bracket being adjusted.

Edge of the box

Ears

A gem box with its L-bracket (ears) reversed so that it could be slid further back from the front edge of the box—to push the box forward towards the room side of the wall.

up on the box's threaded device-mounting ear. This is not impossible, but it sure can be difficult.

Loose Boxes

A box may be loosely mounted for a number of reasons:

- It may be supported by wood lath that has cracked.

- It may be supported by a stud or joist that has deteriorated. When I find this, I suggest that my customers check for water leakage or insects such as carpenter ants and termites.

- It may be supported by a badly-rusted screw.

- Mounting nails may have been partly pulled away because it was stressed beyond its tolerances. For example, a receptacle mounted in it may have repeatedly had cords hit and bent away by furniture, or a ceiling fixture that was too heavy may have been hung from it.

- A gem box stressed in this way may actually have been pulled apart so that its mounting bracket still is attached to the wall, but the other section of the box no longer is attached to the part with the bracket.

- A metal strap that stretched between two joists or studs to support a box used for a light may have been bent by too heavy a light fixture.

There are viable fixes for many cases where a box is inadequately mounted in a wall. (This generally does NOT apply to ceiling boxes, most of which must carry weight.) If the wall is in trouble but not on its last legs, the answer may be as simple as plastering or mortaring all around the box. Boxes that are not reasonably secure in the wall constitute their own problem. If the wall is in good shape, there are two devices most commonly used to secure a box to drywall or paneling, and even to plaster and lath. One is Madison or "hold-it" clips. They are installed, as shown in the pictures on pp. 228 and 229, to keep boxes from pulling out of the wall. Do note three important parts of Madisons' installation.

1. The pair must take opposite positions: the Madison holding one side of the box must have its long leg pointing up (or to the left, if the box is horizon-

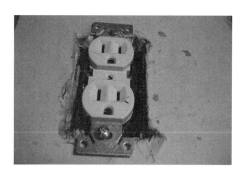

Receptacle ears bear against a wall. These are extra large to augment the box's ears.

Madison clip being folded in. The gem box's ear bears against the wall to keep it from falling in, while a madison on each side will keep it from coming out.

tal); the other must have its long leg pointing down (or to the right, if the box is horizontal).

2. The clips must be pulled firmly forward before, and while, they are bent over.

3. The bent-over arms must sit snugly against the inside walls of the box, so that they don't take up room inside and, very likely, cause a short. I find that pulling them forward with pliers, and then squeezing them tight with the pliers after installation, accomplishes this.

The other helpful device is a flat, squared-off "C" available in stamped metal or plastic, manufactured by Caddy Corp, their "PLC." Normally, the ends of devices' yokes bear against the surface of the wall, keeping the device from falling in if the box is loose or recessed. When the wall is damaged, the PLC can assist in holding the device, and hence the box, forward. When you reinstall the device, start the screws in their holes in the box, but don't tighten them all the way. Then slide the clip along the wall so that it fits under both ends of the yoke. Now tighten the screws the rest of the way. The clip spreads the pressure of the yoke against a larger area of wall (although still small enough to fit under a standard cover plate), so as to give the screw heads and yoke ends something to bear against to keep device and box from falling into the wall. This does not work, though, if the wall has too large a damaged area because the PLC may have nothing to bear against. Of course, if the wall has not just a weakened area but an actual hole around the box, the Code requires you to patch the wall.

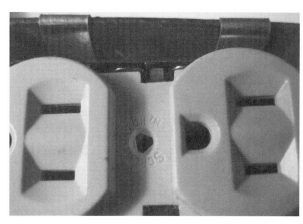

Madison clip—in box, fully installed.

It can be rather difficult to use Madisons or the PLC when the box tries to slip back into the wall before you can secure it. With the PLC, especially, this can be complicated, because the PLC isn't installed until AFTER you have managed to hook the device screws into the box's mounting holes.

There are fixes for the case where a box is recessed further, but space constraints limit what I can cover.

Loose Devices

Even when a box is securely mounted, a device mounted to the box may be loose. When boxes are

One puzzling problem that arises occasionally in dealing with a gem box is that the device is difficult to screw in or out because the mounting screws don't line up with the box's threaded ears, due to the fact that the box has been distorted. If the installer cut out a hole which was a tad short or narrow for the box, he may have squeezed or pounded the box into place. This is found more commonly, but not exclusively, in old work, when outlets were added to an existing wall. The result is that the box and the device no longer match each other. You may be able to enlarge the wall opening enough that you can straighten out the box.

Madison in wood lath, from back.

Madisons in drywall, seen from back.

Madisons going in, front view.

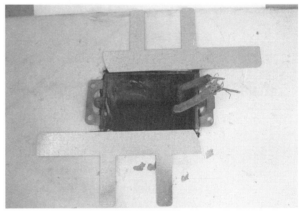

Madisons laid out, long legs facing opposite directions, ready to install.

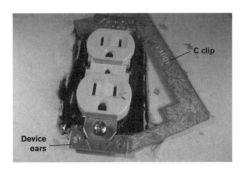

"C"-strap being installed under device ears to help bear against the damaged wall.

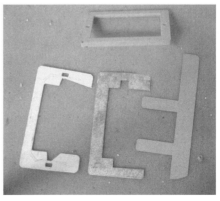

Box support–C clips, Madison, and depth-adjusting spacing device.

recessed, the ends of devices' yokes cannot bear directly against them. They have to bear against the wall for whatever rigidity it can offer. Very often, unfortunately, there *is* no wall between the box and the device yoke because the yoke does not really extend beyond the box. In the past, rubber plumbing washers often were slid onto the 6-32 screw between the yoke and the box to eliminate play without having to recess the device. The spacer shown in the bottom-right photo on p. 229 works much better. Some recently manufactured devices have yokes with extra large (removable) protrusions at their ends, to help with this problem.

Box Volume

The final question to ask about an electrical box is whether it is roomy enough. There are specific formulae in NEC Section 370-16 to use for determining how many conductors, devices, and pieces of hardware may be installed in each size of electrical box. Do read the entire text of Section 370-16 (formerly 370-6), rather than trying to apply the markings inside an electrical box without checking the applicable rules. Box volumes were recalculated in the 1970s; the way we calculate the room taken up by devices changed in 1990. This is one reason that I can say unequivocally that a box that comes out slightly undersized by today's figuring may very well have been perfectly compliant when it was installed, and therefore is not certain evidence of unprofessional work.

Early wiring daisy-chained from the center of each room's ceiling to the next. Then even more cables were added, mostly to feed additional receptacles. The boxes got very crowded. When you find a round or octagonal box has no ears for fixture support, it dates back to those early days. If the fixture is fed by a single cable you're in great luck, as you are when you are dealing with surface wiring and come upon a 17A fed by one cable. If your box is relatively modern, at least having been manufactured with ears, volume problems are less likely, provided that the installer was competent and conscientious.

Sometimes the box is large enough, or nearly so, but the work you are doing requires that you add a cable. One possible reason to add a cable is that you need to run the cable over to a location that is more desirable for the fixture you will install, such as a wall-washer as opposed to a center light. Another reason to add a cable is that the fixture you are installing specifies modern insulation on its conductors, as described below in the discussion of insulation. To solve this problem, you want to run a cable with modern conductors from the old box to a new one. Both installing a new cable and installing a new or replacement box are beyond this book.

Preexisting installations generally are grandfathered, so inspectors and fire investigators generally will not require that you replace slightly undersized boxes, or add surface extensions. And while the rules permit a device ring or fixture canopy to be counted as adding to the available room only when they are marked explicitly with their volume, most inspectors are charitable in allowing an extra conductor or two if the ring clearly adds some extra room to the box.

Use your judgment; without adequate room in a box, you are liable to pinch the conductors when you stuff a device back in. If the box is way overcrowded, you need to do something about that problem rather than try to force a replacement device in. Even a pro can create a ground fault or other short circuit by ignoring this rule. Sometimes I myself have done this by pinching insulation;

other times I've shorted a device's terminals against the side of a metal box; yet other times I've pushed or pulled a wire out of its splice or off its terminal screw to brush against something and short.

From the Box to Your Wiring System

From looking at the electrical box itself, I will move on to indirectly examining the wiring method. First, I'll look at how wiring comes in to a box, and then at the conductors themselves.

Clamps, Connectors, Locknuts, and Bushings

The round knockouts in the sides, backs, tops, and bottoms of most metal boxes allow you to attach conduit directly, and to attach tubing and cables via connectors. Except for a few connectors invented over the last decade or two, each connector is held to a box by a metal locknut on the inside. With a few very rare exceptions, if a box is plastic, it should contain no locknuts, which would indicate that metal pipe or cable was feeding the box. If any are present, the wiring is unprofessional, incompetent and very possibly dangerous. Some plastic boxes, not terribly old ones, come with internal cable *clamps* of metal, intended to secure romex—but not with metal connectors and locknuts.

Here is a tricky part. Certain violations are easy to identify, and should be corrected for reasons of safety. However, they require some investigation; occasionally, perfectly legitimate wiring methods may mimic these unprofessional practices. If a cable is stuck in through an opening in the box, and if by looking through the box past the cable you can easily see the space inside your walls, the cable almost certainly is missing a clamp or connector. Sometimes you can find a replacement for a missing clamp, and hence secure a cable. If the box is metallic, cable never should enter the box without being secured to it by a clamp or connector; conduit always should be attached by a locknut on the outside and a bushing, or locknut and bushing, on the inside.

If a box is plastic, romex could have been installed legally without any clamp or connector as long as it was secured within a few inches of the box. (Secured means stapled, passed through a hole, or strapped.) Nowadays this exemption from the requirement that every wiring method be attached directly to a box is a bit more restrictive than in the past; in recently-wired homes the cable has to be secured within 8″ of the box, and the exemption from the requirements to attach it directly to the box only applies to a single gang switch box.

Another requirement is that no more than ¼″–½″ of the sheath of romex should protrude into the box. This is because the sheath takes up some of the limited interior room, and reduces the amount of free conductor by that portion restrained by the sheath. If significantly more than ½″ of sheath intrudes, it may be worth your while to attempt cutting away the excess. Be especially careful not to cut into conductor insulation as you do so.

The way BX enters and is attached to the box is even more critical. By peeking inside the clamp or connector you should see the very end of the metal armor; in it you should see a red fiber or plastic antishort bushing, or "red hat." Even if you don't see the armor, the red hat is evidence enough. A red hat, a cloth covering over the conductors, or, rarely, a soft layer of lead has been used for many, many decades to protect the conductors from the cut edge of the armor.

"Red hat" antishort bushing.

The variety of clamps and connectors suitable for different cables and boxes is too large to be covered here in any detail. The best way to confirm that connectors were used, in concealed wiring, is to find locknuts in place. Do note that some very old metallic ceiling boxes were manufactured with BX clamps welded onto the *outside.* From the inside, you won't find locknuts, you won't find internal clamps; but you also won't see past the cable into the wall or ceiling, as you would if the requirement to positively attach the cable had been ignored.

Make sure that locknuts are good and tight. If you have any reason whatsoever to question one, try to tighten it further. If it is as tight as can be, fine. If it was loose until you re-tightened it, if it connected BX or a metal raceway to the box, you may have restored a safety ground that had been unreliable. Note that while the sides of round boxes sometimes have knockouts, their sides make intrinsically poor contact with cable and raceway systems because of the basic geometry of trying to mate the flat of a connector's end or of a locknut against a curved surface. There is nothing to do for this other than have an electrician remove the cables—and probably also replace the box in order to be able to secure them properly.

On rare occasions you may see the end of conduit that was not terminated properly. It needs to have a locknut firmly against the outside of the box and a bushing, or a locknut and a bushing, against the inside. If you don't find this, someone has changed what should have been the most reliable wiring system into something that you should have a pro look at.

If you don't see any of these clues to proper entrance, here are at least four possibilities.

- You may be dealing with one of those old, old boxes with the integral, external clamps.

- You may have overlooked the red hat, perhaps because you weren't sure what it (or the end of the armor) looks like from inside a box.

- The cable may be very, very old and have been legitimately installed without a red hat or equivalent.

- Here is the one I worry about. The cable was cut a little too short to reach into the connector or clamp. Either it was left that way, or it was stretched to its limit to make contact, but subsequently pulled out. The gap has interrupted grounding. In this case, you need to treat this box—and any others downstream along this circuit—as an ungrounded box. A better, if quite difficult, solution, if you can manage it, is to bring more cable armor into the clamp, perhaps by moving the box or undoing staples. This probably is a job for a pro; at any rate, it is beyond the level of an introductory text.

Illegal–BX installed without a red-hat, and secured with a romex clamp.

If you find that cables or clamps are missing or have been misused, this is not an omission that would

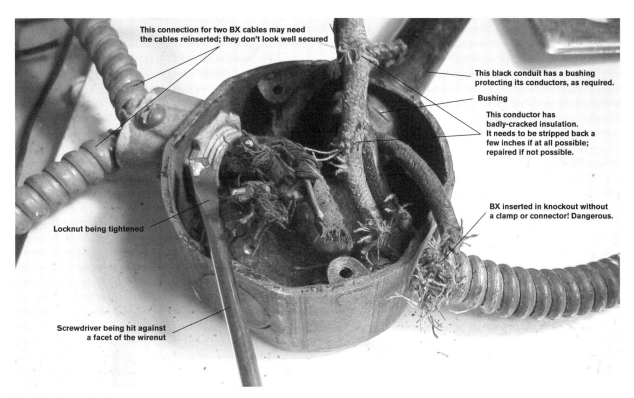

This connection for two BX cables may need the cables reinserted; they don't look well secured

This black conduit has a bushing protecting its conductors, as required.

Bushing

This conductor has badly-cracked insulation. It needs to be stripped back a few inches if at all possible; repaired if not possible.

Locknut being tightened

BX inserted in knockout without a clamp or connector! Dangerous.

Screwdriver being hit against a facet of the wirenut

Tightening a locknut.

result from an electrician's oversight. There are serious problems here, either with the original installation or with subsequent work.

Conductors

Next come terribly important tasks: evaluating conductors and, especially, their insulation. This is critically important not only with truly old wiring but also with wires that may have been overheated, such as those in boxes serving overlamped ceiling fixtures.

As I mentioned in Chapter 12, free conductor length is very important. According to modern standards, each conductor that enters the box should have at least 6″ free, extending beyond any cable sheath in the box and past any clamp or connector. At least 3″ of every conductor should reach beyond the front edge of the box. This enables you to pull out a device and work with it easily. In old work, the conductors may well be shorter, and I don't worry that this may signify a poor installation. But I DO pigtail them.

The Conductors. Next examine the conductors themselves. If they look like extension cord wire or telephone, alarm, or speaker wire, the installation was incompetent, illegal, and HAZARDOUS. Back off. You need rewiring, not just device replacement. Call in a pro.

Rag wiring is legitimate, but dates back at least to the middle of the century. Conversely, thermoplastic insulation with a second covering of clear nylon is the latest type, THHN or THWN. It has a high (90°C) temperature withstand rating, and almost certainly was installed since 1984. There are intermediate levels, plastic with insulation rated at 75 degrees but not 90. In the mid-1980s, and again in 1999, I worked on a system wired in ungrounded romex that had been installed in the 1950s. It was covered with a silver-colored, clearly-marked cloth sheath, impregnated with some flexible compound. Inside it I found not rag wiring but clearly color-coded plastic-insulated conductors in excellent shape. Their temperature rating, though, was 60°C, like many rubber-insulated conductors.

Working on your own wiring probably is a bad idea if you have aluminum branch circuit wiring. However, don't assume your conductors are aluminum simply because they're silver-colored. See Chapter 11.

Color Coding. In the best possible case, inside your romex you find white and black insulated conductors, and bare or (rarely) green-insulated wires. Inside BX you find black and white insulated conductors. There are several situations, however, in which color coding will be far less simple.

- Especially with rag wiring, colors often have faded to the point where all conductors simply look gray or black.
- Occasionally rubber insulation has gotten wet and turned black.
- More often than some might think, wires get unintentionally coated when a wall is painted. You may need to tape them, if you can determine unmistakably which if any serve as neutrals. A lack of adequate or correct color coding is only one reason to do so. See Chapter 11 for a discussion of taping.

Evaluating Insulation

Examine the conductors, one at a time, as discussed in Chapter 11. Any electrical insulation, rubber or plastic, deteriorates over time. This is mainly due to the heat generated by the flow of electricity, but can also reflect the ambient temperature. Kitchens and laundry rooms, for instance, run hot. Air conditioned nurseries are temperate.

Insulation gets brittle and dry, and eventually crumbles. If there are breaks in the insulation, it is in bad shape. If it appears intact, meaning that it has no nicks, cuts, or cracks, bend it sharply. If you can bend a wire double and straighten it out without its insulation cracking, it has retained flexibility. If it is otherwise in good shape but has a small cut or nick, tape it with the appropriate color of electrical tape. If, however, when you examine or test it, the insulation turns out to have deteriorated, you should consider rewiring. If the wiring is severely deteriorated, even if you patch it the disturbance created by your disconnecting and subsequently reconnecting or splicing could be enough to create a short circuit. Even if the insulation survives while you work on the wires, stuffing a wire back into the box may push it over the edge to failure and shorting. This is important to think about in terms of fire safety.

Dealing with Damage

Dealing with Damaged Conductors

Replacement. What are your options when you find deteriorated or damaged insulation? If, as is rarely the case, your outlet is wired with pipe or tubing rather than with cable, it sometimes is possible—quite frequently in surface wiring—to tie new conductors to the deteriorated conductors, and pull in the replacements. The new ones will be good for decades. Moreover, their insulation's temperature rating will enable you to attach modern light fixtures whose installation instructions (sometimes found only on the boxes in which they are sold) limit them to use with post-1985 conductors. Before you do so, though, check for problems by tracing the deteriorated wiring along the raceway back to its origin, or a point where it is spliced to good wiring. Also consider these important questions before deciding to tackle this yourself:

- If the deteriorated wiring runs through many boxes before you reach good insulation, reconsider. You will have to pull from box to box to box, one box at a time, redoing any splices and being very careful of other conductors in these boxes. Can you do this without creating damage to a lot of other wiring along the way?

- Are you prepared to deal with all the conductors in the conduit, if there are more than two? I advise against replacing only some that run through a section of conduit. The pull will be harder than if you replace all, and you risk damaging the ones you do not replace.

- Where's the other end? If you are not sure what path the deteriorated wiring takes along part of the way, stop. Unless you stage someone at the other end, and you see your end move when your helper tugs the other end, you're probably off. Even this attempt may cause damage. A continuity tester will not tell you whether there are junction boxes along the way.

- What if it's too brittle? Unless you can deal with the risk that the conductor might break and prevent you from completing the job and restoring power until you can get a pro in to take over, you are gambling.

- Can you get someone to help, so one person can pull, and the other feed?

- Do you have the length you need? If you don't have enough wire available at one end of each conductor at least to form the nose (which I describe next) leave this to a pro. You need at least two inches of wire to strip.

Here is the procedure for pulling new conductors in, when there are no complications:

1. Strip the insulation way back at the end where you will attach the new conductors. Bare the old conductor at least two inches if possible. Strip at least that much from each of the replacement conductors. If you are not able to strip this much, your old and new conductors may come apart somewhere in the conduit as you pull.

2. Bend the stripped sections of wire in the middle, doubling each one back to form a tight J.

3. If you are replacing multiple conductors, line the Js of the old ones up with the open ends facing in the same direction, alongside each other and do likewise with the Js of the new ones.

4. Hook the old and the new conductors to each other. (Having the Js lined up makes this easier than if the Js face in various directions.)

5. Squeeze the hooks tight, so that the hooks no longer look like loops but each wire just looks doubled. The tighter you squeeze them, the less trouble the "nose" will have fitting through the conduit.

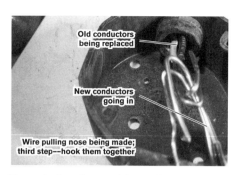

Nose being formed for pulling—wires hooked.

Continuing to form the nose—these hooks are as tight as possible, so the nose doesn't hang on the bushing through which they enter the conduit into which the new conductors are being pulled.

6. Wrap and wrap electrical tape over the mass, forming a smooth "nose," tapered at each end, that will lead the new conductors through the conduit.

7. Slather the nose and the new conductors—along the entire length you are going to pull in—with pulling compound, "wire lube," which is an inexpensive soapy, waxy mixture (nowadays actually containing neither soap nor wax—but with this sort of slippery effect). Squirt or push a little of it into the opening of the conduit the nose will shortly enter, too.

8. Have the person who will be pulling pull on the conductors that are to be replaced just enough to bring the tip of the nose to the conductor opening.

9. Have the person who will be feeding line the nose up with the conduit opening, and if possible push the nose in past the opening.

10. Pull, slowly and steadily, while the person feeding lines up the new conductors with the conduit opening and keeps them untangled and lubricated. The person doing the feeding tells the person pulling how fast to go and when to pause. The person pulling announces to the person feeding that the pull is done, when the nose has come through and brought plenty of new conductor—perhaps as much as a foot of it—into the destination box.

All this gets much, much more complicated if your pull needs to pass through any junction boxes. Usually, it is best to pull it from one box to the next rather than pulling the entire length from the far end. It is wise to pull it out at each box, make sure the nose is intact—retape it if needed—and relubricate.

Repair

This is a completed nose, taped and well "soaped." After taping, wrap nice and tight, but also use extra wraps to make a smooth transition. Then lubricate.

Many times pulling new conductors into old conduit is not an option. Your choices then are repair or replacement. This almost always is the case with cable systems. Some customers ask, "Can I pull new wires through the old cable armor?" No. The ones in

Pulling lubricant.

the old cable are immovable. Many customers ask, "Can't I use the old cable to pull in a new one?" Highly unlikely. Any cable installed during new wiring should be nearly immovable. It is secured at frequent intervals, and can't be pulled out of its staples or out of the small holes drilled for it. It is even more impossible to use it to pull another cable in through those same holes and staples. This also is true of any cable installed during rewiring that took place with walls or ceilings open.

There's only one case where an old cable can be used to pull in a new one. If a cable was fished in—pulled blind through a opening or, even better, tossed across a suspended ceiling—after walls and ceilings were up, it just might be possible to use it for pulling. If cable was fished in, though, it is less likely to be old enough to have deteriorated and require replacement. Replacing cable can be expensive and messy; it certainly is beyond the scope of this introductory book.

The easiest answer to deteriorated insulation is to wrap it with electrical tape. Unlike permanent replacement, this temporary measure may or may not protect you adequately. An intermediate solution—intermediate both in terms of safety and of time, effort, and mess—is to cut back the cable sheath or armor. This is very much at the outer bounds of introductory material.

Here is why cutting back sheaths can be so useful. Insulation deteriorates the most inside boxes, adjacent to splices and terminations. Outside boxes, it usually is healthier, having been further from the source of heat. That's the section of cable you're trying to bring into the box. Whichever way you attempt it, be careful not to pull terribly hard on the cable. I have sprung the armor of BX apart by doing so, and gouged right through the sheath of romex because I hauled it through a staple. In these cases, I worsened the situation and had to open up the wall to remedy it.

Usually bringing more cable in involves some mess. I do not recommend it to a beginner except when working with surface wiring, unless the inside of the wall or ceiling is accessible. For instance, if a ceiling box on your top floor is crowded with conductors whose insulation is deteriorated, you can check in your attic to see whether you have access to the cables. You need at least a few inches of slack; perhaps you can create it by removing a nearby staple or two.

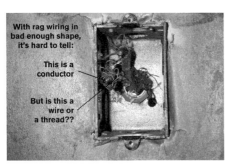

Deteriorated rag wiring can be very confusing; often worth rewiring, by this point.

Then you can temporarily remove the cable from the box, cut back the sheath or armor until you reach a section of cable enclosing conductors with decent insulation, and reinstall it into the connector or clamp. See Chapter 9, Tools, for some guidance about how to cut back sheaths. Be prepared for the possibility that you might fail at this. You may need to call in a pro who will at the least install a junction box or boxes.

On occasion, I am dealing with a box that is not nailed in place, such as an octagon suspended in the ceiling by a screw or locknut that holds it to a strap running between two joists, and I do have access from the side of a wall or ceiling that faces the habitable area. In this case, I temporarily drop the octagon out of the ceiling, as an example, to get at any slack in the cable. This is one way to avoid messing up the wall or ceiling.

Which fix should you choose when conductors' insulation is deteriorated? This depends on the degree of deterioration, the constraints on your time, budget, and ability to put up with mess, and the overall building condition. Rewiring is best. If you just won't be able to rewire, and cutting back seems too much for you to take on, consider using varnish or shrink tubing to repair the insulation; they may be a little better than tape. These are mentioned in Chapter 11.

14

Replacing Receptacles

Joe decided to replace some receptacles in the attractive 30-year-old townhouse that he'd just bought and moved into. He spent well over a month at the job. One of the brightest men I know, he ran into so much baffling miswiring that he realized that he was in over his head.

He called me in after he discovered a deadly case of overfusing. A receptacle was connected to #16 wires (which have no business at all in your branch circuits), fed by a 20-amp circuit, which requires conductors at least two sizes larger. There was no cable, no raceway. The installer had run individual, unprotected wires in the wall, sandwiched between the plasterboard and wood!

The happy-go-lucky remodeler had been working on, or shortly after, October 22, 1986; we determined this by the date on the (nicely combustible) newspaper wrapped around the wires.

Even if Joe had not come up against these dangerous violations, he knew that there would be more to replacing receptacles than, "Black wire to brass screw, white to silvery screw, and (maybe) copper wire to green screw."

While in many cases replacing a receptacle is just about this simple, other times it can be very tricky indeed. And that's even when there's nothing wrong! When the system design is not straightforward, you could be stymied unless you get some guidance. Furthermore, it could be easy for you to not realize that what you're faced with does not match the cookbook model. Every so often I encounter a dangerous situation that a previous worker overlooked, replacing the receptacle without correcting the problem.

You know by now that black is not necessarily hot; that anything electrical requires testing before you can be sure what it is. The information in this chapter does not stand in isolation. Replacing a receptacle probably is the most frequently performed electrical repair. You will call on the information that guides you in evaluating receptacles again and again, as you learn to work with switches, lights, and other outlets. Correspondingly, the information on removing and replacing receptacles builds on information from the earlier chapters. Therefore, if you've skipped ahead, please go back, for your own safety.

I am going to start off by talking about why you might want to replace a receptacle, not how to do it. This is not filler; even if you already plan to replace a receptacle, you will learn about options you may not have considered because

I include some pretty sophisticated material. However, you can skip this section without jeopardizing safety, and proceed to the section where you will learn how to evaluate what you have in place at each outlet. Do not, however, skip past evaluation and pay attention only to the mechanics of replacing a receptacle.

Why Should You Replace a Receptacle?

You may see obvious reasons for replacing receptacles; other good reasons may be much less obvious.

The Two Major Reasons

I'll begin with two very common reasons you might take on this task: a receptacle may have deteriorated, or may simply be antiquated and therefore lacking an opening for a ground prong. After looking at these two possibilities, I'll step back to describe other reasons for replacement, many of which are equally valid.

"Broken" can be a judgment call. A little chip of plastic can break off without reducing the structural integrity, without causing anything to be loose, and without exposing metal. When this is the case, I do not see a need to repair it. On the other hand, I consider repairs very important when parts begin to separate, even without anything visibly being "broken." I have seen half the face of a receptacle lean out from the cover plate, indicating that the face probably was cracked, but more importantly that the face and body were separated or separating. If they fell apart, it is very likely that they would short. A clean short would blow a fuse, but a high-resistance short could smolder.

A Receptacle Shows Deterioration or Damage

All things get old and tired. Receptacles lose their grip. The spring metal inside gradually gives up in response to heat and repeated insertions. Early on, receptacles and switches were held together with screws. Internal components were available separately, so you could rebuild the devices. Now, when a receptacle no longer grips cord connectors firmly, replacement is the only answer. Once the receptacle exerts less than a few ounces of force on each prong of a cord connector, the electrical resistance of the connection becomes too great, and it may heat up dangerously. You need to replace it, just as you need to replace a receptacle that is broken.

How can you tell whether your receptacle exerts, for example, a healthy 10 oz. of pressure, or even a reasonable minimum of 4 oz.? You can't, not without utilizing a tool that is rare for even electricians to own. But you do know when a receptacle fails to hold plugs securely. When this is the case, replace it. Bending the plugs' prongs so they will not slip out of the receptacle is a very temporary, insecure expedient, and it's not very safe.

The Existing Receptacle Lacks a Ground Hole

Grounding problems greatly concern home inspectors, homeowners, and electricians. The most important fact to remember about grounding is that you always need to test if you want to know whether a ground is present or whether you have a grounding problem. The second most important fact is that something may test as grounded but not be a good enough

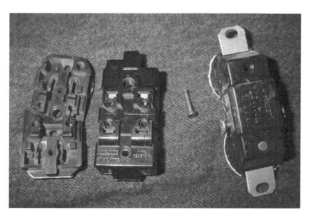

Broken receptacle; of course you want to replace this one.

ground to blow a fuse quickly when something shorts to ground; therefore, it is important to make the best possible splices and the tightest possible screw or locknut connections throughout the grounding path.

Almost all nongrounding receptacles are many decades old, and furthermore they can be inconvenient when you have modern, grounded appliances. This is one reason some people foolishly install three-prong receptacles where there is no ground: not to serve as cheaters, which I discussed in Chapter 5, Safety, but to "modernize"—because they thought doing so was an improvement, and possibly because they thought it would make the outlet safer.

You too might want to replace two-prong, nongrounding receptacles with grounding receptacles, even when they don't show deterioration or damage.

The high quality of some old installations often results in unplanned benefits. I often find nongrounding receptacles in older systems at grounded outlets, those whose wires were run in conduit or BX. Although at the time the systems were installed people did not realize that grounding is important, the metallic wiring methods are intrinsically grounded when they are installed properly and not disarranged. I repeat, however: you never can tell whether an outlet is grounded by looking; you have to test.

Doesn't two-prong *mean* ungrounded? Here is where the difference between an outlet and a receptacle becomes important. A two-prong receptacle (a receptacle being that which you plug a cord into) indeed is not a grounding-type receptacle, but the box enclosing it, hence the outlet, may be grounded. The box then would ground the yoke of the device—in this case, the receptacle—more or less effectively. How effectively the two would bond, tying the receptacle into the system's grounding path, depends on whether the box is flush so that the receptacle's mounting screws pull the two together firmly. This would ground the mounting screw of the cover plate, too, since it is screwed directly to the yoke. The result is that you could ground appliances plugged in by using a three-prong adapter properly. However, if you want to plug in grounded appliances, changing the receptacle is much better than using a three-prong adapter.

Other Good Reasons to Replace Receptacles

Aesthetics

It is surprisingly hard to separate out aesthetics from function. Most of the reasons for replacing receptacles are clearly matters of function, but in many cases old, two-prong receptacles that serve loads such as ungrounded reading lamps are replaced with three-prong receptacles even though the receptacles are in good shape.

Even a more obviously aesthetic preference, such as the desire for a better match between the color of receptacles and rooms, can have a functional aspect. Painting a receptacle is a bad idea, unless you're awfully careful. Paint can make it harder to get plugs in and can reduce the quality of the contact when they have been inserted. If a receptacle is burned or broken in any way, it definitely should be replaced; this is functional thinking. When a receptacle is thoroughly clogged with paint, you are far wiser replacing it than scraping the paint out and hoping that you've restored solid contact; this similarly is functional reasoning.

Since you can get a new receptacle and cover plate—both NRTL-Listed—for a few dollars, why even try painting them? Brown, ivory, and white are standards receptacle colors. Colors that you might not expect to find, such as red and gray, are stock items, though not in the cheapest versions nor at the cheapest merchants. Some colors that are not in stock even at the larger distributors may be available, on special order, at special prices. Listed cover plates are available in a very wide variety.

There are a number of stylistic variations other than color that simply might appeal to your aesthetic sensibilities. Decora style is the rectangular look that puts cover mounting screws at the very top and bottom. Flat face receptacles, where everything is flush with the cover plate, may be chosen as a matter of looks, and make no functional difference.

Aesthetics-Plus

Some options you may consider for receptacle replacement have both aesthetic and functional characteristics. You can purchase a receptacle with a high relief face, meaning that the bevel from the face into each slot is more deeply angled. This is a more "industrial" look than a standard receptacle shape, but it also helps guide a plug into the slots; you've got looks and function.

Other options you might choose have more apparent differences from standard receptacles than do flat face and high relief receptacles.

- An angled receptacle, whose slots are not parallel to the axis of the receptacle face, has an unusual look that might appeal to you. Besides having an unusual appearance, the angled faces make it easier to attach two appliances with 90-degree plugs (the type that hug the wall) that normally would get in each others' way.

- A Quad receptacle accepts twice as many plugs as a duplex, without requiring a multigang box. Besides having an unusual appearance, some quads not only make the task of accommodating angled plugs easier but allow you to safely connect four appliances at a single-gang outlet.

"Clock-hanger" receptacle. The hook is for hanging a wall clock, but you can use this receptacle elsewhere—to protect a plug from being damaged by furniture, for example.

Three-receptacle round cover; no modern close replacement exists. This nonpolarized receptacle may have started out in the days before 3-prong plugs, because the customer suspected that more than two appliances might be needed simultaneously.

• A clock-hanger receptacle is single, rather than duplex, and therefore does not offer the convenience of letting you plug in two appliances at once. However, it is recessed. This means that, for instance, if you want to push furniture flat against the wall, or want to hang a clock or piece of illuminated art, the plug does not get in the way.

Functional Differences

Some receptacles do the same job as what you have, but do it better, or add features. Still, most of them are one-to-one replacements.

Beefing Up to a Heavier Grade. You might consider installing a Heavy Duty or Hospital Grade receptacle, or simply a higher-quality receptacle, at a location with extra-heavy use. Wearing out early and signs of discoloration or distortion all could indicate extra-heavy use. They could have other meanings, though. For example, they could indicate that the existing device was manufactured poorly or was badly installed.

Upgrading an Ancient Floor Receptacle. Floor outlets are very rare outside of offices. Only once in the past few years has a customer asked me to install a receptacle in the floor. However, in this case, a floor outlet was the best choice because this particular location in his entranceway had a brick wall. Besides, my customer was used to floor receptacles.

Floor receptacles have advantages and disadvantages; usually, but not always, more of the latter. If you use a floor receptacle, you know how con-

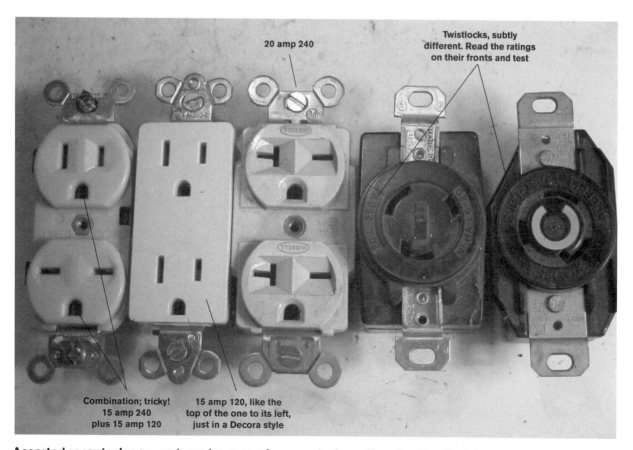

Assorted receptacles to watch out for; except for one pair, these all are functionally different.

Yoke-to-box bonding clip

Yoke

15A-125V

This green dot indicates that this receptacle is hospital grade–a superior design

Hospital grade duplex receptacle, self-grounding. The metal clip ensures that the yoke supporting the receptacle and holding it together, and the ground holes that are connected to it, are bonded to the mounting screw, so if the box is metal, the receptacle is bonded to it without the need for a separate bare wire.

Some very odd older floor receptacles sat under carpeting and required cord connectors with pointy prongs to penetrate the carpet.

venient it can be when you want to avoid stretching a cord from a wall outlet to a desk or a dining room table. However, floor receptacles used to simply be wall receptacles are installed flat on the floor, face up, with standard cover plates. If you have one of these, you probably have seen breakage, or recognized that dust and mopping water normally enter them, accelerating wear—and risk of malfunction, perhaps serious.

If you still have any old-type floor receptacles in place, you may want to replace them. You have two options, aside from moving a floor receptacle to a wall, in the rare instances when that is possible.

- Install a proper floor outlet, which has a heavy, gasketed brass cover.
- Install a "doghouse" or "monument" outlet, which sticks up, resting on rather than in the floor. This is the type I installed near the wall of my customer's entranceway.

Each type has advantages. A flush outlet is much less visually obtrusive than a monument, and, when nothing is plugged in, it should not constitute a tripping hazard. There are several tough noncorrosive floor boxes, with threaded plugs protecting the receptacles. Unfortunately, too often those plugs were and are left out. Some have flip-up plugs, less likely to be lost, but still likely to be left open—and more likely to get broken when they are left open.

The old floor box installations containing these receptacles did not survive intact, because their covers were inadequate. Why, then, is it that I am talking about replacement *boxes*? This has to do with Listing. Testing Laboratories do not investigate the use of the cover part of a floor box assembly by itself, but only the cover as it functions mounted on the box that is sold with it.

Because of this requirement, you cannot replace a cover alone. If you are to install the assembly as intended, you must remove and replace the box in your floor along with the cover. This can be quite a job, especially when there is no access to the box from below, and no slack in the cable feeding it.

Some ignore the restrictions posed by the test standard and fit the cover assembly over the box that already is in the floor as best they can. This means that they void the warranty, and also lose the protection provided by the Laboratory testing the assembly.

This same concern applies to those who skirt the rules in another way, installing covers over the floor boxes that are designed for mounting on walls in wet locations. These do not sit flush with the floor, are relatively unsightly, and can cause tripping and get broken. When they are located out of sight (as may be the case under some furniture), these may not be legal or functionally suitable, but at least they do not get in the way.

Reset-lockout GFCI.

A GFCI lets you safely operate three-prong equipment when grounding is absent.

From an electrician's vantage, the monument wins.

- It can be retrofitted in place of the old receptacle and cover plate over the existing box.
- Whether or not a cord is plugged in, it keeps the receptacle high and dry. It has no plugs to lose, no flip-up covers to break.
- It is highly visible, which arguably means that people are less likely to trip over it, or over cords plugged in to it, than if it were unobtrusive.

It also has disadvantages, starting with the fact that it is highly visible and obtrusive. Its position on the floor also means that it will get in the way. Finally, it could cause quite a stumble.

Overall, though, I favor the monument. If you need a receptacle installed face-up on a surface such as a countertop, I recommend that you use this type. It stands above the floor, which means it can get in the way—but it provides much better protection against water, dust, and "trod-on" damage. A pedestal, whose underside does not even rest on the counter, stands even taller—but the receptacle is further out of harm's way.

While replacing boxes is beyond the scope of this book, deciding whether it is worth having your old floor receptacle upgraded is not.

Adding GFCI Protection. GFCIs, which I discussed first in Chapter 5, Safety, are worth considering strongly, whenever you replace receptacles that lack this protection. This is quite independent of whether the NEC requires GFCI protection at the location.

GFCI protection is required in bathrooms, outdoors, in unfinished areas of basements, and in countertop outlets near sinks. Whenever you replace receptacles in those areas legally you must add GFCI protection. This is an example of a rule with no exemption for grandfathered installations. GFCI receptacles often are the simplest and most economical way to add this protection.

GFCI's were developed, and GFCI installation was first required, not in the U.S. of A. but in the U. of S.A.—South Africa—to reduce electrocutions in wet gold mines. They were installed here in the 1960s, and first required here in 1974. If your house is older, you may very well be due for some GFCI protection, or more of it.

TEST! Look back at the warning in Chapter 5 regarding non-functional GFCIs.

Here's another reason to replace a receptacle with a GFCI. If you have a two-prong receptacle, but no ground is present at the outlet, it can be suicidal to use a three-prong adapter or to install a three-prong receptacle without running a ground wire. Except in very rare cases, you will find it quite difficult to run a ground wire to an ungrounded receptacle correctly, so I will not explain the process.

One popular alternative to adding GFCI receptacles is protecting an entire circuit with a GFCI circuit breaker. However, this is not possible in all cases. Multiwire circuits are one example. You cannot provide standard GFCI protection on a multiwire circuit upstream of the point where it splits off into standard two-wire circuits. Circuits with barely-discernible leakage to ground are another place where you cannot add standard GFCI protection, even though the level of leakage poses no real hazard, at least arguably.

However, you are permitted to install a GFCI to accommodate three-prong cord connectors at a location that lacks grounding. You can do this with a very good degree of safety. In almost all cases, this also can protect similar receptacles downstream. You have a second legal option other than installing a GFCI when a two-prong receptacle needs to be replaced. If you have no need to plug in three-prong equipment at its location, and GFCI protection is not required, you *can* legally replace it with another two-prong receptacle. But I advise that you use a GFCI.

Childproofing. You might wish to install child-proof receptacles or receptacle covers for a number of reasons:

- Children are joining your household;
- you expect young children to be brought into your home without very careful supervision; or
- your home is in transition from infants to toddlers.

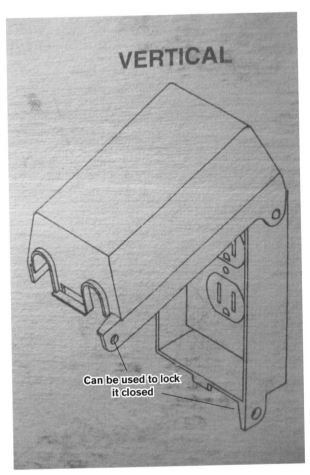

Weatherproof, and child-resistant if locked. Locking it over a plug has the disadvantage that it interferes with your using the cord as a quick disconnect in case there's a problem with what you have plugged in.

Child-protective cover.

Making an outlet childproof definitely means something more than inserting a push-plug or two. These are quite easy to remove, if not by pulling, then by prying.

There are three types of mechanical protection:

- Properly childproof or tamper-resistant receptacles are designed so that it is impossible to gain access to one slot, say with a scissors blade, unless one simultaneously sticks something in the other slot. This just about guarantees that only attachment plugs will be inserted. The plugs, however, do remain accessible.
- Childproof cover plates may work similarly. Warning: some styles cause poor contact between inserted plugs and receptacles.
- High-class, but bulky, cover plate *covers* for outdoor receptacles are designed to keep plugs dry during rainstorms. Some of these lock. While these are not designed for the purpose of childproofing, some people use them to keep children away from receptacles—plugs, too. They are rather bulky.

When you childproof, consider installing not only childproof receptacles or cover plates but also GFCIs. The reason is that childproofing restricts access to the receptacle, but does nothing to protect a child from being hurt by chewing on a cord, or from pulling an attachment plug *partway* out so its live prongs are accessible. The GFCI adds electrical protection—incomplete but superior to that provided by a fuse or regular circuit breaker—to the mechanical when you install it upstream and use a childproof receptacle or "no-tamp" cover as the mechanical protection. With the third option, the outdoor cover, you can fit a standard GFCI receptacle at the location you are childproofing.

> ### ⚙ CAUTION!
>
> Push plugs are not something I'd rely on to keep kids safe. I would use them for one other task: protecting receptacle openings while I paint, if I don't want to mask the whole cover plate, which is preferable.

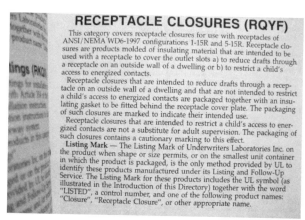

RECEPTACLE CLOSURES (RQYF)

This category covers receptacle closures for use with receptacles of ANSI/NEMA WD6-1997 configurations 1-15R and 5-15R. Receptacle closures are products molded of insulating material that are intended to be used with a receptacle to cover the outlet slots a) to reduce drafts through a receptacle on an outside wall of a dwelling or b) to restrict a child's access to energized contacts.

Receptacle closures that are intended to reduce drafts through a receptacle on an outside wall of a dwelling and that are not intended to restrict a child's access to energized contacts are packaged together with an insulating gasket to be fitted behind the receptacle cover plate. The packaging of such closures are marked to indicate their intended use.

Receptacle closures that are intended to restrict a child's access to energized contacts are not a substitute for adult supervision. The packaging of such closures contains a cautionary marking to this effect.

Listing Mark — The Listing Mark of Underwriters Laboratories Inc. on the product when shape or size permits, or on the smallest unit container in which the product is packaged, is the only method provided by UL to identify these products manufactured under its Listing and Follow-Up Service. The Listing Mark for these products includes the UL symbol (as illustrated in the Introduction of this Directory) together with the word "LISTED", a control number, and one of the following product names: "Closure", "Receptacle Closure", or other appropriate name.

Underwriters Laboratories, Inc. on the limitations of outlet closers.

Child-protective cover.

Unquestionably a wet location; GFCI required.

Can You Convert Your Receptacle Outlet from Single to Duplex? Replacement is not quite so simple a choice in some cases. Suppose you have a single receptacle at a location where a normal, duplex, receptacle would be more convenient. Changing the receptacle to one that can accommodate more plugs certainly is much safer than using an octopus adapter to accomplish this purpose. However, there are cases where this is unacceptable.

One case is where you need a single receptacle at the outlet because it serves an appliance that requires a dedicated circuit. Using a single receptacle, nothing else can be connected at the same time as the appliance. In this case, neither using an adapter nor replacing the single receptacle with a duplex is okay. You will violate the terms of the appliance's warranty, and of the product's Listing, by making such a change. It may be that the appliance would be damaged by the voltage drop or circuit tripping that could result from running a second piece of equipment on the circuit simultaneously.

> Electrocution is too easy in unfinished basement areas, which often are damp and frequently guarantee that your body will be grounded. People have gotten badly shocked using equipment such as power tools in these locations, relying merely on fuse or circuit breaker protection.

A second case is where the outlet feeds a single piece of fixed equipment in an unfinished part of your basement. A single receptacle serving an appliance, such as a freezer that is not easily moved, may be protected solely by a fuse or normal circuit breaker. The NEC restricts the use of any other 120 volt receptacles in such areas to those with GFCI protection. Therefore, if you have such a receptacle, you cannot merely replace it with a standard duplex.

I do recommend—with a certain reservation—that you replace it with a GFCI, even though it is legal not to do so. Besides, the GFCI will give you that extra place to plug tools into, for which you were considering changing the outlet to a duplex. Without the GFCI, you gamble that there won't be any ground faults, at least ones great enough to injure your family, that lurk till someone touches the freezer and is shocked, instead of blowing the fuse.

AFCIs. Arcing Fault Interrupting circuit breakers are available today for most modern circuit breaker panels. What about houses with fuses? AFCI receptacles should appear between the years 2001 and 2002. The receptacles probably will be on the expensive side—but worth the price.

Surge Protection. Surge-protecting receptacles are available to protect electronic equipment. These are worth installing even where you have whole-house surge protection. This type of receptacle can protect not only at the outlet where you install it, but to some extent downstream. Its protection is, however, limited to incoming and upstream surges. (Your heavy loads such as laser printers generate surges—especially following the sags when you switch them on.)

Ground Isolation. One exotic type of modern receptacle, normally found only around serious computer installations, is the Isolated Ground receptacle. Warning: here is a type of replacement that is *not* available as an upgrade to your existing outlet. These receptacles offer an unusual level of data protec-

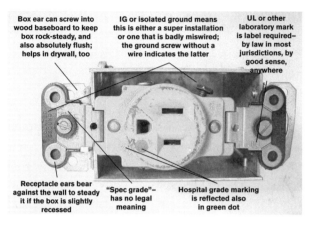

Box ear can screw into wood baseboard to keep box rock-steady, and also absolutely flush; helps in drywall, too

IG or isolated ground means this is either a super installation or one that is badly miswired; the ground screw without a wire indicates the latter

UL or other laboratory mark is label required— by law in most jurisdictions, by good sense, anywhere

Receptacle ears bear against the wall to steady it if the box is slightly recessed

"Spec grade"– has no legal meaning

Hospital grade marking is reflected also in green dot

Hospital grade single receptacle, self-grounding.

tion through isolation of the circuit from surrounding electrical noise.

While the IG has considerable value, I must warn against your trying to install one. I won't cover its installation or replacement because it requires a special type of circuit cabling. If a circuit wasn't wired for an IG installation, the chances are remotely small that an IG receptacle will be safe—or that the ground isolation can be installed as intended. While IG receptacles are quite rare in homes, you should learn to recognize them, just in case one was installed in yours. An older IG's face will be colored orange; a newer one has an orange triangle on its face. Never use an IG receptacle as a replacement for a normal receptacle in the mistaken belief that it somehow is superior in itself; this could create a shock hazard as a result of the IG's special design. If you do find an IG in your home, ask an electrician to check it out the next time you have one over.

Replacing Receptacles

One problem with trying to replace the old connection scheme wire-by-wire is that the coloring on old wires often is faded. You need to keep track of wires that are hard to distinguish. This becomes even harder when they are so short that you need to pigtail them. This is not a trivial problem; you need to deal with it even when you follow the procedures I will describe.

Regardless of whether you are replacing a receptacle with its exact duplicate or choosing one of the alternatives that I just discussed, what you have in your wall will have major effects on how you proceed.

Even when wiring systems are in fine shape, when I look at replacement work that simply tried to duplicate what was there, more often than not it's not as safe as it should be. This is not necessarily because it's been done incorrectly, although this also is common. Simple-appearing replacement may not be quite as simple as it seems.

Don't Decide Prematurely What You Are Going to Do

Before starting the process of replacement, you need to determine whether simply replacing the receptacle will serve your requirements. You might want to consider updating the wiring, in cases where it's not up to the job. The conductors may have deteriorated, but even if they are good, they may not serve your needs. Kitchen and dining room receptacles are supposed to be on 20-ampere circuits serving no other loads; a related requirement is in place for bathrooms. Is it time for you to run new cables to feed your kitchen and perhaps your bathroom to modern standards, so that a toaster or microwave or hair dryer doesn't cause lights to dim or a circuit to trip?

Similarly, if you intend installing a large air conditioner, or a computer, at some location, it might benefit from its own, dedicated line. Sometimes a higher-quality replacement receptacle on the same circuit, as discussed in the section on replacement options, will not survive much better than the origi-

The NEC requires GFCI protection a number of places, noted throughout the Code. You will find most of those that apply to your home early in Article 210, unless you have a pool or hot tub.

nal one. Sometimes, the equipment will suffer or the circuit will trip without a dedicated line. From a legal standpoint, exact replacement of what is there now probably will be considered grandfathered. Where a GFCI is required today, though, you are required to replace it with a GFCI-protected receptacle.

The Simplest Case

Any basic wiring cookbook should show you the simple version of receptacle replacement. Presented properly, even in a cookbook, it should have a few more steps then I gave at the start of this chapter. Here is the same job, with a few critical elements added. This brings it to ten steps, all of which are important. I'll talk about how to determine that everything at the outlet is okay and simple, and about how to replace a receptacle when that is the case. At each step, after I deal with the simplest case, I'll address more-complicated designs, and deal with, or at least alert you to, wiring that has problems. At points, unfortunately, the complications get TOO difficult for beginners to unravel reliably and safely.

If you evaluate your outlet carefully and everything does turn out to be okay and simple, I see no problem with your moving past the subsections that people with tougher situations have to go through, so long as you don't skip any of the steps in the standard procedure.

Despite the fact that in most of the following cases, everything was functioning when you started working, you may feel daunted by what you discover at your outlet. If this turns out to be the case, please respect your feelings and judgment, and don't push your limits dangerously.

The Most Basic Replacement, Step by Step

At its most basic, replacing a receptacle involves these steps: test; gain access; retest; disconnect; repair and identify; connect the new; replace and close; and retest.

The following description starts out by presuming that you need to replace a receptacle that is:

- Standard.
- Grounded.
- Correctly wired.
- 120 volt.
- 15 ampere.
- Duplex.

Each of the numbered steps that follows may take several actions. Some are quite complex and entail "subprocedures," marked with letters rather than numbers. Replacing any receptacle with a GFCI, or converting from 240 volts to 120, takes extra steps; these tasks are in a section for the nonbeginner, called "Advanced Replacement."

Step 1. Test

The very first step of the actual job, the one to take before you dismantle anything, is testing.

A. Begin, as I described in Chapter 3, Voltage Testing, by testing your tester. Make sure that circuit breakers or fuses are on, or you may have no voltage available to test. First I'll describe the tests, and then follow with an explanation of possible results.

B. Test for voltage from hot (the shorter slot in modern, polarized receptacles) to neutral (longer slot), and after this from hot to ground. Just to be safe in case it was miswired, also test for voltage between neutral and ground. If there is no third, round or U-shaped, grounding hole, scratch any paint off the cover plate screw and test to make sure that the screw offers you a ground. If the existing receptacle is duplex, check both the upper and lower (or right and left, if it's horizontal) receptacles, and check for voltage between each hot slot and every other slot.

In the simplest case, your tester indicates that everything is as it should be. You measure

- 120 volts between each hot (shorter) slot and each neutral (longer) slot, and from each hot to ground;

- No voltage from one hot to the other hot, from either neutral to ground, or from neutral to neutral.

  If you get these readings, you're ready to proceed to Step 2. If your readings vary, or if your receptacle is such that you can't perform these tests exactly as I have described them, you need to continue with Step 1.

Dealing with the receptacles whose configuration differs from the standard 15 amp, 120 volt, grounded duplex version often is simpler than tracking down the reasons your readings differ from these. Some variants are very straightforward:

- You may be dealing with a single receptacle. If it is polarized, and you read voltage from hot to neutral and from hot to ground, and no voltage between neutral and ground, you can proceed to Step 2.

- You may be dealing with a nongrounding receptacle:
  - The polarized version has a long slot and a short slot on each face. So long as you measure a ground at the box—at this point you would measure it most likely at the cover plate screw—in all other ways testing and replacement follow the same procedure as in the simplest case.
  - You may be dealing with a nonpolarized receptacle whose parallel slots are equal lengths. In this case, there is no indication as to which slot should be the hot. So long as your readings match those given above—with the hot side determined by reference to the mounting-screw ground—you can proceed, making note for later use of which side is hot.
  - You may be dealing with another nonpolarized variant. It has these same slots, but each has a tail going out to the side from its middle, making it a T shape. I never have found this used for anything other than a 120 volt, 15-amp receptacle, but it was used for 240 volts (D.C., too, once upon a time), so test. If you do get a reading of 240 volts between the two Ts, you cannot treat this as a basic variant on the standard case. Either blank it off, so that no one mistakenly plugs in a 120 volt appliance, or change it to a 120-volt outlet, if doing so is possible, following the amended procedure I describe under "Advanced Replacement."

- You may be dealing with a 120 volt, 20-amp receptacle. When it is mounted with its ground hole down, the left, neutral, slot looks like a sideways T. Confirm that it is on a 20-amp circuit, and it can be tested and replaced the same way as a 120 volt, 15-amp receptacle. The only difference is that it can be replaced with either a 120 volt, 20 amp or a 120 volt, 15-amp duplex receptacle. Even if the present receptacle is a single, you cannot replace a 20 amp with a 15-amp single receptacle, except in the strange case of a circuit feeding multiple single receptacles.

- You may be dealing with a 240-volt receptacle. A 120-volt circuit is protected by a single breaker or fuse. A 240-volt circuit is protected by two fuses, one connected to each busbar, or by two breakers, one of which should be directly above the other, or by one two-pole breaker.

    - A receptacle that has two slots in line with each other, rather than parallel, should be on a 240-volt circuit. It should be protected at 15 amps.

    - The 20 can be tricky. Do not confuse the 120-volt, 20-amp receptacle and the 240-volt, 20 amp receptacle. The latter is a similar-looking receptacle whose T is on the RIGHT side when the ground is down.

- You may be dealing with an antique called the Despard. It has as many as three devices on one yoke. Replacement is simple only when the Despard's devices are all receptacles; if it includes switches or indicator lights, its replacement is a little beyond the scope of an introductory text. Most commonly, but not always, a Despard will have 120 volt, non-grounding recep-

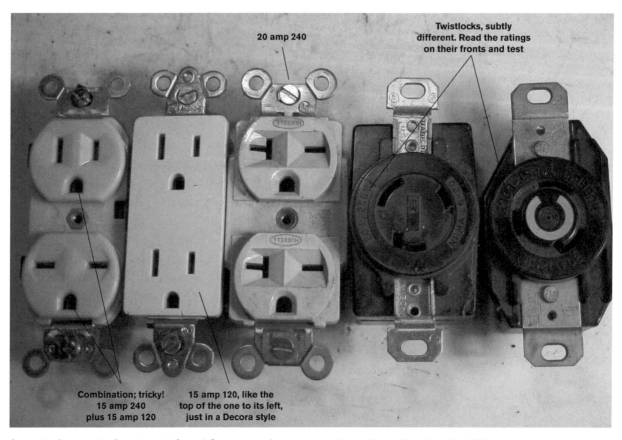

Assorted receptacles to watch out for; except for one pair, these all are functionally different.

> ### 💡 CAUTION!
>
> If your testing does find 120 volts between one slot of any 240-volt type receptacle and the other, call in a pro. The design is unquestionably intended for 240 volts and only 240 volts. Something is quite wrong.

tacles. If you replace a Despard, you will need to replace the cover plate, as the old one will not fit a modern receptacle. It is, however, rebuildable, in that each of the one to three devices on its yoke can be snapped out and a replacement snapped in. If you want to find a replacement, and can do so, you may be able to follow the procedure for daisy-chained receptacles.

- You may have a very much outdated equivalent of a Decora device (Decora being the style that includes modern GFCIs). This fits a slightly smaller rectangular opening than the Decora; as with the Despard, neither the plates nor the devices match anything made today. Aside from this, you can proceed exactly the same as in the simplest case, so long as your tests give the results I described there.

Some more exotic outlets will not accept modern attachment plugs, though they usually can be replaced with normal receptacles. Determining whether this is so may be a bit sophisticated. There is a slight risk of inadvertently converting from 120 volts to 240 or vice versa, so in these cases be careful to test thoroughly before you start, and again when you have finished. Be prepared to open up your outlet and redo your work if it turns out to be necessary.

Receptacles for older electric stoves have this same configuration, in large. Like crowfoots, the NEC has shifted away from these because they do not separate the neutral and grounding functions. This can be dangerous. Older electric dryer receptacles also combine neutral and ground.

- You may have a crowfoot receptacle. This is the same size as a modern receptacle, but it has three slots, in a Y configuration, rather than two slots and a round hole. The two slots forming the arms of the Y should read 240 volts apart, and the third should read like a neutral. If not, something is wrong; you probably should call in a pro.

- The old four-slot receptacle is rarer. Its two sets of slots are parallel, with their outer (likewise their inner) ends forming a diamond shape. The slots that are in line, one with the other, may have been used for a 240-volt circuit; the slots that are parallel, one with the other, may have been used for a 120-volt circuit. If you find ground present at the cover plate screw, replacement may be straightforward. Check for voltage between the in-line slots. If you read 240 volts there, treat replacement the way you would the replacement of any 240-volt receptacle. If you read 120 volts there, something is nonstandard even for this style of receptacle; call in a pro. If you get no reading between the in-line slots, treat testing and replacement as though this were a nongrounding, unpolarized 120-volt receptacle.

- An even more exotic-looking and rarer—but modern—receptacle is the twistlock. Its slots are curved instead of straight or round. It rarely is found in residences because it is used only with an appliance whose cord connector matches the twistlock's configuration. Like the last two varieties, replacing it can be simple. Test it; if you seem to have hot, neutral, and ground, you can proceed, but check again inside the box.

Nonstandard test results can be a bigger problem. The possibilities branch considerably.

- Your tester may give no reading, now or as you retest during subsequent procedures. When this happens, as well as in many other cases, you will need a reliable external reference. Plug a three-wire extension cord into a

One probe is inserted into an extension cord that you have just tested; the other into the one you are checking. Most of the time you will not be checking from the extension cord's hot to the receptacle's hot, as pictured.

receptacle that has tested as normal and live, and that remains live when you kill any circuit you are in the process of testing. Test the female end of your extension cord to make sure you're getting normal readings between its three openings. Now you can proceed.

• Test for voltage between the ground or neutral of the extension cord and all three slots of the receptacle. If you confirm that there is no reading, test for voltage between the hot of the extension cord and the receptacle's ground and neutral. If you do find 120 volts now, there are a number of possibilities:
  • The circuit may be off.
  • The receptacle may be controlled by a wall switch.
  • Power may be interrupted by a bad splice or termination or a broken wire somewhere upstream.
  • The receptacle may be protected by a tripped GFCI upstream.
  Check the circuit breaker or fuse, look for a tripped GFCI, and try wall switches. If none of these measures restores power, call in a pro.

• Your tester may show voltage between hot and neutral, but no voltage—or only a momentary flicker of a reading—between hot and ground. If you subsequently test between hot and neutral, using this neutral or that from an extension cord, and find no hot, you almost certainly have a receptacle that is protected by a GFCI upstream. Find the GFCI, reset it, and you should again have 120 volts between hot and neutral. When you install the replacement, attach a "GFCI-protected" sticker to it.

• You may find 120 volts to ground and neutral at one receptacle (one "face") and zero at the other. If all your circuits are working, the part that is dead almost certainly either is controlled by a wall switch that is in the OFF position or is making such poor contact with your tester that it does not register. In the former case, you have a split-wired receptacle, which is explained immediately below. You will determine whether this is the case after you have opened the box. You will be retesting after you have opened the box; that will let you know whether there is any reason for this reading other than poor contact.

• You may find a hot but no neutral. In this case, there are a few possibilities:
  • Your receptacle is controlled by a wall switch that was miswired.
  • A wire is broken.
  • It was wired at 240 volts and has power coming out of the panel from only one busbar.
  • Power is interrupted by a problem somewhere upstream.

One wire broken off at terminal screw—fatigue.

Due to space constraints, I do not lay out the troubleshooting required to correct these problems.

- Each hot could show 120 volts to neutral and to ground, but 240 volts to the other. Most likely, this represents a multiwire circuit.

- You could find voltage from neutral to ground, but not from hot to ground. This almost certainly means reversed polarity—the wires were attached to the wrong terminals.

Split-Wiring

Split wiring is a technique allowing one duplex receptacle to serve two different purposes. To split-wire, you break the link on the hot side of a receptacle, or rarely, the links on both the hot and the neutral sides. This allows you to treat the duplex as though it were two single receptacles:

- One may be switch-controlled and the other not; this is convenient when you want to turn off a floor lamp from the wall switch in your living room, but leave the clock radio powered. In this case, you would break the link on the hot side and leave the link on the neutral side. Feed one terminal on the hot side with unswitched power and the other with switched power.

- One may be on each leg of a multiwire circuit; the advantage to this is that the location can serve much heavier loads than if only a single 120-volt circuit were available. Again you break only the hot-side link, and bring one hot leg to each hot-side terminal.

- Finally, and most rarely, you may have two fully independent circuits. In this case, you break the links on both sides, and bring one circuit's hot and neutral to the top side terminals (assuming you have mounted the receptacle vertically) and the second circuit's hot and neutral to the bottom.

Split-wiring is advanced circuiting, beyond what I can cover in this book in adequate depth.

Step 2. Kill power

A. Look on your panel directory for the outlet.

- A 15 or 20 ampere fuse or circuit breaker may be marked, clearly identifying this receptacle as part of the load the fuse protects.

- If two fuses or circuit breakers are marked as controlling this receptacle, or if a two-pole breaker or a pull-out fuse block is marked, if you are replacing a 240-volt receptacle you can proceed comfortably. If it is a 120-volt outlet, it may be part of a multiwire circuit. This could complicate your job, and add an element of risk.

- If the circuit is not clearly marked, you need to return to Chapter 8 and label your lines.

B. Unscrew the fuse (some very old fuse boxes will have a switch handle to flip to turn the fuse on or off) or flip (in some rare cases, push) the circuit breaker handle, and return to the receptacle. Use your tester to verify that all parts of your outlet are now dead. If it isn't, you need to correct the labeling. Return to Step 2A. That should solve your problem, except in the very rare case where there is something wrong with the feed. In the latter case, you urgently need a pro.

Step 3. Expose the Wiring

A. Open the outlet box holding the receptacle. In the simple case, the cover plate screws come out easily and the cover plate pops right off revealing an intact plastic single outlet box secured flush with the wall surface. If this happens, proceed to Step 4.

Any of the following variations require additional attention:

- If you have difficulty with unscrewing or removal, return to the discussion of these tasks in the previous chapter.

- If you have an intact metal box, your task will be the same as if you were replacing a receptacle in a plastic box, except that you will need to bond box, receptacle, and grounds all together. See the discussion of grounding screws and clips in Chapter 12, Terminations.

- If you have a multigang box, metal or plastic, your task will be nearly the same, so long as the box contains just two conductors from outside plus some connected receptacles. Additional conductors add complication, as I discuss below.

Step 4. Test Again

Now that your voltmeter probes can reach the wires and terminals, test again to make sure power is indeed off. Unfortunately, receptacles that are due to be replaced often make unreliable contact with anything stuck into them—including the leads of your tester. Often this is why receptacles need to be replaced! Scratch, scrape, and prod wires and terminals, and make sure. If you have any doubt, use an extension cord reference to give yourself a reliable ground to use in confirming that no terminal, wire, or surface shows voltage with respect to ground. Once you are sure that the terminals all are dead, you can proceed to Step 5.

When you try to remove a receptacle or receptacles, you may discover that the conductors are all too short. I say "try to remove" from experience. I have had to cut wires in order to get a receptacle out.

If the conductors you need to work with are short, you will need to splice pigtails on for adequate length. Whether or not pigtailing is necessary, I have found that it is very easy to lose track of which conductor went where. Mark the conductors; I sometimes use colored tape. As you work inside the box, you may choose to, or need to, undo soldered-and-taped splices that are so far back that you have to reach way inside the box to add your pigtails. In this situation, if you do not proceed methodically and carefully, you may well lose track of which wires came from which splices.

Step 5. Remove, Retest, and Examine

A. Remove the receptacle, as described in the previous chapter.

B. Test again now that you have yet better access. If everything tests fine, as before, proceed to Step C.

If you do find power at a receptacle terminal, or at a wire that is or was attached to the receptacle, stop. Identify the circuit involved and do not proceed until your voltage checker confirms that you have killed it.

If you did not find a good ground when you tested from outside, now is the time you may find one. If you have a metal box, the box itself or a clamp or connector may test as grounded. Whether your box is metal or plastic, you may discover evidence of a ground wire that was clipped or left outside the box, and reach it with your meter.

If the receptacle you removed was a crowfoot, now is the time you can determine whether you have both a neutral and a separate means of grounding.

C. Look for additional insulated conductors besides the two needed to bring power to a receptacle (and, in rare cases, a green grounding wire). If there are none, proceed to Step 6.

Additional insulated wires may merely pass through your box, spliced or, much less commonly, unspliced. If this is the case, you can proceed to Step 6; however, you may want to back off on the chance that this is a multiwire circuit that could cause shock even though you tested for voltage. At the least, don't touch any splices in the box that are not connected to the existing receptacle or receptacles.

WARNING! Beware of Bootlegs

A "bootleg" involving grounding exists in four cases:

- Someone may have joined the neutral and the ground; they may have literally spliced a bare or green wire to another, insulated conductor from a cable. (Their reason doesn't matter.) If you see this, get the installation checked out. If the insulated conductor does not enter from a cable or raceway, it may possibly be legitimate; this is unlikely, though.

- The neutral may have inadvertently contacted a ground—a ground wire, for example, may have rested against a neutral terminal. Clear away the contact, and make sure that this does not happen when you install the replacement receptacle. Taping the terminals helps, but is no guarantee.

- Where no ground is available, a neutral may have been spliced or daisy-chained both to the location intended for a neutral and to a grounding

terminal or a metal enclosure. You need to undo this, and treat the outlet as ungrounded. Beware of other dangerous make-dos.

- Finally, a ground may have been connected to the neutral terminal or splice to replace a missing neutral. This is not easily fixed. The outlet must not be used until the source of the problem has been tracked down, and it has to be tracked down; it could cause a fire.

Any of these arrangements can cause shock, and any of them most certainly can confuse electronic equipment installed on their circuits. Any of these can trip a GFCI. Very often this type of mistake is due to a misunderstanding of switching theory, sometimes of grounding. Finding any of this, except for inadvertent contact, suggests that someone has played very loose with your electrical system, and you should be on the lookout for booby traps.

On the other hand, conductors that pass through may be connected to the receptacle.

- If the links on the sides of the receptacle are intact, it merely is being used to pass power through to another device or devices. You will be able to connect your new receptacle the same way, or, better yet, pigtail all these wires.

- If the receptacle is split-wired, replacement can be more complicated. If there are only three insulated wires, replacement is straightforward, but evaluating what you have can get complicated even when that is what you find. The various split-wiring designs are beyond the level of this book. You are best off calling in a pro. Despite the sophistication that may be involved, split-wiring should be old hat to a master electrician or experienced journeyman. There is nothing about it that is uniquely associated with old wiring; I do not even cover it as a separate subject in *Old Electrical Wiring: Maintenance and Retrofit*.

- If the box is multigang, the receptacles may all be wired from one hot and one neutral, in which case you can proceed to Step 6. However, they may be wired in a way that is similar to a split-wired receptacle; they can share a multiwire circuit, or they can include both switched and unswitched receptacles. If they are not all wired from the same hot splice and the same neutral splice, you are best off leaving this to a pro, for the same reason as applies with other split-wiring.

Step 6. Check the Box and the Cable Entrance

Make sure that there are no problems with the enclosure or the wiring method, as I explained in the previous chapter.

If you find a problem, the previous chapter offers guidance for remediation. If you need repair beyond that which I can give in this book, you will need to call in help.

If what you discover next, or have discovered to this point, makes you feel that you are in over your head if you try to complete the receptacle replacement, you may need to call in a pro. In certain cases, though, there will be another option for making your house safe. If you are sure that you don't need the use of this receptacle, and the outlet box and conductors are intact, you probably can eliminate the outlet safely. This presumes that you have not encountered evidence of illegal work—a case in which you probably *should* call in some help. Once you have disconnected the wires as described below, and confirmed that they are dead, cap them with wirenuts, or, if conductors are daisy-chained, splice them. Then put a blank cover over the box.

Step 7. Disconnect, Testing As You Go

Disconnect the wires from the receptacle. In the simplest case, you'll find one conductor whose insulation clearly is black, attached to a copper-colored screw. The insulation of the other conductor clearly is white, and it is attached to a silvery screw. A third wire is bare, and it is attached to a green screw. All the wires are solid copper, and all the insulation is flexible and intact. If this is exactly what you find, go on to Step 8. Otherwise, continue here.

If polarity seemed correct from the outside, but the white conductor is on the copper-colored terminal and the black on the silvery terminal, polarity probably is reversed somewhere upstream; you should track down the location.

If white, or possibly white, conductors are attached to both sides of the receptacle, and black to (and only to) the side with the coppery screws, it is conceivable that a "switch leg" is attached—legally. The white wire on the coppery screw ought to have been taped black, but this may not have been a legal requirement at the time that the receptacle was wired. When the conductor connections depart from "black (or blacks) to coppery side, white (or whites) to silvery sides," the setup quickly departs from anything a beginner can be expected to untangle.

If insulation color is not clear, whether due to fading, to taping or shrink-tubing, or to paint, now is the time to address this problem. If your testing has indicated a correctly-polarized 120-volt receptacle, tape the conductor or conductors attached to the hot terminal or terminals black and those attached to the neutral terminal or terminals white. If your testing clearly indicated that you have a simple, two-wire 240 volt circuit, tape both conductors black.

If two or more wires are attached to a side, especially the neutral side, beware of voltage appearing as you disconnect wires on the same side. I discuss "inadvertent multiwire circuits" in Chapter 5, Safety.

Step 8. Confirm Ampacity

Check that your replacement receptacle, wires, and circuit protection match in accordance with the NEC. Check the insulated conductors by comparison with a

CAUTION!

If anything about the wiring is unclear, you may choose to restore power at this point so that you can test the wires. If, however, all you are trying to identify is a neutral or ground, an ohmmeter may suffice.

sample or with a wire gage. If the fuse or breaker protecting the circuit is marked 15, the receptacle must have the 15-amp configuration.

Wire gage may not be unmistakable; quite often it can be confusing. Older insulations were thicker than modern, and stiffness varies from one to another. If you have a raceway carrying stranded conductors, this can be even more confusing. You have to judge by the wire used, not the complete conductor. Once you're sufficiently used to working with conductors, you'll be able to gauge a wire's gage by flexing it, judging by stiffness—differing materials or not.

There are arguments for and against daisy-chaining. First, it may save a little time and some materials, compared to pigtailing, especially if you quickwire. This argument, however, should be irrelevant. Working on your own home, a few cents and a few minutes is less important than doing the job better. The other argument in favor of daisy-chaining is that splices take up extra room in sometimes-crowded boxes. This is more worthy of consideration; note, though, that really crowded boxes may be worth replacing. The strongest argument against daisy-chaining is numerical: measurements have shown that circuits with daisy-chained splices tend to have higher resistance than those using pigtailed connections, even when both are installed by experienced electricians.

A mismatch between circuit protection and wire size is not always evidence of illegal work. Some old cables were manufactured with #12 hot and neutral wires, but #14 grounds. Installing this cable on 20-ampere circuits no longer would be legal, but it was considered fine at one time. I am quite sure that your local inspector would judge it to be grandfathered.

A 15-amp duplex receptacle generally is fine on a 20-amp circuit. The reverse, however, is illegal; if you found the latter, someone installed it ignorantly. Beware of what else they may have done wrong.

Step 9. Reconnect

The replacement may be a different color or shape, but so long as its terminals are equivalent to those of the old receptacle, connection should be straightforward. Secure the wires to the new receptacle, properly, following the same color code as used by the old receptacle.

The replacement receptacle can vary quite a bit from the old one and still be connected the same way. A hospital grade, flat-face, childproof, angled, high-relief, or Decora receptacle certainly can replace a standard, two-prong, or uncomplicated Despard receptacle. Even modern floor receptacles and quad receptacles are connected electrically the same as others, although you need to follow their instructions, or have good mechanical sense, to mount them properly.

When you have multiple receptacles that were connected to the same conductors, the best way to connect all hot and all neutral conductors to corresponding terminals on the new receptacles is by pigtailing, even if the old receptacles were daisy-chained. This is even more true if you have converted a box from single-gang to two-gang and added a second receptacle.

If your replacement is a 240-volt receptacle, you should have two conductors that are black or are taped black (at any rate, some color other than white or green), as noted in Step 7. Whether or not the old receptacle was a standard 240-volt receptacle, installing the replacement will be simple: connect one of these conductors to each copper-colored screw, and a ground wire to the green screw.

Grounding and Bonding

Most old receptacles are not grounded and bonded in accordance with modern standards. Grounding and bonding are highly important for safety, so you will be upgrading these almost every time you replace a receptacle in a metal box. Splice all ground wires together; include a wire bonded to the box and one going to the grounding screw of each receptacle in the box. Bond to the box with a ground screw or ground clip, as described in Chapter 12.

Suppose your tester indicates that a box is grounded, and despite this fact no ground wires enter it. What could this mean? Normally, when no ground wires enter a metal box that is fed by a metallic system, the box is grounded by the metal raceway or cable. When this seems to be the case, do what you can to make sure that locknuts are tight, make sure that wiremold is intact and fits tightly, or make sure that BX is solidly wedged into its clamp and the clamp is screwed down securely. This way, when you bond the receptacle to the box, the ground path back to the panel will have the lowest possible resistance.

☀ CAUTION!

If a box is connected both to a metal raceway or BX and to a plastic cable containing a ground wire, be sure to splice the cable's ground wire to the wires bonding box and receptacle.

This cable's ground wire was wrapped under the cable clamp; you need to unwrap the wire and ground and bond in one of the more reliable, low resistance ways I describe in Chapter 12, Terminations.

It is very, very hard to get at the ground in a cable like this. You may not even realize that it was a grounding-type cable until you remove it from under a clamp or inside a connector, and see the stub of truncated ground wire.

If you determine that a metal box is grounded, and yet observe that it was wired using a nonmetallic cable and no ground wire enters the box, there are a few possibilities. The original installer may have used a cable containing a ground wire, but wrapped the ground around the cable sheath under the clamp or connector. In this case, try to unwrap the ground wire, draw it into the box, and splice it to the wires bonding the receptacle and box together.

In other cases, you may find ground wires twisted together without a wirenut, perhaps wrapped around a cable clamp, and clipped short. This is one of the nastiest situations. You have to straighten them out enough to splice a grounding pigtail onto them in order to reach the receptacle. At worst, you may find that the cable has a ground wire that was just clipped because the installer knew no better. While it made good enough contact, somehow, for your box to register as grounded, you are going to have to get more sheath into your enclosure so as to get hold of enough of that ground wire to pigtail. See the previous chapter for a discussion of how you may, just possibly, manage to do this. Finally, some old romex was manufactured without a ground wire.

"Self-grounding" receptacles can save you some time and some box space. They bond to metal boxes without your having to attach bonding wires to the receptacles. This is useful when, and only when, all ground wires entering or passing through the boxes are bonded to the boxes. One is shown on p. 244.

The reverse is unsafe; never ground a box by bonding it to a grounded receptacle. Here's the reasoning. Suppose you don't ground the box, but rely on its bonding to the receptacle to ground it. Eventually, when the receptacle is replaced, both box and receptacle will become potentially dangerous if the replacement receptacle is not grounded properly. On the other hand, suppose that you ground the box but rely on bonding the receptacle to the box to ground the receptacle. When the receptacle is replaced, if you don't ground it properly, it will be at least somewhat safe because its mounting will more or less bond it to the grounded box.

Step 10. Prepare to Replace the Receptacle

A. Complete any necessary splices.

B. Attach conductors to your receptacle. At this point, the receptacle should be ready to function.

C. Before stuffing the receptacle into the box, you may wish to restore power and very carefully test it in accordance with Step 1. If anything does not test as it should, you will need to back up.

D. If you do test, be sure to kill power again before touching the receptacle.

E. Now wrap all live parts of the receptacle with electrical tape. One wrap around the screws will minimize the chance of their shorting against a ground wire or a screwdriver when power is restored. (Don't wrap the tape around your 6-32 mounting screws.)

Step 11. Stuff the Receptacle Back Into Its Box

A. If the box is crowded, arrange wires and wirenuts so that they do not obstruct the space into which the receptacle must fit. Make sure that the locations where you stuff the wires back in the box do not put them at danger—right where long mounting screws will penetrate, or where they will be pinched by the devices they feed when you push them back in. Finally, dress (shape) the receptacle's conductors so as to minimize the chance that a terminal screw will be loosened as you push the receptacle back into the box.

Dressing Wires

"Dressing wires" is the trade term for training conductors into shapes that serve the principle of "neat and workmanlike installation." Adhering to this general-duty precept enhances safety. When it comes to replacing a device in its box, this has two applications. First, in putting the receptacle back, be careful not to pinch any conductors. Tuck them in as far back as you can before pushing the receptacle in; then stuff the receptacle in before engaging and tightening the mounting screws. Second, bend each conductor attached to a terminal screw in the direction such that pressing the receptacle back against the conductor will curl it further around, tightening, rather than backing it off the screw. This is an extension of the idea that wires always should wrap around screws clockwise, so that tightening the screws doesn't push the wires out.

To my mind, the ideal sequence for stuffing wires in a box when there are splices present is to stick the grounding wires in deepest, then the returns/neutrals, and finally the hot wires that are spliced together. The reason is that it is safest to push hard against the ground wires because I can't do mischief by cutting through insulation. I even can push them back with my lineman's pliers, so long as I am not putting stress on any conductors. I am not concerned about forcing them against metal clamps or screws at the back of the box.

B. If the receptacle is going into a standard box that is mounted in a wall, or into a wiremold box, seat it flush against the box or the wall so it is not bowed out by crowding behind it.

If the receptacle is going into a surface-mounted box, or into a round or octagon box mounted in a wall, mount the receptacle to the cover plate and then carefully fit it back into the box and screw the cover plate down against the box.

C. Mount the receptacle to the box; level or plumb it before tightening the screws.

D. Attach the plate; level or plumb it before tightening the screws.

Hints on Mounting Receptacles

• Force

As you push the receptacle back into its box, be sensitive to the risks of pinching insulation on the wires behind it, and of forcing wirenuts off wires. I've done both, myself, as I was learning this trade.

If you are working with a multisection switch box, be careful not to force the sections apart with your pressure as you push the receptacles back in.

• Receptacle Orientation

This high-up receptacle's orientation assured that when the cord that usually is plugged in pulls out, the ground comes out after the hot and neutral— arguably a safety advantage.

This receptacle serves a sump pump that has an angled plug. The receptacle should have been installed with the reverse orientation. The way it is installed, the plug is stressed.

The receptacle has been reinstalled so that its orientation no longer stresses the sump pump's plug.

Depending on your aesthetic preferences, and on the orientations of other nearby receptacles, you may want to install your receptacle with the ground prong either up or down, or to the right or the left. It matters little.

If you intend the receptacle for plugging in a particular appliance, one that has a right angle, wall-hugging plug, install the receptacle so that the plug and cord hang against the wall without making a loop. Otherwise, there is a slight safety advantage if you orient low vertical receptacles with the ground up; if horizontal, with the ground to the right (neutral up). The logic of these choices is that if something slides down along the wall between the plugs and the receptacle, it will not hit the hot prong first. Also, if the cord is yanked up, the hot prong will lose contact first. Similarly, high receptacles (for instance, those above a countertop) should have their grounds or neutrals on the lower side, other factors equal, because cords are more likely to be pulled out of them downwards.

• Mounting in Surface Boxes

If you are installing a receptacle in a surface box, break off the tabs from the ends of the receptacle's mounting yoke so that it fits better. In flush installations, when the box is slightly recessed the tabs help hold it firmly against the wall.

• Cover and Wall Issues

The cover plate also goes on with one or two 6-32 screws. If the plate is to be painted or wallpapered, do so before putting it back. However, do not wrap wallpaper around the plate to the point where paper will be inside the electrical enclosure.

Finally, if the wall finish shows signs of your struggle, you can avoid repainting by using a goof (oversized) cover plate. Note that these will only serve to conceal spoiled finish; actual holes need to be repaired. The NEC insists that no gap greater than ⅛" be permitted around an electrical box.

Step 12. Power Up and Retest

A. Restore power.

B. Test your tester again, and then the receptacle. If the tester says the receptacle is okay, both in terms of grounding and in terms of polarity, you're done. And you haven't even gotten dirty.

Advanced Replacement

GFCIs

If the wires bringing power in from the electrical panel are connected by mistake to the terminals marked, "Load," the test and reset buttons will work, but you won't be protected. This can be deadly.

Where a receptacle is split-wired or part of a multiwire circuit, replacing it with a GFCI is a little too complicated for this book.

The Simplest Installation. You should have no trouble installing or replacing a GFCI in a box containing only two insulated conductors. The only difference from the basic, simple version of receptacle replacement is that the conductors must go to LINE terminals, not the LOAD terminals. The latter have tape over them to help prevent confusion.

Basically Straightforward Installations. You also should have no trouble installing or replacing a GFCI in a box containing more than two insulated conductors, if and only if three conditions are met:

- All the additional conductors were attached to the old receptacle, whether directly or by pigtails.
- The previous receptacle was not split-wired.
- You bring all the conductors to the LINE terminals, which usually means that you pigtail them.

Protecting Downstream Outlets

Considerations. Protecting outlets at additional locations downstream can be so useful that it is worth explaining; even though when you don't bring all the conductors to the LINE side, it is easy for you to make a dangerous mistake. There are pluses and minuses to providing downstream protection:

- The most important advantage is that you can protect locations where it is difficult or impossible to install a GFCI. An example of the latter is a receptacle that is an integral part of a bathroom light fixture.
- A lesser advantage is that you can create increased safety at other locations, ones where you could install GFCIs, without having to go to the trouble or expense of putting in multiple GFCIs.
- Another advantage is that you can install a GFCI where you don't have a ground, as described in NEC Section 210-7.
- The most serious disadvantage is that you could miswire it in such a way that it does not protect you, as described in Chapter 5.
- A much lesser disadvantage is that you could miswire it in such as way that it simply does not work.
- Another disadvantage is that whenever power goes off downstream, you must remember to check your GFCI receptacle as well as the panel. Often people forget about the GFCI receptacle that might be responsible for an outage at other outlets. I had a customer who owned a large place who ran into this problem. The employee showing me around remembered that they had a GFCI protecting an outlet I was investigating, but forgot where the GFCI receptacle was located. The hunt took quite some time. (I discovered a GFCI receptacle that protected outlets in the main hallway, located outside the building. Whoops.)

Procedure. Remove the old receptacle, following Steps 1–5 of the general procedure I laid out in this chapter.

You have to identify which conductors are which. This requires live testing, which is inherently dangerous. I will use letters rather than numbers, as a reminder that this augments rather than substitutes for the basic 10 steps for replacing a receptacle.

A. Carefully separate the conductors so that they are easy to get at and hard to nudge into contact with anything, and temporarily restore power.

B. Use your voltmeter to determine which conductor is at 120 volts with respect to ground. This conductor is your incoming hot wire. Fix its position in your head or identify it in some other way that does not involve contact; you don't want to handle it until you have killed power.

C. Kill power.

D. Tape or otherwise identify the incoming, or LINE side, hot conductor, to distinguish it from the other hots. Any other conductor or conductors that were spliced to it, or daisy-chained to it, are outgoing, downstream, hots. The outlets they serve will be the LOADs protected by the GFCI. If color coding is unclear, now is a good time to tape these black.

E. Move the wires out to where they are easy to get at and clear of each other again, and well away from grounded surfaces, and restore power.

F. Now use your voltmeter to find out which conductor is 120 volts with respect to the incoming hot wire; this is the incoming neutral. Fix its position as well.

G. Any wires that were connected to it are outgoing neutrals. Confirm that no other wires show voltage either with respect to ground or with respect to the incoming hot. If any do, stop. You probably cannot install a properly-functioning GFCI here.

H. Kill power.

I. Connect the incoming, upstream neutral and incoming hot to the terminals marked LINE. Connect the other, downstream conductors to the terminals marked LOAD.

Now return to the basic sequence of replacing a receptacle, beginning with grounding and bonding in Step 9. The only other modification is that testing will not quite follow the pattern of Step 1. Any hot to neutral test should give you the standard 120 volts, and any hot-to-hot or neutral to neutral test should give you zero volts. However, any test to ground with a wiggy should trip the GFCI, causing its RESET button to pop out. (Some testers will not produce this effect when you test from neutral to ground; that's all right.)

Size and Shape. There are physical differences between GFCIs and standard receptacles that will affect your work. GFCIs are considerably bulkier than standard modern receptacles, so be prepared to discover that you have insufficient volume to get your GFCI back in place of an existing non-GFCI receptacle—especially in boxes that have additional conductors. GFCIs manufactured in recent years all have rectangular Decora faces, so you need to use matching cover plates. If you are installing the GFCI in a surface-mounted box, the fit may be very tight—too tight for some utility boxes, and too tight for any round or octagonal box. If you had a decorator cover plate, you may have to hunt for a satisfactory replacement.

CAUTION!

I do not rely on GFCIs' TEST buttons. Experienced electricians believe that inserting a wiggy or equivalent in the hot and ground slots is *the* way to find out if it will trip. The TEST button is valuable to confirm the device's calibration (and if the TEST button won't trip it, you needn't bother using your wiggy).

Why Won't It Work?

You may have done everything right in replacing or installing a GFCI and find that it still won't RESET when you restore power. There are any number of possible reasons, quite aside from a defective GFCI. Most of them will be reasons to call in a pro.

- A bootleg neutral.
- A bootleg ground.
- An informal multiwire circuit, which really is another type of bootleg.

- Miswiring downstream, if you included downstream GFCI protection.
- A ground fault downstream, such as an old light with some worn fixture leads near the lampholder—just not enough of a fault to blow a fuse.
- A fault at the location where you're working, such as pinched insulation or a ground wire nudging a GFCI terminal.

Voltage or Ampacity Conversion. Converting a 120-volt outlet on a dedicated circuit to 240 volts or vice versa is fairly easy. Unfortunately, it is as easy to do wrong as to do right. Converting a high-ampacity outlet, say a 30-ampere outlet, to feed a lighter-rated receptacle, say a 15- or 20-ampere outlet, is similarly easy to do, and even more easy to do wrong rather than right. It is very, very rare for a lower-ampacity outlet to be legally and safely convertible to one with a higher ampacity.

Which Conversions I Cover. If you have very good reason to be sure that you know how to work safely, you may want to attempt one of the following conversions. The conversions I describe here are the simplest on this list, but still can be quite dangerous.

This is how you might be able to convert a properly-grounded 240 volt, 15, 20, or 30 amp outlet, served by a circuit that unquestionably feeds no other outlets, to a 120 volt outlet of the same or (in the case of a 30 or 20) lower ampacity. The following applies only when you have a *Main Circuit Breaker* or *Main Fuse* panel, or when you have a split-bus panel and the circuit originates in the *branch circuit* section, *not* the Main section.

Procedure. Most of the steps match those listed earlier for replacing a 240 volt receptacle. However, there are a number of steps you must take, before replacing the receptacle, to convert from the 240 hookup to 120, and, when you are changing ampacity, from the higher-ampacity fuse or circuit breaker to the lower. These are identified by letters to remind you that they do not substitute for the basic steps. I follow these with a more detailed, but more general, description of the procedure for removing or replacing conductors on circuit breakers, or removing and replacing circuit breakers.

If you plan to change current rating, you need to buy a new circuit breaker or deal with a fuse changeover. The latter can entail complications that make you change your mind about attempting the conversion. If, for instance, your 30-amp circuit is protected by a cartridge fuse in a fuseholder that will not accommodate a 15-amp fuse, or if it is a fustat, which means that it will only take the size that it now serves, you need to call in a pro. (There is a slight chance that you can find a replacement for a fuse block that has the wrong size clips.) If you don't have any of these problems, you can proceed.

A. Identify the two-pole circuit breaker or the two fuses or breakers feeding the old 240-volt receptacle.

B. Kill power.

C. Open the panel.

D. Identify the wires feeding the circuit.

E. If one is black and the other white, remove the white one from its fuse or circuit breaker terminal and attach it to the neutral bar. If both are black, or both white, stop and leave this to a pro.

F. If you are lowering the circuit rating from 30 to 20 or 15, or from 20 to 15, replace the circuit breaker with one of the new rating.

G. Close the panel.

H. Restore power.

I. Go back to the outlet and retest. One of the hot slots in the receptacle you are removing should now be a neutral, the other remaining at 120 volts to ground, as well as 120 volts to the first one. If this is not so, call in a pro. Soon. And leave the circuit off.

J. If the outlet tested correctly, reidentify the two lines on the circuit directory. You now should have a spare fuseholder or circuit breaker, or a two-pole circuit breaker that only is being used on one leg. Go to Step 2.

K. When you reach Step 4 of the basic procedure for replacing a 120-volt receptacle, confirm that the white wire indeed is the neutral. If it isn't, call in a pro. Soon. And leave the circuit off.

Here is a procedure for installing a conductor on, or removing a conductor from, a circuit breaker, or replacing a circuit breaker. It presumes that you know, perhaps from instructions accompanying a new circuit breaker or printed on your panel's schematic diagram, how circuit breakers are connected to your panel. Most residential circuit breakers simply plug in (with a bit of force).

Unless the manufacturer's instructions recommend otherwise, you would be wise to follow this sequence:

Breaker Removal

1. Make sure that the circuit breaker is in the full OFF position. Test its terminal; it is not live.

2. Remove the circuit breaker from the panel.

3. Pull the conductor out of the panel as far as you can without significantly disturbing the other conductors.

4. Now, disconnect the wire from the old circuit breaker and connect it to the new.

Breaker Installation

1. Make sure the new circuit breaker's handle is in the OFF position.

2. Install or reinstall the circuit breaker on its busbar.

    (If you are inserting a wire on the neutral bar, now is the time to loosen the screw you will use, insert the wire, and tighten the screw.)

3. Making sure that the breaker's handle has remained in the full OFF position, and being careful not to brush against any of the other terminals or pull on any of the other conductors, dress any conductors that are attached to it, making sure that they leave its terminals straight, and do not interfere with access to other terminal screws.

4. Check the tightness of the terminal screw, because the twists and pulls involved in reinserting the breaker or in dressing the conductor often loosen the screw.

5. Replace the panel cover.

6. Turn the circuit breaker to the ON position, unless it now has no wires attached.

7. Reidentify the circuit breaker in the panel directory if you have removed its wire and it is now spare or if you have changed its function.

15

Replacing Switches

Normal wall switches are used for two purposes, but they are intended just for operating loads such as your home's lights. (I will not directly address switches used as controls for disposals, exhaust fans, and furnaces.) However, in addition to their use as lighting controls, sometimes wall switches are relied upon, inappropriately, as shortcuts to safety. Using a switch to kill power to a lighting outlet when you work on its wiring, or even relamp, can be risky. We have wiring conventions intended to ensure that the power to a light fixture is off once the switch darkens, but these rules were (and are) not always heeded. This is particularly true in houses wired long ago, and in systems wired by the ignorant or indifferent. The failure by anyone who worked on your system to heed these conventions can cause something as simple as changing a light bulb to be dangerous.

The theory needed to troubleshoot a lighting system regardless of what's wrong, and even—and especially—to identify and correct hidden threats to your safety, is a very big bite. It is well beyond the scope of an introductory text. Much of this is covered in *Old Electrical Wiring: Maintenance and Retrofit*. (The latter is more of an advanced than an intermediate text; some of its information will be of limited use to a nonelectrician.) For this reason, I will simply not address the question of whether your switching circuitry is correctly and safely designed. I won't even discuss lighting circuits to help you decide when you should look at your switch as the possible culprit if there's a problem. I'll focus on two things. One is how to test a switch to determine whether the physical device itself is internally damaged. The other is how to replace a switch, whatever your reason for doing so. (The benefit to your ability to end up with a properly-functioning light will depend on the presumption that you have a lighting system that was working and has not been modified since it last was working.)

There are a number of reasons to replace a switch: aesthetics, damage, and malfunction are three. When I say aesthetics, I include quiet operation. An old snap switch that still may be in use literally makes a snapping sound when flicked.

When I say malfunction, I should point out that one of the first things I look for when a light misbehaves is a dying or dead switch, or one with bad connections at its terminals. The most common problem with a switch is that it stops conducting as it should; the wires seem to be disconnected. There are other failure modes besides this open "circuit," though they are far rarer.

- A switch can fail closed, meaning that the handle does not interrupt continuity through the switch as it should.

- It can short to the yoke, and thus to a metal box or cover plate to some extent. Your circuit breaker may trip—or you may get shocked.

- It can develop high resistance, even internal arcing, with the result that the light does not operate at full brightness, or, in some cases, come on at all. (This cause of trouble is hard to determine, except by a "not right" look or smell, or an ohmmeter reading.)

- It can exhibit delayed action, waiting—usually just for a moment—after you flip the handle.

- Any of these problems can be present intermittently, rather than continuously.

- It can come apart, most likely the face of the switch separating from its body. This mechanical failure may not become evident until you remove the cover plate, but it can result in one of the electrical modes of failure.

- Internal damage can become even more evident, in the way the handle moves or doesn't move.

I will not go into detail about these modes of failure, due to space constraints.

There are some exotic variants that can be worth substituting for standard switches for reasons of aesthetics or convenience. These include lighted-handle switches, dimmers, timers, remote-controlled switches, and occupancy sensors. Most of these are one-to-one, wire-for-wire replacements for standard, simple, ON/OFF switches. Some of them are available in variants that substitute for other types of switch, too. You may well be able to use the same procedures listed in this chapter to install these as replacements, if the instructions that come with them say so.

Switching Patterns

There are three basic patterns of switch control. You can control a light (or a group of lights that go on and off together) from one location, from two locations, or from multiple locations. Both of the latter sometimes are called, "multipoint switching." Except for exotic variants that I will mention briefly towards the end of this chapter, but not explain—I don't have room to go into thorough enough detail—these utilize three types of switch. If your lights were set up with control from:

- one location, you use one "single pole" switch (technically, any that you are likely to work with will be a "single pole, single throw," or "SPST");

- two locations, you use two "three-way" switches;

- more than two locations, you use two three-way switches, plus one four-way switch for each additional switching location beyond two. It is relatively rare to control a light from more than two locations. Still, if this is what you have, you may want to evaluate and, if necessary, replace these switches.

Single pole switches have marked ON and OFF positions; the other two types do not. There is, however, no such surface marking you can use to differentiate three-way switches from four-way before you remove the cover plate. I will describe three-way and four-way switches further, later in this chapter, when I explain how to remove and evaluate them. Four-way connections, however, can be so confusing that you may be better off getting a pro in to deal with them.

Safety

Switching designs can be very complex, well beyond the layouts suggested in many basic wiring texts, and well beyond the designs you may find in the

> **None of this discussion, to reiterate, involves changing circuiting; even converting from three-way operation to single-point is beyond what I'll explain.**

manufacturer's literature accompanying switches such as three-way dimmers. Because of this complexity, and because of the potential shock hazard, I have chosen not to describe live testing techniques that might come in handy on those occasions when you are trying to solve problems (as opposed to replacing switches for reasons such as aesthetics, when everything is working). Instead, I suggest that you use one of two options:

- simply replace switches and see whether the new ones work; or

- test your suspect switches with an ohmmeter or continuity tester.

Before working in switch boxes (and, in the next chapter, in light fixtures), take special heed of the need to turn off circuits, and confirm that they are off.

The Basic Procedure for Testing and Replacing Switches

If you don't have a clear recollection of the procedures I refer to, please reread the chapters, or at least those sections of them.

There are many similarities between the ways you remove, test, and replace various types of switches, and a few differences. Therefore, I will first describe the simplest case, the single-pole switch. Most people will be much less comfortable with the steps needed to evaluate other types of switches, most especially four-way switches. I will deal with them following this first case.

Step 1: Remove the switch from the box, following the procedure for removing devices that I detailed in Chapter 13.

Step 2: Before doing anything further, test again for voltage, following the procedures in Chapter 3.

Step 3: Remove the conductors from the switch, following the procedures that I describe in Chapters 12 and 13. Note whether any of the conductors' terminations on the switch seem poor. This very easily can cause a switch to work unreliably or to stop working.

Step 4: Evaluate the box, conductors, and cable connections, considering the issues I discussed in Chapter 13. If the light was not working right, check especially for a broken wire or bad splice.

Step 5: If you will be testing the existing switch (rather than simply replacing it), position the switch so it will be convenient to your tester.

Mercury switches can be identified by three characteristics:

- They are absolutely silent in operation; no click.
- They don't give the "click" *feel* of most other switches.
- They function only when plumb or very nearly plumb; tilt one and it doesn't work until you restore it to vertical.

If the switch seems to have shocked people, test for two things: continuity between either (non-grounding) terminal and the yoke—there should be no continuity—and box grounding. Sometimes a switch is unfairly accused of being dangerous because someone feels a shock touching it—a shock resulting from the discharge of static electricity to the grounded device

Unless you hear distinct snaps when you operate the switch's handle, keep it quite vertical, in case it turns out to be a mercury switch.

Step 6: Test the switch terminals, ignoring the grounding terminal if present. To evaluate a single-pole switch, touch the terminals with a continuity tester or ohmmeter, and compare readings in the two handle positions. OFF should show no significant continuity between the two terminals—the needle shouldn't move, or the number change; ON should show no significant resistance between the two terminals, meaning good continuity. If either reading is different, you need to replace the switch. When you replace a switch, it does not hurt to test the replacement, just in case you got a dud.

Step 7: Attend to grounding. If you are going to reconnect the old switch, and it does not have a ground screw, skip this step. Similarly, if there is no ground present, skip this step. If the switch you are installing has a ground screw, and there is a ground present in the box, ground the switch, following the procedure I described in Chapter 14.

Considerations Regarding Grounding and Bonding

What happens if your switch does not have a ground screw, or your switch box is not grounded? Switch grounding is less serious an issue than receptacle grounding because a switch isn't going to have three-prong equipment plugged into it. Still, should some mishap cause a switch yoke to get live, it could pose a shock hazard if it does not trip the circuit—especially if it is located in a bathroom.

New switches and even dimmers do have ground screws. When an older switch is reinstalled by screwing down on a grounded surface-mounted utility or wiremold box, you need not worry about its grounding, as direct metal-to-metal contact adequately bonds such switches. Other situations are more problematical.

This leaves other cases, though, where grounding is not possible. You have four choices. First, you can ignore the issue. I advise against this. Second, if you are working with a flush installation, you can install a nonmetallic cover plate; this is the minimum solution most inspectors accept. Third, with a flush installation, you can install a nonmetallic cover plate and hold it on with nylon 6-32 screws. Fourth, you can protect the location with a GFCI. This will provide full protection at the switch location if you install a GFCI upstream, and some protection if you install a GFCI switch, as briefly described later in this chapter. (A GFCI does cost much more than a replacement switch, but there may be additional safety reasons to add a GFCI.)

Step 8: Reconnect the conductors. You can attach either conductor to either terminal.

Step 9: Reinstall the switch, following the procedure for installing devices I described in Chapter 13. Make sure that ON is the up position, OFF the down.

Step 10: Restore power and make sure that your light or lights work right. If they don't, and your installation—including your evaluation of the wiring you were dealing with, and your switch—was good, the problem is elsewhere.

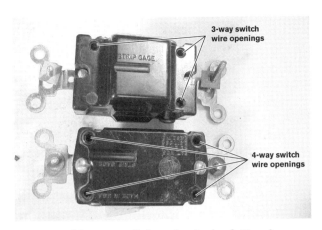

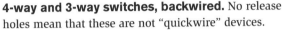

4-way and 3-way switches, backwired. No release holes mean that these are not "quickwire" devices.

Testing a 3-way switch.

Modifications to the Basic Procedure for Three-Way and Four-Way Switching

When you work with three-way or four-way switches, the terminals are not equivalent, as they are with single pole switches. This is why several of the steps are a good deal more complicated. As you look at the information on four-way switches and see the complexity of the process, it wouldn't be at all unusual to decide to leave troubleshooting or upgrading a four-way switching system to an electrician.

With three-way switches, one terminal is a "Common," and two are "Travellers." The Travellers are interchangeable with each other, but the Common is different; and it is marked and treated differently. The switch will have the Common terminal one color, perhaps black, or brassy, and the Traveller terminal screws another color, perhaps coppery. The switch may also have the words, Common and Traveller, embossed on its back by the terminals.

Four-way switches are still more complicated. These have four terminals, consisting of two pairs. Usually, one pairing will consist of the terminals on one side of the switch, and the other on the opposite side. I also have encountered four-way switches that had one pairing across the top and the other pairing at the bottom. If a four-way switch is old enough, you may not be able to tell which way the four are divided up into pairs.

Usually, however, the switch is marked. The system will more or less correspond to that used with three-way switches. Most likely, one pair will be one color, for example black, or coppery, and the second pair another color, for example brassy. There are no terms for the terminals on a four-way that correspond to the terms "Traveller" and "Common." The switch still may have markings embossed on its back (or on the package, when you buy a new switch), though, indicating its layout. They usually will show a set of parallel lines, plus diagonal lines. Each of the parallel lines connects the terminals making up one of the pairs.

Substitute Step 3: Three-way and Four-way Switches

Steps One and Two are the same for three-way and four-way switches as for single-pole switches. Once you have removed the switch from the box, follow-

A Most Peculiar Arrangement

I have encountered two switching arrangements, using three-way switches, that don't fit the designs I have described. In both cases, despite being a three-way switch, the switch does *not* seem to operate in conjunction with a second switch. In each of these cases, the installation of a three-way switch was inappropriate, albeit not necessarily illegal or dangerous. You should consider replacing it, unless, on further investigation, you discover that it is indeed working in conjunction with a second three-way switch.

In the first of these two cases, conductors are connected to only two of the switch's terminals. (This could be done with a four-way switch as well.) Once you open the box far enough to see the switch's terminals clearly, you will know if you have encountered this. It could be that a single pole switch needed replacement, and this was all that the installer had on hand. There is nothing terribly wrong with this, so long as the switch was installed so that flipping the handle up turns the light on. You can easily replace it with a single-pole switch following the switch instructions above.

In the second case, however, all three terminals have wires attached. Check carefully to make sure that you have not overlooked a second switch that is paired with this one. If you find no such switch, there is a good chance that one was bypassed by having the Common spliced to a Traveller. This could be quite legitimate, so long as the connections remain accessible. They might be in some electrical box serving another purpose, or one with a blank cover. If, however, you conclude that they were buried inside a wall, you have uncovered evidence of dangerous work.

If you find no evidence of improper wiring, you could replace this switch with a single-pole switch. However, leave this to a pro unless you are absolutely certain that you understand the deadly, nonstandard switching designs that may have been used. There are descriptions in my advanced text, *Old Electrical Wiring: Maintenance and Retrofit.*

ing the procedure I described in Chapter 13, check to make sure no voltage is present. Then identify which conductor is attached to which terminal.

(a) First of all, differentiate among the terminals. For a three-way switch, this means picking out the Common. For a four-way switch, it means identifying one pair, if the existing switch gives you an indication as to how the terminals are paired. If it doesn't, it means identifying each terminal uniquely.

(b) Second, make sure you have a means to distinguish the conductors. Are the conductors' colors clear and distinct from one another? Cable with a black, a white, and a red conductor is used for three-way switching quite commonly. If the colors are not distinct, you need to do something that will let you distinguish the conductors. One of the best approaches is to use electrical tape, as described in Chapter 9, Tools.

(c) Third, distinguish among the conductors by what they do. If the conductors had distinctive markings, make a record of which conductor, in a three-way, is the Common; which conductors, in a four-way, are paired together. This probably will be unmistakable if you have just marked the conductors yourself. If you have a four-way switch that gives no indication as to which conductors should be paired, make an initial assumption that the conductors on each side of the switch are paired.

(d) Now remove the conductors in accordance with the procedure described in Chapter 12. If you are comfortable doing so, and you know that power is off and will stay off, leave one conductor connected—the Common, in the case of a three-way switch. This reduces or removes the risk that you will mix up which conductor was which.

Some readers might be tempted to leave a switch fully connected as they test it. Rather than mark the conductor corresponding to each terminal and then remove it, they reason, why not avoid the need for this identification? One of the problems with this model is that instead of the tests indicating the status of this switch, they now are affected by the entire lighting arrangement; when this is the case, it can be harder to know whether tester readings indicate something about the switch itself or about the rest of the circuit.

Steps Four and Five of the procedure—evaluating the box, conductors, and cable connections, and positioning the switch for testing—are the same for three-way and four-way switches as for single-pole switches. Step Six, however, is different for each of the three types of switch.

Substitute Step 6

Three-Way Switches

Touch one lead of your ohmmeter or other continuity tester to the Common terminal, and the other to each of the two Traveller terminals in turn. Then flip the switch's handle and repeat. The Common should alternate being connected with one Traveller, *only* one, and the other as you flip the switch handle. If this is what you find, the switch is good. If you find anything else, the switch is bad.

Four-Way Switches

The switch is good if and only if you find one type of arrangement. Here is the generalized form of Step 6 for four-way switches.

With the handle of a four-way switch in one position, you find continuity between the members of one of the possible pairs of terminals (A with B or A with C or A with D, to use arbitrary names). You find continuity between the remaining two terminals, but not between either of them and either of the first pair of terminals.

Flip the handle. Now you find continuity between the members of two *other* possible pairs of the four terminals (A with D and C with B, for example, given that your first pairs were A with B, and C with D), but no continuity when you test any other possible pairing. *Note that each terminal must connect with one and only one other terminal at a time.*

If the switch has markings on the back, they will indicate which two sets of pairings are supposed to alternate connection as you flip the handle. However, if you do not find such markings, if you are not reinstalling the present switch, you will have to rely on the results of testing if the old switch is good. Record which conductors functioned together as pairs. You will need this information when you connect the replacement switch.

If the old switch is not good, and not marked, you will be stuck with guesswork. Here is a pretty good basis for a guess. If two of the conductors attached to the switch enter the box in one cable, and two in another, assume that the ones from one cable belong on one side of the new switch and those from the other cable belong on the other side.

Substitute Step 8

Step Seven of the procedure I described earlier, attending to grounding, is the same for all switches. The next step, installing or reconnecting a switch, is

WARNING

With one illegal old design, a wrong guess will result in a bolted short. With power restored, instead of turning the light on, you'll blow the fuse. See *Old Electrical Wiring* for further details.

straightforward when you are dealing with a three-way. The Common on a new three-way switch is clearly identified, and if you are reinstalling an old switch you have confirmed which terminal is its Common. If the old switch was good, and you could not make out its markings, you know which conductor corresponds to the Common from your testing in Substitute Step 3. Terminate the Common conductor on the Common terminal, and either Traveller conductor to each Traveller terminal. The only case that should be up in the air is where the old switch was both defective and unmarked. You have to guess; there is one chance in three that you will pick out the Common with your first guess.

Reconnecting a four-way switch usually is not much harder. If the terminals on the old switch were marked, it is straightforward. If it was not marked, but it was functional, you differentiated the terminals when you tested it in Substitute Step 3. Whichever of these was the case, you know which pairs of conductors go together. If so, reconnect them to terminals that go together.

If you are installing a replacement switch, attach two conductors that you have marked as being paired to one of the pairs of terminals indicated on the switch. Then attach the other two conductors to the other terminals.

If the old switch was not working, and not marked, treat the conductors that were on each side as a pair—unless you have a reason, or even a hunch, suggesting that you do otherwise. However, hold on to your record of which conductors corresponded to the screws on the top and the bottom. You will need to take the connection apart and try this latter pairing if your initial hookup does not work right.

Substitute Step 9

> There is no right way up for most three-way and four-way switches. If yours is any different, the manufacturer's instructions will say so.

If you are installing a four-way switch that replaces an unmarked, defective, switch, your hookup involves trial-and-error. Therefore, while you need to attach the conductors to the terminals securely, it makes sense to leave the mounting screws in the switch and the cover plate loose until you have finished testing.

Substitute Step 10

You will use the same testing approach for multipoint switching arrangements whether they are three-ways or four-ways.

Restore power, and try the switch you have just connected. If it works the lights as it should, leave its handle in one position and try each of the other switches in each of their positions. Repeat this process with the handle of this switch in the other position. Once you have checked every possible combination, you know that your installation is fully functional.

Tracking Down Causes of Failure

Some mistaken patterns of termination will enable you to control the light from this location, but will not let you turn it on from the other locations with the handle of this switch in one of its positions. With some patterns of misconnection—or of switch failure—one switch operates as a master on-off switch. The other switch, or switches, have no affect on the light when the handle of the "master" switch is in a certain position.

If the system does not work as it should, you have some choices. If it did work all right before you replaced the switch, I certainly would recommend that you recheck the installation you just completed. It could be that an old wire or splice broke, or it could be that a termination was not good enough.

With multipoint switching, it also could be that you lost track of which wire went where. If this was the case, and you kept notes, check them. At worst, you may have to swap wires and test, and swap and test, until you get it right. Don't forget to test the system as a whole at the end, trying every combination of switch handle positions.

If the system had stopped working properly before you started, and you found this switch and your reinstallation okay, the problem may well be with one of the other switches. That can be the case even when this is the one that seemed to have a problem, and the other switch or switches operated the lights as they should. This is a peculiarity of multipoint switching arrangements.

If the system had stopped working properly—or at all—before you started, and none of the switches nor their wiring appeared to be at fault, there are several possibilities. (Further troubleshooting is beyond the scope of this introductory book, though.) Some possible locations for the problem include:

- The light.

- The wiring between the light and the switches.

- The wiring bringing power from upstream.

- A four-way switch that you overlooked.

Alternates to Standard Types of Switch

Dimmers

Dimmers suffer from one special problem when wiring is unreliable, whether because of its age or because of misapplication. They are vulnerable to voltage aberrations such as may be caused by electrical faults. If a dimmer stops dimming but continues to work as an on-off switch, either it has succumbed to age or, more likely, it has suffered an electric shock such as may be created by a short on the circuit, or a sudden utility surge. You can use it as is, as a switch, or replace it with a new dimmer. If this problem repeats, find out if there is some dangerous problem that you have overlooked.

You probably will not find a standard four-way dimmer, but some exotic, relatively expensive electronic dimming systems allow you to dim from three or more points. These follow their own logic. Whether you use them to control your lighting from a single point or many, you need to rely on the manufacturers' instructions.

Aside from this sensitivity to wiring problems, dimmers often can be used to replace standard switches, in older houses as well as newer. Standard single-pole or three-way dimmers normally follow the same logic as single-pole and three-way nondimming switches. This means that in dealing with normal types of dimmers, you can remove, evaluate, or replace a single-pole or three-pole dimmer as you would a nondimming single pole or three pole switch.

There are functional disadvantages to using three-way dimmers (except in the case of some electronic versions). These inconveniences get worse if you substitute three-way dimmers for both members of a pair of three-way switches. The problem

CAUTION!

Do not substitute a dimmer for a switch that controls a receptacle. Dimmers are for hard-wired lights, exclusively. Also, most fluorescent lights don't work with dimmers.

is this: suppose you turn one dimmer way, way down, to the point where the light appears to be fully or nearly off. The reason for having a second switch, dimmer or no, is to be able to turn the light on and off from the other location. Unfortunately, most people will forget to turn a dimmer back up when they leave the area, especially if it is the type that has a knob that you rotate to dim, but can push to turn off. When this is the case, the other three-way switch is useless for turning the light back on. You have to walk across the dark area to the dimmer—presuming that you remember that it is the reason the light is off. This problem is doubled when you pair three-way dimmers. In some cases, you may have to walk back and forth.

One more point with regard to three-way dimmers. Some variants, the toggle-type for example, operate as plain on-off switches some of the time. A toggle-type three-way will dim only when its handle comes up from the full-off position. This strikes a trade-off between its dimming and its three-way benefit.

If you go for "the high-priced spread," a premium electronic dimmer, you may get a device that is better at withstanding shock, and can select other features as well. These include multipoint dimming, remote operation, and assorted presettings.

Differences between Evaluating Ordinary Switches and Dimmers

Much of the procedure for evaluating possibly defective dimmers, and for testing replacements, is the same as that for switches. There are, however, a few changes to ensure that your tests give you correct information.

Dimmers do not have handles that you simply flip between two positions without any change of procedure. For accurate testing and wire identification, you need to proceed as follows to mimic flipping a normal switch's handle between its two positions:

- The dimmer with a handle that you flip or slide through a range of positions: make sure that you toggle it or slide it to the full-ON and full-OFF.

- The dimmer that simply rotates from completely off to completely on: rotate it to full-on and to full-off, when it clicks.

- The dimmer that you push-on and push off; (rotate it to vary brightness; if it's giving light, click it and the light goes off): make sure that this dimmer stays rotated fully clockwise. You will have to push even a single-pole push-on rotary dimmer twice to determine whether it works, since there are no markings to tell you whether any push-on dimmer is in the ON or the OFF position.

- The dimmer that has a small slide that varies brightness, and an on-off switch: slide to the position that should be brightest, which should be all the way up or over.

Steps 3 and 8   Removing Conductors from Dimmers and Reconnecting Them

Many dimmers have leads, permanently connected small gage flexible conductors, rather than terminal screws. Therefore, you will follow the procedures for splices, rather than terminal connections. The leads of a modern three-way dimmer are color-coded, enabling you to differentiate the Common from the Travellers.

Steps 1–10   Replacing a Dimmer with a Switch. To replace dimmers with switches, follow the procedures I detailed earlier in this chapter for replacing switches.

Variants on the Basic Type of Dimmer

There are several types of dimmers that not only operate differently than others, but may be hooked up differently, too. This means that you may not be able to evaluate or replace them by following the procedures I described. These include:

- The dimmer that you tap repeatedly to move it through its range. (You may be able to test this type after tapping it till it has adjusted as much as it can.)

- The dimmer that you keep a finger on to make it move through its range. (You may be able to test this type after holding a finger on it till it has adjusted as much as it can.)

You can run into difficulties with these dimmers. Try to obtain manufacturers' instructions, if you're having trouble with one. If these are not available, examine the connections; if they seem at all different than those I have described, you may have to replace the dimmer with an identical substitute if you want it to function the way the present one is wired. You may well want to call in a pro, but even a pro may be at a loss without manufacturers' instructions, and may be unable to give you similar control except by installing an identical or very similar unit as its replacement.

If you want to install one of these in place of a more-standard type of switch or dimmer, be aware that the manufacturers' diagrams may well not reflect the way your system is set up, which means that the instructions may not work with your system. The most common variation from how you install a standard dimmer is that these fancier electronic ones may require you to differentiate between the upstream conductors feeding it and the downstream conductors going from it to the leads or terminals of the light fixture. Another variation may require you to attach a neutral.

GFCI Switches

GFCI switches are used for the same purpose as GFCI receptacles: to protect from shock. All the installations I have heard of use them to protect downstream loads, though they do provide some protection at the switch locations as well. I am familiar only with single-pole versions.

If you install one of these to replace a switch, you need to identify the Line wires, those coming from upstream, following the procedure I described in Chapter 14. Then follow the manufacturer's instructions. Unfortunately, in some switching arrangements, you may not *have* a pair of Line wires, in which case you may have to find another location, perhaps further upstream, to install the GFCI protection you seek.

Other Less Common Devices and Arrangements

Stacked Switches

It's not rare for multiple lights to be controlled independently from one location. Usually this is done using a multigang box containing several switches side by side. Whether a conductor coming in to a switch you want to replace comes from a pigtail, or is daisy-chained from a second switch; whether it leaves the box in its own cable or raceway, or shares one with another switch; whether you are replacing only the

> Stacked devices incorporating receptacles often present a special problem. See "Receptacles at Switch Locations," later in this chapter, for details.

one switch, or several, you can follow the procedures I explained earlier in this chapter.

Sometimes, though, a "stacked switch" is used. You may recall Despard devices, which allow one yoke to hold a combination of up to three devices—switches, receptacles, or indicators. You may have occasion to replace one of these. Nowadays, combination or "stacked" devices are used to perform the same function. They normally contain only two devices, but you can obtain a stacked device containing three. Both the two-switch version and the three can be purchased for use with a Decora cover, the type of cover with the same rectangular opening that covers a GFCI receptacle. The more standard stacked device design, though, is covered with a standard duplex *receptacle* cover.

Wiring Stacked Switches

A stacked switch normally has one side with independent screw terminals, and one with two linked screw terminals. I have installed stacked switches incorporating combinations of

- Two single-pole switches.

- Two three-way switches.

- One of each.

I also have replaced Despard devices that incorporated three switches.

The loads are connected to the separate terminals. (In the case of a three-way switch, a pair of Traveller terminals takes the place of one of the separate terminals.) If the two switches on the device are fed by the same conductor, this is terminated on the side with linked terminals, and they are left connected. On the rare occasions when unrelated conductors feed the two loads, the tab linking these two terminals is broken, and each of the two switches on the yoke is connected separately.

The procedures for removing, evaluating, and replacing stacked switches are essentially the same as those for individual switches, except that everything is more crowded. However, if you are replacing a stacked switch, check to see whether the link is broken out. If so, should you be installing a replacement, break out the link on your new switch. Be especially careful not to mix up the conductors as you transfer them from the old switch to the new.

Switches Incorporating Indicators

Yet another type of stacked device consists of a switch and an indicator light on the same yoke. The light is used to remind you that the load controlled by the switch is turned on. This is convenient in instances such as where a switch controls a remote load. If a switch located in the hall below controls the attic light, noticing that the indicator is lit could save you from leaving the attic light on for weeks or months. However, to install one in place of a standard switch may require that you have suitable circuiting, a task that I will not cover for the reasons I explained at the beginning of this chapter.

Lighted handle switches, which come in both single pole and three-way versions, work the opposite way. Their handles are lit when the loads they control are *off*. This can be convenient to help you find a switch's handle when the room lights are off. These are wired the same as normal single-pole and three-way switches, and so can be substituted readily.

Timers

You may need to replace a timer, or you may choose to use a timer to replace a switch or a switch to replace a timer. Many timers fit in outlet boxes and are wired like standard switches. Do check the rating when you choose a replacement timer, however, to make sure that it is suitable for the load it will control. A heavy-duty type, typically used to control motorized appliances and even water heaters, may sit on the surface of the wall rather than in a switch box, and its wiring may differ considerably from that of a standard switch or timer.

If a heavy-duty or "time clock" type timer goes bad, you may well be able to purchase a replacement mechanism from the manufacturer. This would enable you to restore operation with little mess. The enclosed instructions will tell you how to remove the old mechanism, and then you usually will have little more to do than swap the wires, one at a time, from screw terminals on the old one to the equivalent terminals on the new one. Be very careful to make sure you have killed power; these may carry a greater risk of shock and other electrical hazards.

Sensors

Generally, you can install a light and motion sensor, or a sound-activated switch, as a direct replacement for a standard single-pole switch. This can be handy in an older home where the builder did not install three-way switches.

Occupancy sensors also are touted as energy-savers; lights are on only when someone's in the room. The most simple-minded one is sound-activated. Its sensitivity can be set so that the lights will go on or off if you clap your hands when you enter the doorway, or simply in response to your footsteps. Oversensitivity could be irritating—the sensor might go on and off repeatedly as you sit there, or not go on in response to a reasonable level of sound. A more sophisticated occupancy sensor combines a light sensor with an active infrared or ultrasonic transceiver. Both versions have manual overrides. Infrared sensors, by the way, can't see around corners; ultrasonic can. Consider that factor if there are architectural barriers in the space whose light you're controlling. Note that occupancy sensors may turn the lights off while you're in the room, if you don't shift position for a few minutes.

Humidistats are relatively rarely installed, but easily available and straightforward substitutes for single-pole switches. A humidistat will automatically turn on a load such as an exhaust fan when air gets too moist.

⚡ WARNING

The older a fuseholder, the more likely that you may be exposed to live parts when you approach it, either because it was manufactured before modern designs developed or because safety barriers are missing.

Fused Switches

Horsepower-rated switches designed to control equipment such as air conditioning compressors rarely go bad. Some, however, contain fuses. When the fuses blow, be prepared to replace them in accordance with the procedure in Chapter 7.

Other Control Systems

The remaining types of home switching vary in terms of how complicated they are, how easy to troubleshoot, and how dangerous. While generally they are not difficult to repair or replace, I do not have the space to cover them adequately. Given that work on each one can expose you to 120 volts, and, in four cases, to 240 volts, I recommend that you do *not* deal with them by extrapolating from other repairs.

- A knife switch, a switch that was designed with exposed live parts, normally is used as a disconnect for 240 volts. Even if this were not the case, it is a dangerous item. It is highly unlikely that you have one, even in an old house, except perhaps as the disconnect of an old appliance such as a furnace. There are special issues associated with knife switches' wiring, but, basically, they are too hazardous to keep around, and need to be replaced. I recommend against even trying to replace a fuse if you have a fused version. If you even suspect you may have one, check with a pro.

- There's another switch besides the "four-way" switch that has four terminals: the "double-pole, single-throw" switch, used to control 240-volt loads. It has marked ON and OFF positions; a four-way (or three-way) does not. It is not likely that you will find one in your house. Still, if you open up a switch box expecting to find a single-pole switch, because it has ON and OFF markings, but you find one with four conductor terminals, that's what it is; replacement is just slightly beyond the scope of this book.

- Although most thermostats serve power-limited systems that should not be able to hurt you, some thermostats carry 240 volts.

- Power-line carrier technology is the first of two switching systems that allows you to control any outlet you wish from a master control or controls. It superimposes control signals on the regular power line, somewhat like a broadcast engineer imposes a music signal over what otherwise would be a "dead air" hum. An X-10 installer (X-10 is the best-known brand) uses rather expensive modules in place of switches or receptacles, to receive these signals. Usually the modules go bad. Warning: These are 120-volt equipment.

- The second unusual switching design uses low-voltage wiring. Skinny, lightly-insulated wires run from push-button switches to relays. These relays go bad. Warning: While the relays are operated by low-voltage, they switch 120 volts.

Receptacles at Switch Locations

There are a number of designs where it is safe and legal to add a receptacle at a switch location, or vice versa. Unfortunately, bootleg grounds, which I mentioned in Chapter 14, are very common where a receptacle has been added to a switch. For this reason, I suggest that beginners don't touch stacked switch/receptacle arrangements, unless it is pretty clear that they are original, or at least unmistakably professional, installations. Combinations in ganged gem boxes, and switches and receptacles combined in Despard devices, are far less likely to have been installed by someone ignorant of safe wiring rules. With exercise of due care, you can evaluate and replace switches at these locations the same way you would switches at any other. You may find side-by-side switches and receptacles in jury-rigged "two-gang" boxes, by which I mean single-gang boxes haphazardly wedged together with their wiring illegally joined; dealing with anything jury-rigged is well beyond the scope of an introductory text.

16

How to Replace a Light Fixture or Small, Hard-Wired Appliance

Replacing lights and small appliances carries its own dangers. After most consultations where I identify errors involving receptacles, I warn most strongly about shock hazard; when I see problems with lights, I talk about fire. (Small appliances? Usually shock.)

Replacing lights definitely tends to be more difficult than replacing receptacles. This is because:

- Their wiring tends to be more complicated.

- Their boxes tend to be far more crowded. An electrical box at the center of the ceiling is conveniently located to serve as a junction box for lines going out to other outlets and to switches. If your house dates back to the early twentieth century, the ceiling light may have started out as the only electrical box—since its switch was a pull chain—in the room.

- The wiring at ceiling lights tends to be the most worn of all, especially in old houses. This is because:
  - It is far more difficult and messy to run cables through ceilings than through walls at renovation time, so a lighting outlet in the center of a ceiling often is left out of an upgrade.
  - Heat rises, as a result of which the conductors tend to suffer accelerated aging.
  - Every error that could result in overheating at a receptacle has the same result at a light—and lights suffer from a number of additional, very common sources of overheating such as overlamping.

- The fixtures themselves are mounted in quite a few ways, and replacement fixtures very frequently are mounted differently from the old ones. This chapter has many, many more pictures than the chapters on replacing switches or receptacles, simply to illustrate a few of the widely varying systems you may encounter.

- The fixtures themselves constitute the "cover plates," forming part of the enclosures, and the designs of replacements very frequently change the shape of and, more importantly, reduce the room in the wiring compartments.

And then there is the simple fact that most lights are 8–10′ above the floor.

Some Dangers of Ceiling Work

There are a number of special risks associated with ceiling work, and hence with the installation and replacement of most lights. These include the dangers associated with ladder work. Protect your body! Besides avoiding falls, beware of the risk of being struck by pieces of the light or of the wiring. All sorts of rubbish may drop on, poke towards, or fly out at you, so wear goggles and clothing that covers you. Cover the floor and furniture so that you don't damage them or leave debris that could injure someone later.

Other risks are especially associated with older ceilings. Ceilings have two reasons to suffer more physical deterioration than walls. First, gravity is working against ceilings far more than against walls. Second, except on the top floor, people are walking above the ceilings, shaking them.

Types of Lights

Before I begin to talk about how to replace lights, I will go over some basic design concepts.

- A normal incandescent fixture puts 120 volts through a light bulb. Feed it AC, DC, odd wave forms such as dimmers produce, even somewhat higher or lower voltage; it still will glow. This is one of its main advantages over other light sources. It is the oldest type of fixture you might find in service in your home. The principle does not change whether the bulb is the standard pear shape (the "A" bulb); round (the "G" shape), a spotlight, a floodlight, or other shape; a full-size ("medium base"), larger ("mogul," for example) or smaller ("candelabra," for example).

- A halogen lamp that screws into the same socket differs minimally, except that it gives whiter light and operates slightly more efficiently. It is heavier, though; make sure your fixture can handle the weight.

- A low-voltage fixture incorporates a transformer to produce 12 volts, which it puts through a corresponding light bulb. This is used indoors to operate some very narrow-beam spotlights, such as occasionally are used for theater, or to highlight art. Outdoors, it is used to provide landscape lighting that doesn't pose the shock hazard associated with 120 volts.

Here's a previously unmentioned factor in any work on older plaster walls and ceilings, but far more of a concern in the case of ceilings. Plastered surfaces consist of three elements. Directly under the primer and paint is a thin coating of plaster—actual Plaster of Paris (the mineral gypsum). This is held in place by a thicker layer of sand mix—a material with a high proportion of sand, making it coarser and less hard. The sand mix is laid over, and pressed into, lath— either a metal mesh or, in older times, strips of wood. It is held to the lath by "keys," the sand mix that oozed through as it was trowelled on. Over the years, the keys break. When too many keys break to support the weight, chunks of plaster come down. You face this risk whenever you work on an old, plastered, house. Hammering a nail can cause this; tugging on cables, or removing or replacing boxes, can make it even more likely. If you're lucky, of course, you'll find gypsum wallboard (drywall). It was used as early as 1921.

- All standard nonincandescent fixtures utilize "electric discharge." Rather than heating wire filaments until they glow, they illuminate by using electric arcs—yes, quite like the early 20th century arc lamps—to stimulate chemicals to phosphoresce. Electric discharge fixtures are far longer-lasting than incandescents, and more economical in their use of electricity. Fluorescents are the least efficient of these. However, the other types tend to have components operating at more than 1000 volts, and therefore are not permitted at your house.

Reasons to Replace a Light

There are several good reasons to replace a light fixture. One is purely aesthetic; if you don't like its looks, the only reason not to replace it is the one given at the beginning of Chapter 13. Doing anything at all to wiring that's on its last legs can push it over the edge to the point where you need rewiring. Rewiring is a mixed blessing: you're not relying on marginal wiring, but you are facing mess and expense. One reason is somewhere between aesthetics and function: if a plastic lens or cover has yellowed over time, it may be not only visually unappealing, but also less useful, because it will let less light, and less white light, through. Also, that cover may be on its way to cracking—and coming down.

There are plenty of reasons to replace a light that have to do with functionality. The first and most obvious is that a fixture has stopped working. It is due for replacement or rewiring—*if* it is the source of the problem. You need to check the circuit and the switch or switches. First of all, confirm that the bulb or bulbs are good. With most older multitube fluorescents, you need both tubes of a pair to be functional for the light to work at all; and confirm that the branch circuit's hot and neutral conductors are securely spliced to the ballast's leads. You also have to decide that the fixture is not worth rebuilding; most fixtures can be, if the only problem is that standardized parts, such as sockets or ballasts, have deteriorated.

Safety

There are many reasons for replacement that are related to safety. Some take a little investigation.

Examine all visible parts of the fixture, most especially the fixture wires. If you find brittleness or darkening, it's time to put in a new fixture. Any visible damage at all may mean that it should be replaced.

Missing Parts

If a part is missing, whether it's a lens or an apparently decorative part, the fixture probably should go—unless the part can be replaced. Two examples have to do with outside lights. If either is incomplete, rain can get in to shatter the lamp. A "carriage light" type fixture, consisting of a metal frame and three glass panels, often loses a panel. A jelly jar type fixture has a canopy on top and a glass cylinder (the "jelly jar") below that screws into it or, less commonly, is held to it with three setscrews. Occasionally the jelly jar goes missing, leaving the lamp wide open to the outdoors. A more subtle problem is that

A 4-inch round cover rosette-pendant adapter, from below. Sliding the slotted hole over one of the box's screws helped support the fixture temporarily while the conductors were spliced. Then it was swiveled so that the other opening lined up with the second threaded hole and the second 8-32 screw inserted; then both screws were tightened. To remove, reverse this process.

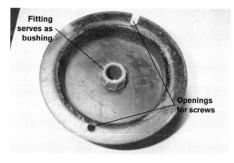

A 4-inch round cover as a rosette-pendant adapter, from above. Note that the box's 8-32 screws directly supported the pendant (cord-hung) fixture this mounting served.

this type of fixture often has a finial or cap nut at the very top. That can go missing easily without anyone's noticing. Unfortunately, this means that water and bugs can enter.

Adequate and Suitable Light

Another reason to replace a light is to cut down on glare. As we get older than 40, than 50, 60, and 70, we need even, bright light, and less glare. "Even" light means light that doesn't leave shadows.

Another reason you might want to replace a fixture is to get more light. If it's not bright enough, though, why not just screw in brighter lamps? Wrong move: A brighter light bulb, drawing more watts than the light fixture is designed for, could start a fire.

How Can You Avoid Overloading a Fixture?

What's too many watts? All light fixtures are marked with the maximum wattage and the types of lamp for which they are designed. Usually this marking is right at the lampholder itself, so you can find it readily when you replace a lamp. If the marking is missing, you have to guess—unless the manufacturer can be contacted. If a lampholder has a standard ("medium base") diameter, I generally guess 40 watts, or 60 watts maximum. If its lampholders are some reduced size ("candelabra" size for example), I guess 25 watts maximum.

Safe Ways to Get More Light. Sometimes you can get more light from the same fixture without inviting fire; wattage is not the same as brightness. All of the following solutions are based on three presumptions.

- The fixture can take the additional weight to which they subject it.
- The fixture does not have a marking forbidding their use, such as "Use 50 Watt PAR 20 floodlamps only."
- The fixture can accommodate them physically, and can function with them.

If these presumptions are valid, consider these options:

- Use a halogen lamp. Halogens are brighter than standard incandescents, and are available in almost all popular configurations.

A turn-knob lampholder, one lead frayed bare—no wonder the fixture died. The support area took the heat and the twists every time a light bulb was replaced, because it was above the fixture.

An old but functional turn-knob lampholder; the leads look OK.

This lantern-style outdoor sconce is missing glass, which means that rain hits the lamp, so it gets wet and shatters. If the glass cannot be replaced, or exposure has deteriorated the fixture, you will need to remove the fixture.

As with most fixtures lacking separate canopies, the back wall of the fixture covers the outlet box, completing the wiring enclosure. To change the fixture, unscrew the single cap nut, which holds the fixture to the outlet box. Pull the fixture away, and you have access to the splices.

- Cast more light down, at the same wattage, by using spotlights or floodlights.
- Install a compact fluorescent adapter, which could save electricity and still be brighter than what you had. Compact fluorescents, however, have a drawback that may or may not be important in your case: unlike the preceding varieties, all of which are incandescent, a compact fluorescent generally cannot safely be controlled by standard dimmer switches. (If it can, it will say so.)

Replacing the Fixture

A surface porcelain with compact fluorescent adapter. This lampholder is one of the very, very few fixtures that unquestionably can safely support these adapters.

You can choose from an incredible variety of light fixtures. If I had the room, I could go into considerable detail about how you can install and repair a representative assortment. Here I must limit my explanations to some general principles that will help you to remove and replace fixtures and, most importantly, to perform some very basic evaluation of lighting outlets to complement what you already have learned about the mechanical aspects of wiring systems. You may find that you need to review the chapters containing some earlier material, such as that on evaluating wiring, on safety testing, or on connections.

Running power to new lighting outlets in older houses, like running power to new receptacles, runs into complications so often that it doesn't belong in an introductory text. For the reasons I listed at the beginning of this chapter, even attempts to replace existing light fixtures result in calls for help much more often than attempts to replace receptacles.

Approaching the Fixture

Before I talk about probing a lighting outlet, and then opening it up, I need to reiterate the importance of safety issues.

I talked about killing power before taking wiring apart in Chapter 5, Safety. If possible, use insulated tools; gloves, too. You are far more likely to be exposed to shock when you have reason to believe you are safe as you work at lighting outlets than as you work at receptacle outlets. I detailed some of the reasons for this in *Old Electrical Wiring: Maintenance and Retrofit.*

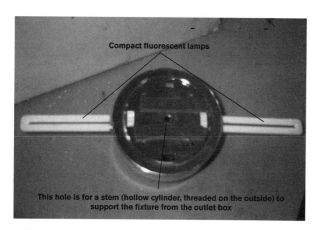

A fluorescent substitute fixture. These compact fluorescent lamps are part—the essential part—of the fixture, rather than a modification such as compact fluorescent adapters installed in fixtures specifying incandescent lamps.

Fixture Support

In order to gain access to an outlet box, you need to deal with the fixture that is in place. I will give you a brief overview of how that is likely to be supported before giving you step-by-step instructions for its removal.

There are two common ways of supporting a fixture from an outlet box or attached fitting. In the first method, either two screws or two inverted or headless screws descend from the box through holes in the fixture or its canopy and are secured by nuts, usually decorative cap nuts. In the second method, a threaded hollow cylinder, or "stem," descends from the box through the center of the fixture or canopy and is held on by a bezel ring or by a large-diameter nut. This stem may continue down, after holding the canopy in place with a bezel ring, to support the fixture shade by a cap nut. Be careful not to turn the stem when you unscrew the nut that holds the shade on, or the fixture may suddenly come down in your hands.

Before describing how you can proceed further, I again urge you to use safety equipment, especially goggles.

Accessing Ceiling Outlets

To gain access to the outlet box in order to perform your evaluation, you need to at least partly remove the fixture. Here is the sequence you will follow.

1. Kill power.

2. Remove any parts associated with the fixture that you can clear away by lifting out or similar simple means. This includes light bulbs, shades, and loose decorative pieces. This makes it safer for you to work by reducing clutter and extra weight, and reduces the risk of damage.

3. If you were able to gain access to lamp sockets at this point, recheck there for voltage.

Here is an important rule regarding all light fixtures: don't twist them. Four and only four fittings accessible from outside a fixture's outlet box are intended to be screwed and unscrewed:

- Screws, including setscrews.
- Nuts, including cap nuts and finials.
- Circumferential rings, including bezel rings and locknuts.
- Some globes, shades, and other lamp protectors that are not held in place with spring clips or setscrews.

Twist anything else and you are likely to twist the wires inside, very likely damaging them.

Straightening a sconce is likely to be harmless. Twisting a fixture, or part of a fixture, up to 180° probably won't hurt it. More, though, gets chancy.

Caution

When you have to tighten or loosen, install or remove, any fitting that needs to be screwed or unscrewed, restrain anything that its motion might cause to rotate. The most common reason that I am called in to replace or rebuild a fixture is that screwing and unscrewing light bulbs has twisted the conductors feeding them repeatedly and eventually caused them to short.

Removing Broken Light Bulbs

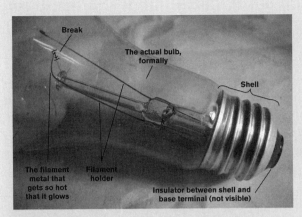

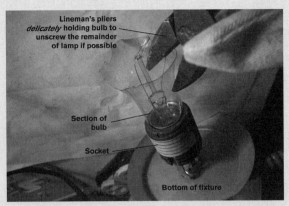

Type A lamp, removed very successfully. A break in the filament is the source of the problem. The shell *should* be connected to the neutral. It *should* be dead when the light is off; if not, touching it as you unscrew the bulb could shock you.

How do you get a lamp out after its bulb has broken? (For clarity, I'm using the formal term, "lamp," with "bulb" as the glass part, "shell" as the metal part of the lamp that you screw in, which forms one electrical contact, and "base" as the shell plus the remainder of what gets screwed in, which includes an insulator and the lamp's bottom center electrical contact.) Recognize the likelihood that glass will fly if you follow these suggestions, and protect yourself.

Whatever approach you take, make absolutely sure that power is off before you start. The only way to be certain of this is to use your voltage tester to confirm the presence of power, kill the circuit and then use your tester again to confirm that power went off. Before working on the fixture, especially before touching it, check for voltage from the ground

What the task generally involves—a broken lamp, pliers, and gloves.

prong of your extension cord reference to any part of the filament that your probe can reach to, and then from the ground reference to the shell. If you find voltage, hunt for the right circuit or circuits until you have killed power—until it's no longer possible to get a reading anywhere.

I have heard of many approaches to the problem of how to remove a broken light bulb, including sticking a cork in and twisting. Here is what I do.

1. The first thing I try is to twist out the base by delicately grabbing the glass—using my lineman's pliers, with which I have some finesse—and gently twisting; this works for me, but not very often. It doesn't cost extra effort to try this, given the next step.

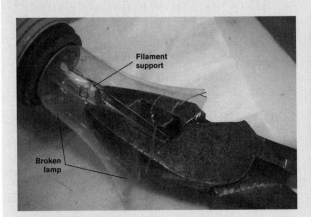

Using lineman's pliers on the filament.

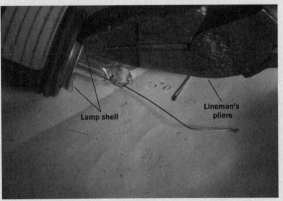

Lineman's pliers successfully twisting the shell by grabbing the little bit of lamp shell that was not blocked by being inside the socket.

(Continued)

Removing Broken Light Bulbs *(Continued)*

Needlenose pliers twisting the shell.

2. What works for me most frequently is to break away any remaining glass and then grab the shell and twist it directly.

   When I'm lucky, I can get my pliers on a bit of the lamp's shell that extends beyond the lampholder's "shell." The lineman's pliers flatten the shell slightly where I grab it, pulling its outside away from the inside of the lampholder's shell, thereby reducing friction.

   When there's nothing to grab, I insinuate a pair of needlenose pliers in between the two metal shells, trying my best not to distort the lampholder's shell when I wiggle the tip of my pliers in.

   With the pliers in place—not necessarily all the way down to the bottom—I grab the lamp's base, without grabbing the socket, and twist. A teasing rather than a brute-force approach tends to succeed more easily and lessen the risk of damage to the socket.

3. If upon doing that I can't unscrew what's left of the lamp right away, I squeeze the lamp's shell away from the socket with pliers. If necessary, I twist the pliers counterclockwise until I wrap the lamp base around them, torturing it unmercifully till it breaks free from the socket.

4. If it still doesn't unscrew, I repeat step 3 with my pliers at another point along the shell. On rare occasions, I've used a narrow chisel (or a screwdriver that I can safely employ as a chisel) to drive the lamp base in, away from its contact with the socket.

   At the worst, this distorts the lampholder to the point that I have to rebuild or replace the fixture. This may be necessary anyway, in cases where the sticking that made it hard to unscrew was caused by corrosion.

A ceiling box with a strap held by the box screw, conductors entering through a stem supported by the strap. Note that the ground wire is an intrinsic part of the strap.

4. Break any adhesion, such as is often created by even a single layer of paint, between the fixture or canopy and the ceiling, using a knife (or other tool that works for you).

5. Now free the fixture or canopy, without letting it drop or hang from the wires.

In two common designs, the fixture is attached directly by screws. Supporting the fixture, loosen each screw a turn or two. Now, if the heads are in keyhole slots, rotate the fixture a few degrees so that the heads are in the wide part of the slots, and let the fixture down enough that you can see the inside of the outlet clearly. If the heads just come down through the fixture, without keyholes, remove one screw and put it in a pocket, pouch, or other place where it is safe but can be retrieved easily. Then, supporting the fixture, remove and store the other screw, and let the fixture down enough that you can see the inside of the enclosure clearly.

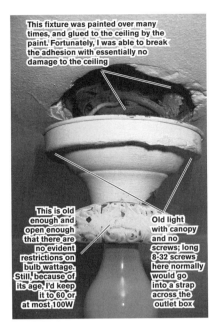

This fixture was painted over many times, and glued to the ceiling by the paint. Fortunately, I was able to break the adhesion with essentially no damage to the ceiling

This is old enough and open enough that there are no evident restrictions on bulb wattage. Still, because of its age, I'd keep it to 60, or at most 100W

Old light with canopy and no screws; long 8-32 screws here normally would go into a strap across the outlet box

A fixture canopy separated from the ceiling, giving access to the wiring in the outlet box.

If spliced branch circuit conductors separate from each other, or any conductor breaks at any point during these procedures, you need to check each conductor for voltage against a known ground.

In other cases, the fixture is suspended from a stem that descends through a canopy. There will be a bezel ring at the bottom of the canopy. Being careful not to unscrew the stem—but being prepared to catch the fixture in the event that the stem does unscrew and let the fixture down—unscrew the bezel ring as far as you can. Normally you can back it right off the stem. If you are working with a chain-supported chandelier, bring the bezel ring further down, over the top link or two, being careful not to scrape conductors' insulation. Now drop the canopy away from the outlet box, over the stem and first few links of chain, down to the bezel ring. The stem should continue to support the fixture.

If what you see in the outlet box at this point suggests that further work on the fixture is beyond your present skill level, you now can reverse steps 5 and 2, unless what you find looks so hazardous that it seems safer not to disturb the fixture further until you've had a pro evaluate the wiring. If you are ready to continue with removal, this is how you proceed:

6a. Look at the fixture wires. Normally they are color-coded, the hot one black and the neutral white, and the rare ground green or bare. Sometimes, though, they are coded by "polarity" more subtly: the hot is smooth and the neutral has a raised ridge, or one has printing on it and the other doesn't.

6b. Now look at the conductors they are spliced to. If the fixture conductors are coded, by color or otherwise, and the colors of the branch circuit conductors don't match the fixture conductors' coding, someone made a mistake. Most likely, the mistake was on the part of the fixture installer, who ignored polarity; when I find this, it does not dismay me, so long as the splicing was done well.

6c. If your fixture conductors are coded by polarity, but you can't distinguish any colors on the conductors the fixture wires are spliced to—usu-

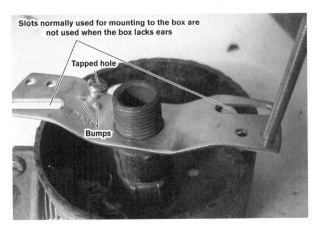

Slots normally used for mounting to the box are not used when the box lacks ears

Tapped hole

Bumps

A fixture support strap being installed over the center fitting of box lacking mounting ears. Note that the long screw is threaded into strap backwards (with its head hidden, facing towards the box). It will be used for fixture mounting. This old box lacks ears; extensions of its center fitting will be used to support the strap, which has a tapped hole for the ground screw and bumps to help keep a ground wire in place.

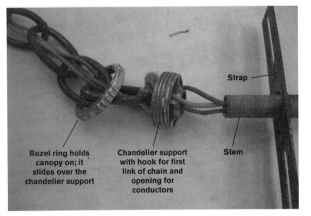

Bezel ring holds canopy on; it slides over the chandelier support

Chandelier support with hook for first link of chain and opening for conductors

Strap

Stem

Chandelier support (canopy left out).

I find channels to be the most useful tool for unscrewing nuts and rings that are on too firmly for my fingers to spin. Whatever tool you use, be careful not to scrape the finish of any fixture that you plan to keep.

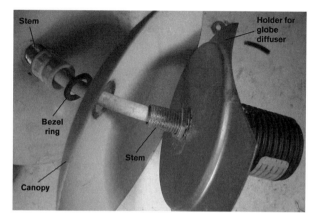

A pendant, modern, opened to show stem, bezel, and so forth.

ally this is the case because the branch circuit conductors are old and faded—make a mental or paper note of which splice is connected to which fixture conductor.

7. Unless the conductors seem too fragile or too tangled for you to do so safely, move the splices connecting to the fixture conductors—leaving alone any splices at the outlet that are not connected to the fixture—to where access to them is as clear as possible.

8. At one of these two splices, unscrew the wirenut, or unwrap the tape, far enough to expose copper, and test for voltage. Then undo the second splice and test for voltage. If you find voltage, stop, go to your electrical panel, identify the circuit, kill power, and confirm that you have done so by rechecking at the splice.

9. Now, if the polarity of the fixture conductors was color-coded but the branch circuit conductors that they were attached to were not, you can code the branch circuit conductors, one with black electrical tape and the other with white. Warning: if either the circuit or the fixture was miswired, this taping may not result in correct coding.

10. If the fixture included a ground wire, undo its splice or termination.

11. Take the fixture down.

11a. If the fixture was held in place with a pair of screws or nuts, you now are ready to put it down somewhere safe.

11b. If the fixture was supported by a stem, straighten out the wires coming up through the stem from the fixture. If you can do so, pull them back down till just a little, maybe half an inch, remains peeking out of the stem. This is to prevent the wires from causing problems as you proceed. Now and only now, unscrew the strap, if the stem was threaded into a strap, to disengage the fixture. Otherwise, now and only now, rotate the stem and fixture together, counterclockwise, until the stem has unscrewed. Having totally disengaged the fixture, lower it and put it down somewhere safe.

Evaluating Replacement Possibilities

Is the lighting outlet properly installed and in good condition? Is it suitable for the type of fixture you want to install?

The first questions to deal with are how the fixture is going to be fed and what's going to hold it up. Answer these before you buy a new fixture. I'll talk about aspects peculiar to lights and some issues regarding ceiling outlets supporting paddle fans. This adds to rather than replaces the information in Chapter 13 about how to evaluate, and when possible, repair boxes, cables, clamps, and connectors.

Structure

What's going to hold the fixture up? First, I'll talk about the support provided for outlet boxes by your structure—usually by joists. I'll talk about supports in the fixtures themselves later in the chapter.

Is the box (or the fixture, in the case of a fixture designed for direct support) that now is in place screwed to the underside of a joist? Great. Unless the hardware or the structural member is very shoddy or rotten, support will be no problem, at least if your replacement fixture is designed to take the same support hardware. In an old, plaster-on-wood-lath ceiling, sometimes the lath itself—narrow, skinny, dried-out wood—is used for support. That's far less trustworthy. You will have to judge how much weight the outlet can support and choose a replacement fixture accordingly—or have the box replaced or its support improved, perhaps with some aftermarket device.

Very commonly, a steel "hanger bar" (most commonly stamped out of sheet metal) is used to support a box. This is fine for supporting most lights—*although paddle fans require more*—so long as the bar is nailed or screwed to the joists on either side. Sometimes, though, these hanger bars, or even more marginal hardware, are simply laid on top of the ceiling drywall or the lath. This is inadequate; boxes supported this way should be replaced or remounted.

The Box

Is the box suitable to support the fixture? A switch box, for instance, is not intended to be supporting any fixtures below it; it might do for a wall sconce, which doesn't pull straight out from the box, but not for a ceiling fixture. If a switch box supported your old ceiling fixture, have the wiring checked out by a pro; you have found evidence of ignorant work, which might or might not be hazardous.

Paddle fans can be wired the same as lights, if manufacturers' instructions permit, with one exception. Do NOT use a dimmer to feed a fan motor. If you want to vary its speed, use a fan speed controller. It may look quite the same as a dimmer, but it is designed to handle motors. If you install a paddle fan with a light kit, you cannot dim the light unless you have installed a system that allows you to control it separately from the motor.

No ceiling box not explicitly manufactured to support a paddle fan is considered suitable for the purpose, so if you are considering replacing a light with a paddle fan, or, in order to maintain general room lighting, with a fan-light combination, you need to find some better support. Fans have fallen. If your present box is right under a joist, though, you can run screws right past or through the box from the fan's hanging hardware into the joist. Otherwise, you have three choices: have the box replaced; securely extend wood from your joists over the box so you can screw into it; or investigate aftermarket fan supports. The latter two options, I should note, become difficult to employ when cables emerge from the back of the box. The cables may block the space where you want to run the support.

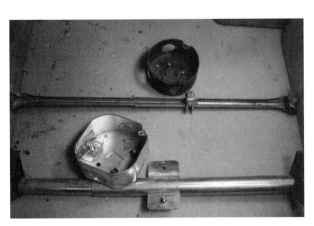

The stamped metal adjustable hanger bar traditionally used to support boxes for lighting, such as the round black enameled box, are much less sturdy and rigid than the type of system required for paddle fans, shown in front.

Is the box not only properly supported, but also flush with the ceiling surface (the wall surface, in the case of a wall sconce)? If it is significantly recessed, you may need to move or replace it, or to add a box extension, as discussed in Chapter 13. Slip-in extenders now are available for ceiling boxes.

Does the box have enough volume, from the looks of it? If your old fixture had a canopy that contributed to the volume of the overall enclosure, will the new fixture reduce it? Look through NEC Section 370-16 and calculate how safe what you have in the ceiling will be if you install the fixture you want.

People often overlook one nuisance of a requirement when selecting fixtures. When the fixture's canopy is wider than the outlet box, the wiring enclosure is incorporating some ceiling or wall. Therefore, if that surface is combustible—like wood paneling or fiberboard indoors, or wood or vinyl siding for outside lights—it has to be protected with something noncombustible. Protecting the transition space usually means cutting a piece of sheet metal to bridge between the outlet box and the edge of the canopy.

Surprisingly, UL Lists some foreign fixtures whose canopies or cover plates are inadequate to cover the outlet boxes. It is your responsibility to bridge between the box and the canopy/cover plate with sheet metal or with a patent adapter. Reducers are available for the most common fixture support box, the 4″ round or octagonal box, bringing its opening down to a 3″ diameter. Even if you find a suitable reducer, so you're not filling in with sheet metal or something equivalently non-combustible, you may find the transition from the edge of the canopy to actual ceiling ugly, even when painted. When it is necessary to use sheet metal, moreover, I don't like the sharp edges that can cut my fingers, or cut the insulation on the conductors, and short them out.

To improve the looks of these adaptations, you can install ceiling medallions of plaster or plastic that fit around the fixture and hide the metal—or any other unsightliness. Note that using one of these does not exempt you from the requirement to repair ceiling damage around the box or to bridge across any combustible material—it just takes care of the aesthetics, when your repairs would be unattractive.

Wire Insulation

The next questions have to do with the wiring. Insulation temperature rating can be a real problem, even beyond the question of insulation condition, which I addressed in Chapters 12 and 13. Most conductors manufactured before 1975 or even 1985 had insulation rated for 60 or 75°C. Almost, but not quite all, surface-mounted incandescent fixtures being sold today require 90°C-rated insulation. Pendant fixtures such as chandeliers are less likely to have these restrictions because their lamps' heat does not affect the wires in their outlet boxes. Fluorescent fixtures usually carry this restriction only within the vicinity of their ballasts, whose leads connect to the branch wiring. The restrictions are permanently marked inside fluorescent fixtures, but may appear only on the packaging of incandescents.

The temperature restrictions on installation instructions refer to the *building wire,* the permanent wire of the branch circuit that feeds power in from your house's system. It is not at all unusual for the fixture to have *internal* wiring that is marked 105°C. Some non-electricians are misled by this. The rating of the fixture's own wiring does nothing to help you meet the restriction on the wiring to which you may connect it. (I talk more about the issues regarding the temperature ratings of conductors, and describe a specialty product for upgrading their ratings, in *Old Electrical Wiring: Maintenance and Retrofit.*)

Mechanical Connections

The next issue is how the old fixture was designed to be fed, whether this aspect of its installation was correct, and how replacement fixtures you are considering are designed to be fed.

Some fixtures are designed to be fed directly through knockouts, with power normally arriving through cables that come from the switches controlling the fixtures. Others are designed to be installed over outlet boxes. Some are adaptable for either use. Be sure to buy a replacement fixture that is appropriate for the arrangement in your ceiling; changing over from one system to another is a task beyond the scope of an introductory-level book.

If you buy a fluorescent fixture, for example, that is designed to be wired directly with a raceway or cable, and only directly, you will have a tricky time installing it under an outlet box and feeding it from the box. Even if you manage to do so, it certainly won't be safe. If you need to install a fluorescent fixture under an outlet box, choose a fixture designed with a knockout in the back that is nearly big enough to fit a hand through, so that you can mount the fixture under, and feed it from, your box and retain access to the box interior. If you buy lighting track, you can get either a cable or an outlet box feed fitting. (Some tracks are sold with both fittings.)

Supporting Your Replacement Fixture

The final issue is whether what you have in your ceiling outlet matches up with how the fixture you want to install is designed to be supported.

All modern ceiling boxes (and box fittings such as mud rings) designed to support light fixtures have two ears diagonally across from each other, tapped for 8-32 screws. Some fixtures simply are designed to screw directly to the tapped ears of a box.

Most likely, your new fixture will come with a strap, a piece of roughly rectangular flat steel about 5″ long, designed with two purposes. One is to allow the conductors to get past it without obstruction. A second is to straddle the two tapped ears so it can be screwed to them, attaching it to the box, and in turn to provide a means for the fixture to be attached to the strap. The strap will have lengthy openings that allow the screws passing through it to be varying distances apart, depending on whether it straddles a 4″ octagon box, a 3¼″ or a 3½″ box, or a mud ring.

The strap's design will correspond to the fixture's mounting system. The strap may have two or more 8-32 holes lined up opposite matching holes on the canopy. This strap is designed to hold headless screws or reversed screws.

If the fixture uses a stem for mounting, the strap provided has a tapped hole in the center whose diameter corresponds to that of the fixture's stem.

Sometimes a modern strap also has a tapped hole marked "ground," intended for terminating a grounding conductor rather than for mechanical support.

Sometimes, though, the box wasn't designed to take a strap.

Some very old round boxes, usually pancake boxes, have no ears. Instead of ears, this type of box may have a threaded, essentially solid metal fitting coming out of its center, sometimes with extenders screwed onto it to bring it forward.

In boxes lacking mounting ears, the support systems can involve more pieces. In some cases, a stem was threaded onto a hickey, a fitting that was screwed onto the solid center fitting. The hickey both adapted the diameter of the solid center fitting to the generally narrower diameter of the hollow stem and allowed any con-

Insulated Spaces

When installing an incandescent fixture under an outlet box in a cavity containing thermal insulation (usually, but not always, this means an attic or the space over a cellar) keep the insulation away from the outlet box so heat from the fixture can dissipate. Some do this by cutting a rectangular strip of sheet metal and forming it into a large circle around the box. Others surround it with a large, empty can such as a five-pound coffee tin with the bottom cut away.

A reminder: Beware of any very old electrical box that is supported on a capped pipe. It might be a gas pipe that originally fed the old gas light, and it might have gas in it. Don't open it. If you do, inadvertently, close it again tightly. House gas is not under very high pressure, but if allowed to leak, it can explode. If in doubt, call a plumber or your gas utility.

As I mentioned earlier in this book, any pancake box, and almost any round box whose diameter is smaller than 4″, has to be significantly undersized if it is fed by more than one cable. This does not necessarily mean you must replace it, but be especially alert to the condition of the conductors, and be especially careful not to pinch or crush them during reinstallation.

ductors coming through the stem to exit through a side opening before they ran into the solid metal of the center fitting. Less commonly, a strap that had a large-diameter opening in the center and 8-32 holes toward the outside was secured over this center fitting with a locknut or locknuts.

If the strap or stud is appropriate for your new fixture, you certainly can keep using it as long as it is securely mounted.

If the box has tapped ears you can use, and also center fittings or extenders that you can remove without chafing the insulation of old conductors, without removing the support for the box itself, and without risking other damage, remove them to free up the room they occupy.

Always read the manufacturer's instructions, especially those that come with larger or heavier fixtures. In rare cases, heavy fixtures require support independent of the outlet box: that's a Code requirement for all fixtures weighing over 50 pounds. "Independent of the outlet box" means that some hanger must go from the fixture into a joist or the equivalent.

Grounding/Bonding. A new fixture installed with older wiring can present a problem because modern fixtures assume the use of romex. Therefore, they make provision for and expect a grounding conductor. What if your conductors are in BX or conduit? Even if the manufacturer's instructions for the fixture don't explicitly demand that you install a grounding pigtail, I advise that you ground any time a grounding means is present. Technically, the NEC requires it. Incidentally, with a replacement fixture that has a long pull chain (a paddle fan is an example), when the outlet is not grounded you should use nonmetallic chain, or at least insert a nonmetallic link.

When your test says that the box is grounded, but the design provides no solid metal-to-metal contact to ground the fixture, run a grounding/bonding wire from the box to the fixture's ground screw.

So long as your fixture is not too heavy to be supported from an outlet box, a 2-part fitting such as this one can make it unnecessary for you to splice while supporting the fixture.

Pointers on Hanging Heavy or Awkward Fixtures

A heavy fixture can be a brute to hang. First, you have to lift the fixture up to the outlet box. Then, you have to do three things simultaneously: with one hand, support the fixture; with a second hand, hold the wires together; with a third hand, screw on the wirenuts. Then, finally, you need to support the fixture with one hand while you attach it to the outlet box with your second and third hands.

Various companies offer specialty fittings like the one shown on the bottom of p. 294 to solve this problem. Using these fittings, you do six things sequentially rather than having to do any steps simultaneously.

1. Splice the leads of one part of the fitting to the wires in the outlet box.

2. Screw this part of the fitting to the outlet box.

3. and 4. Repeat steps 1 and 2 using the other part of the fitting and the fixture. (Sure, you can reverse the order.)

5. Lift the fixture, including this second fitting part, to meet the first fitting part, the one attached above.

6. With a single push-and-twist, connect the two. This joins the fixture and the outlet box electrically and mechanically at the same time.

(See my discussion of grounding/bonding in Chapter 14.) A track light, for instance, may be fed through a fitting contained within a plastic surface enclosure that fits over the ceiling box. The fitting will have a grounding screw; since the wiring system does not include a ground wire if the box is grounded by being fed by BX or a metal raceway, you need to run a bonding wire from the fitting's ground screw to the metal box.

Pointers on Securing Diffusers

Most incandescent fixtures, especially those using "A" lamps, use diffusers—light-softening shades, globes, or other glass or plastic shapes between the lamp and the room. The shades' mounting arrangements are described in the fixtures' installation instructions. There are two finicky considerations that may not be clear from the instructions, but that you need to keep in mind. First, make sure that the support is snugged up sufficiently that the glass won't drop when you let go of it. If it seems advisable, before removing your hands completely, jiggle the diffuser and make sure that it stays put. Second, don't snug the support too tightly.

Both considerations can be particularly a source of "Whoops!" in the case of semiglobular diffusers. I'm referring to those that have lips that snug up just inside the fixture bodies (in what is sometimes called the fixture's "belly band"—like a cummerbund) through which three setscrews are tightened to catch beside the lips. If two setscrews are tightened enough to grab the lip, but the third is just hanging by a little friction, the diffuser can slide out and crash down. This sometimes happens when the installer doesn't push the diffuser up into the fixture enough while tightening.

Conversely, glass has no elasticity, and the metal of the rim may have little. Therefore, if you tighten just a little bit too much, the glass may crack. For this reason, some of these diffusers have rubber

You must tighten a diffuser screw till it touches, then stop—glass has no give.

bands literally in the setscrew channel. Besides protecting the glass directly, this rubber has an additional benefit; it lets you feel some resistance when tightening the setscrews, with much less pressure against the glass.

Pointers Regarding Chandeliers

Chandeliers can be some of the more awkward fixtures to install. The first thing to do with a chandelier, as with any fixture, is to make sure it's Listed. Some non-Listed antique, imported, and artist-designed chandeliers are notorious for having undersized wire, or even underinsulated wire designed for stereo speakers, and canopies that are undersized or that do not permit safe access to the wiring compartment.

Concerns Associated with Closet Lighting

NEC Article 410 gives requirements for fixtures in closets. The options for closet lighting are very, very limited. Too many fires have been started by lights that were left on out of sight but within reach of combustibles.

Pointers Regarding Vanities

Some incandescent fixtures, for example those intended for installation above or alongside bathroom vanities, are designed rather like fluorescents in some ways:

- Some are supported directly from the wall studs, drywall, or lath, rather than the outlet box's screws, or in addition to them. You still are required to have access to the splices. Usually this feature means a two-part fixture, with the visible section discreetly but reversibly secured to the wall-side section.

- Some have long, rectangular wiring compartments.

- Some have larger, totally-enclosed wiring compartments and are designed with knockouts for cable entry instead of being designed for mounting over outlet boxes.

- Another type does mount over an outlet box, but is readily removable, and therefore doesn't require a large knockout in back for access. Instead, such a fixture may have a front and a back section with a small hole, normally protected by a grommet (a rubber or plastic bushing to protect the wires from the sharp edges of openings in sheet metal) between the back section and the outlet box (like the Circline fluorescent shown on pp. 300 and 301). WARNING: don't mistake this hole for the knockout required for direct connection. It's a different-size hole, and the fixture itself would be deeper if it were intended to serve as the sole splicing compartment.

  Grandfathering. On the subject of replacing bathroom fixtures, here's a caution about grandfathering and, more importantly, life safety. Many bathroom fixtures, both those that are integrated into medicine cabinets and stand-alones, incorporate receptacle outlets. You—or a child, or an elderly or feeble adult—are at as grave a risk from shock from such a bathroom receptacle as from a stand-alone bathroom receptacle. For this reason, when you replace one you are required to provide ground fault protection. If you cannot install a GFCI upstream, protecting the outlet, you must leave the receptacle's leads disconnected. As long as you provide GFCI protection, you can hook up the light with the receptacle leads attached.

While I don't have room to talk about outdoor wiring explicitly, it is worth mentioning one problem regarding siding, one that now may have a simple solution. If you leave a gap between the fixture's flat back and the siding's lapped and angled front, you are subjecting the fixture and outlet wiring to the same problem associated with incomplete enclosure that I mentioned above in the discussion of missing parts. Fortunately, this problem no longer requires that you cut the fixture into the siding. Adapters now are available that match the shape of siding, to bridge from the siding to the fixture.

Generally—GFCI requirements being an important exception—if a once-legal fixture is rebuildable, it is legal for you to do so. (I have left repairing fixtures for a slightly more advanced text, in part due to space limitations.)

If, on the other hand, a fixture needs replacement, a greater degree of responsibility is demanded. For example, if you had a bare-bulb incandescent fixture in a clothes closet, you can't legally replace it with a similar fixture. I would advise against even rebuilding it. On the other hand, if, by a strict reading of the rules, a lighting outlet box is slightly crowded, most inspectors will allow you to replace the fixture it supports without replacing the outlet box.

Recessed Lights

Recessed lights are an attractive option for people who don't want their fixtures to be obtrusive. When basements are converted to habitable spaces, recessed lights often are chosen because the ceilings are relatively low. Unfortunately, the people doing the upgrade or conversion may not have had your particular needs in mind when they installed the recessed fixtures. Their biggest disadvantage is that if you don't like the illumination they give, there are limits to the changes you can make.

A recessed light fixture consists of two parts. The first, the "frame-in kit," is permanently installed in the ceiling. It includes mechanical support and a junction box that is connected to a lamp socket by flexible metal conduit, rather like BX. The second part is a "trim kit" that literally hooks into the frame-in kit from below. The trim kit can be removed and replaced without damage to the fixture; it has to be designed this way because the Code requires that all junction boxes remain accessible. Several trim kits generally are Listed for use with each frame-in kit. These differ in appearance and in the type and amount of light they throw into the room.

If you don't like the light your recessed light gives, you may be able to replace the trim kit. You need to determine the manufacturer and the model of your light.

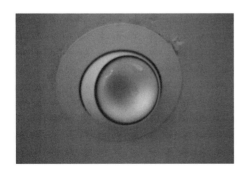

An aimable light. This trim gives an aimable recessed light; you can substitute other trim kits that would look different and give different light patterns.

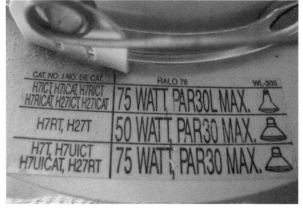

Sample lamping restrictions found inside a trim kit. The wattages depend on the specific types of shapes of light bulb. They vary by trim; that's why they are listed on the inside of the trim itself.

Unscrew the lamp, cut away any paint adhesion, and then literally pull the trim kit down a few inches till you can reach in and release a support spring, as shown in the lower right illustration. Then look inside the frame-in kit for the information. If the fixture is not too old, you can investigate other trim kits that the manufacturer's catalog says can be substituted. If you find one that will work better for you, you should be able to install it, in accordance with the accompanying instructions, without touching a wire.

If a recessed light occasionally goes off, and then comes on by itself, it's likely that a feature of the frame-in kit is protecting you by causing this—the alternative being fizzling its way to a fire. Most frame-in kits have thermal protectors that act like self-resetting circuit breakers; these are triggered not by high current but by overheating. If you do notice periodic blinking, it may be that something about the installation has changed. Have you insulated recently in the area above the recessed light? Unless the fixture is "Type IC," it needs to be kept away from thermal insulation.

Your recessed light probably constitutes a fire hazard if the frame-in kit was installed in such a way that when the trim kit is installed, the part containing

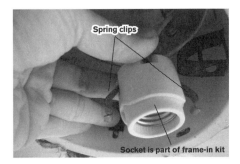

The frame-in kit in place, the lamp socket awaits trim kit. The black spring snaps into the trim kit's reflector. To insert or remove the socket from the reflector/lower trim, spring clips need to be compressed by finger pressure.

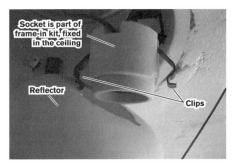

The lamp socket partly hooked into the trim kit's reflector. The reflector is the lowest part of this trim kit. It hooks to the socket with the two clips. I find hooking in one at a time easiest.

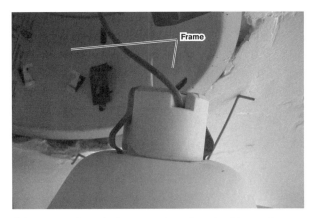

A lamp socket, fully inserted in the trim kit's reflector.

A trim kit. Note how wide the reflector/trim spring is before you squeeze it for insertion; also when up in the ceiling holding the trim in place.

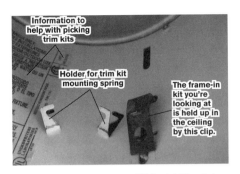

The spring clip that will hold the trim kit goes into one of these paired holders on each side of this frame-in kit.

the socket sticks up through it so that it is within ½″ of wood—even the subfloor above. If you suspect that it may have been an amateur installation, and it is located in a shallow joist space, get this checked.

Track Lighting

Tracks are frequently chosen to add light, or at least move lights, without opening up the ceiling. If you are satisfied with their looks, you gain a great deal of flexibility by installing them. You can replace the actual fixtures you screw light bulbs into, called "cans," (for canisters), based on manufacturers' instructions. "Twist and pull," commonly, is all it takes; occasionally a can is designed with a lever that latches it in place.

There's nothing special about moving or replacing cans in older buildings. Installing the tracks themselves, as replacements for other types of fixture, is a little different than installing them in new buildings. First, there are electrical issues. You need to perform the same sort of evaluation you would when replacing any old fixture with any new fixture. In this case, the canopies are quite flat, so you need to check carefully for adequate box volume, as I discussed earlier in this chapter. Second, there are structural issues. Is the ceiling in good enough shape, both figuratively and literally? Does it have the structural integrity to hold the track, and to continue to hold it securely if lights are added or removed or re-aimed? Is the ceiling flat enough for the track to lie against? Is the room square enough for a track array to look good?

By squeezing together the springs that hold the trim kit in place, you insert (or release) them.

The trim kit spring started into its holder in the frame-in kit, shown from below (room side).

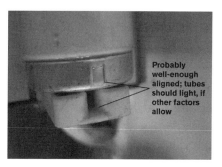

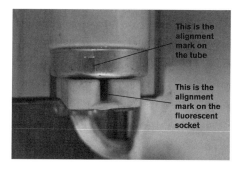

This fluorescent tube probably is well-enough aligned in its terminal.

This fluorescent tube and terminal are clearly misaligned—reason enough for the light not to operate normally.

Fluorescents

Tube-type fluorescent fixtures have been around long enough that I find them more frequently in older buildings than in newer. If you should ever notice an unfamiliar smell near one, or see a black excrescence of pitch, turn the fixture off and keep it off until its ballast, or the fixture itself, has been replaced.

Aside from this, which may or may not be accompanied by malfunction, if a fluorescent malfunctions, and power is getting through to it, check its tubes. If any one of them is going bad, it may malfunction. If you replace tubes and it still malfunctions, make sure that you have installed them correctly; see the illustration. If it has a starter, a shiny metal cylinder whose end pokes out behind or near a tube, replace that. To remove a starter, twist it counterclockwise while

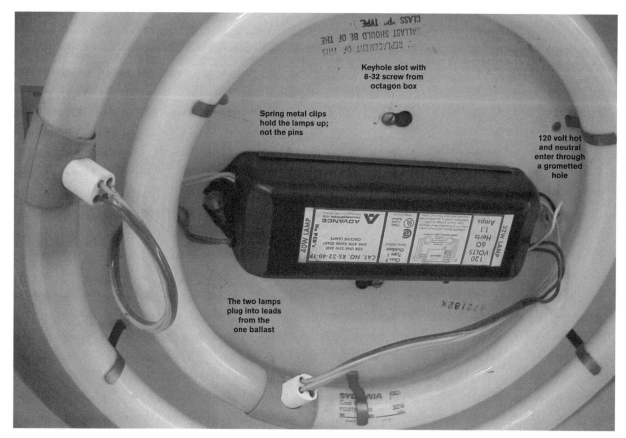

Circline, showing the keyholes and screw heads you address for fixture removal; the tubes, and mounting clips for their removal; and the ballast, with the mounting nut for its removal.

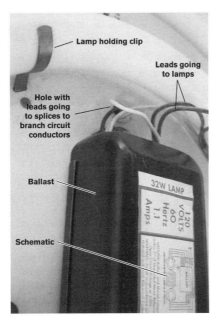

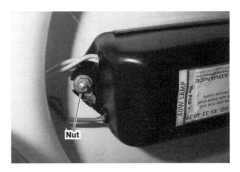

Circline, showing the hole passing leads through to the outlet box.

Ballast mounting nut. To remove the ballast, drop the fixture to undo the splices. Then unscrew the mounting nuts and the ballast can be removed.

pushing up against it. After it has rotated a few degrees, it should drop out of the fixture when you pull on it.

When you make sure that power is getting through to the fixture, be sure to check its grounding, too. This is not solely a matter of safety; fluorescent tubes rely on the close proximity of a grounded fixture surface for proper function.

Repairing fluorescents in old buildings is no different from repairing them in new buildings; replacing them, likewise. There are two important issues. First, almost all fluorescents have restrictions regarding the temperature ratings on the conductors entering them, at least near their ballasts. Second, in the normal course of operation some very old fluorescents have enough ground leakage current to trip GFCI's. GFCIs can't be used upstream of these antiquated fluorescent ballasts.

Circular fluorescent fixtures are installed and replaced much like standard incandescent fixtures. They are attached to ceiling outlet boxes, and they have to be dropped down from the boxes for access to their splices. They are, if anything, easier to rebuild than incandescent fixtures.

Illegal Fixtures

I'll lump together a number of fixture types. Antiques, home-created fixtures, artists' designs, field-built neon fixtures and unListed imports all probably are illegal. It costs many thousands of dollars to get Listing for a line of fixtures. Artists very rarely bother. Even getting a single fixture Listed by a laboratory specializing in that is expensive. Hardware stores and antique emporia doing rewiring that involves changing a fixture's design don't bother. Many importers, including some posh designer-goods stores, don't know or don't care, because their customers don't know any better.

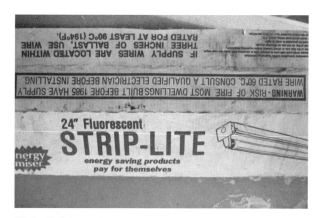

Strip light–90 degree wiring warning.

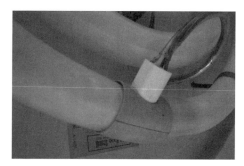

The terminal of a circline's lamp lead, plugged in. Unlike straight fluorescent tubes, the lamps in circlines cannot be installed misaligned. Also, the terminal/socket plugs into the lamp at one point, rather than the two ends of a lamp plugging into terminals, with careful insertion and twisting.

The worst problem with these fixture categories is that the designers and sellers often haven't a clue as to what makes a fixture safe.

What are some common mistakes?

- Thin conductors. Imported fixtures designed to European Union standards tend use wire with good insulation, but wire that may be too thin for the amount of current those 220-volt fixtures will draw when working on 120 volts.

- Thin insulation. Hardware stores, artists, and antique restorers sometimes use stereo speaker wire, in which the conductor may be thick enough, but the insulation is not designed to protect against anything approaching 120 volts.

- Unprotected conductors. Freelance manufacturers and restorers frequently neglect the need to protect conductors from abrasion and pinching.

- Inadequate grounding, mounting, and enclosure. The interface with houses' wiring systems is standardized. Canopy and box designs for European fixtures are different than ours. Artists and restorers sometimes don't even think about how you're supposed to gain access to and protect the splices, or how the system is to be grounded. As a result, boxes may not be fully covered, or fixtures may not attach solidly to your outlet box. Grounding may be interrupted, and if the fixture does not meet UL standards for protection of its internal wiring, a short could result in a live fixture waiting to shock someone. Some fixtures contain combustibles. Some canopies mount to the outlet boxes by screwing into place, resulting in twisted, eventually damaged wires.

A lamp terminal, unplugged for lamp removal. Plug and socket on circular ("Circline") fluorescent disengage with a tug. These use only one point (with four pins) for their electrical hookup.

None of these warnings are about your wiring being old; they all are about the fact that the older a home, the more common it is that someone along the way did ignorant work.

Small Hard-Wired Appliances

Some hardwired small appliances regularly wear out and need replacement. Aside from discomfort associated with physical access, you should have little difficulty evaluating, and perhaps taking care of, a few problems of garbage disposals and bathroom ceiling exhaust fans.

Bathroom Exhaust Fans

There are four things worth knowing about a modern bathroom exhaust fan, the type located behind a grill in the ceiling.

- Some people install them—as well as other exhaust fans such as range hoods—without ducts. Dumping the fan's output into the ceiling space can cause various kinds of damage and safety hazards. (The same is true of duct-type kitchen exhaust fans or range hoods.)

- The fan motor/blade assemblies normally are secured to their enclosures by two screws set in keyhole slots. Tightening those screws sometimes reduces noise and vibration.

- The assemblies plug into single receptacles located in their enclosures. Therefore, if a fan assembly fails, you can unplug it, loosen two screws, turn it a bit so the screw heads clear the keyhole slots, and drop it out. Use your tester on its receptacles to make sure the problem *was* with the fan itself. If it's not terribly, terribly old, you can easily buy a replacement.

- These fans are sound-rated. An old one might be rated at 3½ sones, one sone being about as loud as a refrigerator. You may be able to get a new replacement—because the old one has died or simply because you want less noise—rated at one sone.

Garbage Disposals

Disposals are spliced in just like fans or receptacles. They are not permitted on kitchen receptacle circuits, and their nameplate and package instructions need to be respected. For instance, they might say, "protect at no more than 15 amps," or "supply with a dedicated circuit."

You should look for two items on the underside of your garbage disposal. One is a reset button. When its motor overheats, usually as a result of stalling out, a thermal protector will interrupt power to it. Clear the blockage that caused over-heating, push this button, and you will solve the problem of "the disposal doesn't go on; it doesn't even make a sound like it's trying." The other item—which you *shouldn't* see—is a conductor or conductors. Unless it's a plug-in unit, you should find cable, but it should be attached by a connector to a covered splicing compartment. In either case, the conductors should be hidden. If you can see wires or wirenuts, you see evidence of illegal, incompetent work—this is true even if the disposal was recently replaced by a licensed plumber.

For technical terms, please first look in NEC Article 100. Also try your dictionary.

8B A 4″ round or octagonal box.

11B A 4-11/16″ square box.

17A A 3-1/4 or 3-1/2″ round or octagonal box.

90 A right-angle turn, or a piece of hardware enabling a right-angle turn.

1900 A 4″ square box.

AFCI Arcing Fault Circuit Interrupter.

AHJ Authority Having Jurisdiction—the local electrical inspector.

Air splice A splice that is unenclosed; illegal and dangerous, in 120- and 240-volt wiring, except for one very outdated system, and another that is so rare that I've never seen it used.

Al/Cu A marking on terminals of heavier-duty equipment, such as electric dryer receptacles, indicating that the equipment terminals are suitable for use with either copper or aluminum conductors.

Alligator clip A spring-loaded, toothed, temporary connector used at the ends of the leads of some testers.

Amp, Ampere Measure of current flow.

Ampacity Current-carrying capacity.

Analog For the purpose of this book, a meter that provides a reading by pointing to a position on a scale.

Anchor Hardware designed to hold a screw in a wall or other surface, to support equipment mounted to it.

Antishort (Slang) Synonym for Bushing, antishort.

Apprentice See Electrician.

Arc   To draw current through the air, often spattering metal in the process.

AWG   American Wire Gage; the smaller the one- or two-digit number, the thicker the wire and the higher its ampacity.

Backfeed   A situation where power enters at what is normally the Load side, where it normally exits. Backfeeding is dangerous, except in one or two rare circumstances where it is legal and intended.

Bakelite   An early plastic—black or dark brown where used as a construction material—that is more brittle than modern plastics.

Ballast   The part of a fluorescent or similar fixture that raises the voltage and restricts the amount of current that will flow through the lamps.

Bezel ring   A ring that is knurled on the outside and threaded on the inside, to screw over a threaded cylinder.

Black Iron   Usually refers to enameled (painted) steel pipe—perhaps electrical, perhaps plumbing, perhaps gas. Rarely used to refer to enameled electrical boxes.

Blow   For fuses, to operate so as to interrupt the flow of power.

Bond   To join parts electrically so as to maintain the same voltage level; in normal use, ground.

Bonding strip   The thin aluminum wire inside modern armored cable, designed to overcome an undesired electromagnetic effect of the armor's configuration; not intended for use as a ground wire.

Bolted short   A very solid, positive, contact creating a low-resistance short, as when incorrect wires are spliced together. If power is on, Bang!

Bootleg   (Slang) An unsafe substitute.

Box:

Gem   A sectional switch box, 2″ wide × 3″ high by a choice of depths.

Junction   A box used to enclose connections rather than to contain devices.

Mud   (Slang) An enclosure, usually octagonal, capable of being embedded in concrete, used most commonly for junctions or for light fixtures. Apartment houses with poured cement ceilings employ mud boxes, though they are not restricted to this use.

Pancake   A round box, 1/2″ or 5/8″ deep, 3–4″ in diameter.

Sectional   An enclosure designed to be taken apart and combined with its fellows as needed.

Switch   An enclosure designed to support a device via 6-32 screws. Most commonly used for switches and receptacles.

Utility   A rectangular enclosure 4″ long and 2″ wide, nongangable, by a choice of depths.

Bra cup   (Slang) A cheap paper breathing mask. These are not intended as protection from hazardous dust or fumes, but can keep nuisance dusts such as plain sawdust from causing you to cough.

Brownout A reduction in voltage, normally to your entire system, from the level that ensures proper functioning of electrical equipment.

Bug For the purposes of this book, a splicing device used to connect larger sizes of wire than you would or could splice with a wirenut.

Bugeye tester (Slang) A specialized receptacle tester that purports to tell you, by combinations of colored LEDs, the polarity and grounding status of a receptacle.

Building wire See Wire.

Bus, busbar A solid, rectangular metal bar that distributes electricity to fuses or circuit breakers.

Bushing

A small protector, such as the plastic or rubber grommet around the edges of a hole between one part of a fixture and another, to protect the fixture wires passing through; see also the following entries.

Bushing, antishort A bushing inserted in the end of armored cable to prevent the cut edge of the armor from damaging the conductor insulation.

Bushing, conduit A fitting that screws over the end of a piece of conduit to protect conductors passing between the conduit and the enclosure from abrasion. Sometimes a steel bushing is used to hold the conduit to the inside of the enclosure, in place of a locknut.

Bushing, reducing

1. A very shallow cylinder, threaded inside and out, designed to adapt a larger threaded opening, as in a weatherproof enclosure, to a smaller raceway or connector.
2. See also Donut.

BX (Slang) Armored Cable, Type AC. The present, most-modern version is type ACHH, because of the High Heat temperature rating of the conductors it contains.

Cabinet An enclosure for mounting equipment such as heavy-duty switches, or busbars and the associated fuses or circuit breakers.

Cable A flexible assembly of conductors inside a covering or sheath. The term can refer to BX, romex, and other types.

Caddy clip (Trademark, slang) Spring steel clamps for conduit and tubing.

Can (Slang for canister) In track lighting, the fixture itself, the movable part of the system that holds and aims the light bulb.

Cancer boxes (Slang) Proliferated subpanels. Generally, you are better off with fewer panels.

Canopy The part of a fixture that bears against the ceiling or wall and covers the outlet box feeding it.

Cap nut A nut that is open only on one end. Usually, rather than being hexagonal on the outside, it is round and knurled.

Carbon monoxide An odorless, deadly gas created by incomplete combustion, for instance at some furnaces and water heaters. Carbon monoxide detectors can be an excellent protection.

Carriage light A sconce-type fixture, usually used outdoors, consisting of a metal frame and four glass panels.

Cartridge fuse See Fuse.

Cash sale (Slang) A transaction at an electrical supply house via cash, check or plastic—anything other than using a credit line. When the clerk asks you, "Is this a cash sale?" he is not expecting you to pull out $300—or, necessarily, even $3—in greenbacks.

Chase nipple See Nipple.

Cheater (Slang) An adapter used to bypass a safety feature.

Circuit The path current takes starting from your electrical service, passing through the loads it feeds, and returning back to the service.

Circuiting, circuitry Circuits considered collectively, or the way your outlets are divided up into those fed by your various circuits.

Clamp (noun) A fitting that takes the place of a connector (but that is considered as occupying interior space in an enclosure, which is not true of a connector). A clamp is designed to secure one or two cables of a specified type or types to an enclosure. Clamps found in metal boxes are embossed with letters to indicate what the clamps are suitable for. An "A" indicates BX, and an "N," romex.

Classified Investigated and certified by a Nationally Recognized Testing Laboratory as suitable for use with equipment whose manufacturer has not necessarily authorized it.

Clearance, working You are legally required to leave a space 30″ wide and 36″ deep unobstructed from floor to ceiling in front of your electrical panel, air conditioning disconnect, or other electrical equipment that might need to be opened by an electrician while live.

Close-in (Slang) An inspector's approval to go ahead and hide wiring by covering walls, ceilings, or trenches.

Close nipple See Nipple.

CO Symbol for carbon monoxide.

Co/Alr A rating marked on lower-power devices such as 120-volt receptacles that are suitable for connection to either copper or aluminum branch circuit wiring.

Code The National Electrical Code, or NEC, NFPA 70. Caution: Your town or county may not have adopted the NEC as written. Check for local amendments.

Common In three-way switching, where an outlet (usually lighting) or outlets are controlled from two locations, the Common is the conductor that is switched between one Traveller and the other.

Compact fluorescent adapter A fluorescent fixture that screws into an Edison-base incandescent socket. The ballast is contained in its base and the lamps are narrow, folded, or twisted tubes that take up not a great deal more room than a normal incandescent light bulb. It is sufficiently heavier, though, to damage some fixtures.

Compressor In residential wiring, the part of a central air conditioner that sits outside, usually alongside the house on a concrete pad.

Conceal inspection, Concealment Inspection Synonym for Close-in.

Conductor
1. In this book, a wire plus the insulation covering it.
2. Any metal, in fact anything, that conducts electricity fairly well.
Caution: Something termed a nonconductor, such as concrete, may nevertheless conduct well enough to cause injury or death in some circumstances.

Conductor, hot See Hot.

Conductor, identified A conductor whose insulation is white or gray, either solid color or, very rarely, striped. In some cases the insulation is taped white or gray, or taped only near the ends.

Conductor, return The conductor that is eventually grounded near the source of the circuit. It completes the path for electricity from the hot wire through a 120-volt load. I refer to it in this book as the neutral, for convenience. It should be white.

Conduit A pipe designed to contain wiring.

Conduit body, condulet A fitting inserted in a run of conduit to allow access to the wires inside it.

Conduit bushing See Bushing.

Connector A fitting used to attach a cable or raceway to an enclosure.

Connector, wire Synonym for Splicing device.

Connector, cord (male) That which you plug into a receptacle.

Contaminant Anything that can intrude into electrical equipment and degrade its operation by causing deterioration, like solvents, or caustic cement dust; or causing mechanical interference, like paint or dust; or pose the risk of fire, like insects, some paint, or sawdust.

Contractor See Electrician.

Counter The section of an electrical supply house where parts other than decorative lighting fixtures and fans are purchased. The other main public area is the Showroom.

Coupling A fitting that attaches two pieces of raceway.

Cross-listed Designed to work properly with parts of another (specified) brand, one that may or may not be produced by the same manufacturer.

Current Rate of flow of electricity, closely related to risk of fire.

Cut-in notice An inspector's notification to the utility providing power that your interior wiring is safe and they can proceed with their end of the job.

Daisy chain (Slang) To connect, each to the next.

1. When devices are daisy-chained, conductors loop directly (with or without being cut) from the terminals of one device to the terminals of the next, rather than going from each to a splice.

2. When outlets along a circuit are daisy-chained, cable or raceway goes from the panel to one outlet, and from that one to the next, rather than radiating separately from the panel to each outlet, as in "star" or "radial" wiring.

Decora (adjective) A device whose face is relatively smooth and rectangular.

Despard A fixture yoke into which can be fitted any combination of one to three switches, receptacles, and indicator lights.

Device, wiring Equipment, typically mounted in a switch box, that helps you use electricity but does not itself use an appreciable amount. This excludes lights and heaters, but includes switches, receptacles, thermostats, timers, dimmers, and free-standing sensors.

Diffuser Something inserted between you and a light bulb to reduce the discomfort associated with a "point source" of light impinging on your eye.

Digital For the purposes of this book, a meter whose readout is in numerals.

Disconnect (noun) Anything designed as a means of interrupting power. Disconnects generally are not intended for regular use as on-off switches, nor is a switch—especially a dimmer!—intended for use as a disconnect. Some circuit breakers, though, are embossed with the letters, "SWD," for Switching Duty.

Service disconnect The switch(es), circuit breaker(s), or fuse(s) you need to operate in order to interrupt all power to your premises.

Distributor An electrical supply house or wholesaler. Distributor personnel tend to be more knowledgeable than those at home centers, and distributors tend to have better assortments of electrical supplies.

Doghouse (Slang) Synonym for Monument.

Donut, doughnut (Slang) A stamped reducing washer that reduces opening sizes in enclosures; used in pairs, with the concave sides facing each other for springiness.

Downstream Further from the source of power.

Drop light A portable lampholder on a flexible cord; it may include a receptacle.

Dual-element See Fuse.

Duplex

1. A device that doubles up, offering two switches, or two receptacles, on one yoke.

2. A connector that accepts two cables.

Ear A protrusion at or near an end to help bear against an underlying surface or to allow equipment to be mounted.

Earth British for Ground.

Edison base   Having a threaded base whose diameter is that of a standard light bulb.

Electrician:

> **Apprentice**   Someone engaged in the 4–5-year process of becoming an electrician, which includes classroom training, supervised installation, and testing.
>
> **Class A**   An electrician qualified to perform commercial and industrial work, which are considered more valuable; not necessarily a superior electrician for residential work.
>
> **Contractor**   Someone licensed and insured or bonded, as may be required, to do business as an electrician, rather than merely to perform the work as an employee.
>
> **Handyman**   No, a handyman is not an electrician. Even a "certified home improvement contractor" does not need electrical study, training, or experience to earn his or her license.
>
> **Helper**   Sometimes used as a synonym for Apprentice; sometimes indicates a lower level of qualification, such as a laborer who is not engaged in formal apprenticeship training.
>
> **Journeyman**   An electrician who has completed apprenticeship training and can work without close supervision, but who has not moved on to being authorized to take sole responsibility for electrical work.
>
> **Master**   The highest level of qualification to perform and supervise wiring and electrical repair. Warning: Even a master electrician may not have experience dealing with your particular wiring system.
>
> **Mechanic**   (Slang) Synonym for Journeyman, usually, but also used to include Master.

Electrode   For the purposes of this book, synonym for Grounding electrode.

Element   The part of a heat-producing appliance such as a toaster that heats; the part of a fuse that melts.

EMT   Electrical Metallic Tubing, a nonthreadable raceway that is vulnerable to moisture.

Enamel   Black, oil-base paint, normally. An old-fashioned means of protecting steel from corrosion.

Enclosure   Any electrical box, which may include circuit breaker panels, junction compartments that are part of an appliance, even meter bases and lampposts.

FD, FS   Weatherproof rectangular (parallelopipedal—rectangular in three dimensions) enclosures, traditionally made of cast metal.

Fault   An unintentional and potentially hazardous electrical connection or contact.

Final   (Slang) The final inspection approving electrical installation for use.

Firestop A cross-brace between studs. It can play hob with attempts to fish cable without opening a wall.

Fish
1. To push or drop cable through a space you can't actually reach into.
2. To pull conductors through a raceway.

Fixture wire See Wire.

Flush Even with the surface of the wall or ceiling.

Frame-in kit The part of a recessed light fixture that is permanently installed in the ceiling.

Fuse:

Cartridge A cylindrical fuse, usually held by a clip at each end.

Dual element or time delay A fuse that permits a very short-term overload, as is required for motor starting. Not normally used for internal protection of electronic equipment.

Fustat A time delay plug fuse that cannot be interchanged with one of a different ampacity.

No tamp Synonym for Fustat.

Plug A fuse that screws in.

Gage See AWG, Strip Gage.

Gem box See Box.

GFCI Synonym for Ground Fault Circuit Interrupter.

Gang The number of device yokes a box or device ring is designed to support.

GEC Grounding Electrode Conductor.

Grandfather To permit wiring that does not meet the latest safety standards, because it was in compliance when originally installed.

Green sticker, green tag Notice that an installation has been approved.

Ground, grounding
1. Connection to the safe voltage level represented by the earth.
2. A wire or terminal intended for use in grounding.

Grounding electrode Something in solid electrical contact with the earth that is used as a stabilizing voltage reference for nonlive parts of your electrical system.

Ground Fault Circuit Interrupter (GFCI) A device designed to interrupt the flow of power within 4–6 milliseconds if it detects a minuscule imbalance between the amount of current flowing out along a circuit and the amount returning.

Halogen An incandescent light that runs more efficiently because chemicals allow the filament to run hotter without burning out prematurely.

Handy box, Handi-box Slang for utility box.

Handyman See Electrician.

Hard wire To attach directly to the circuit wiring, as opposed to plugging in to a receptacle.

Heavy-up A service upgrade.

Hickey A device for mounting a fixture to an outlet box. Also a pipe-bending tool.

Hot (Noun, adjective; slang) A live conductor; in house wiring, normally at about 120 volts to ground.

Identified conductor See Conductor.

Incandescent Producing light by being heated till it glows.

Inspection, final See Final. Compare with Rough-in, Close-in, Red-Tag, Green Sticker.

Insulator or insulation, electrical Something that prevents current from flowing through it when intact. Undamaged conductor insulation prevents the wires inside it from shorting out; undamaged insulators keep busbars from shorting to the cabinets of panels.

Interchangeable See "Will fit"; not reliably the same as "Cross-Listed."

Intermittent (Adjective, noun) A connection or action that is maintained inconsistently. Intermittent problems are extra dangerous and hard to trace.

Inverse-time A standard type of fuse or circuit breaker; the higher the overload, the faster it interrupts the circuit.

Inventory engineering. Synonym for "Will Fit."

J-box (Slang) Synonym for Junction box.

Jelly jar A sconce-type fixture, usually used outdoors, with a canopy on top and a glass cylinder (the "jelly jar") below.

Joint
   1. (Archaic) An electrical connection.
   2. A mechanical connection, often pivoting, between a fixture and its outlet box.

Journeyman See Electrician.

Junction box See Box.

k, kilo (prefix) 1000; kWH is a measure that represents the number of volts times the number of amps you've used times the number of hours you've used them divided by a thousand.

Knockout A semiprepunched, normally round, opening in an enclosure.

Knockout seal A component used to close openings in metal enclosures.

KO or K.O. Synonym for Knockout.

KO seal Synonym for Knockout seal.

Labeled Marked UL (including 🆄🄻) or FM; probably okay also if marked

MET, APL, CSA, Battelle, or AGA. Beware of European Listing marks. Not every esoteric emblem means a fixture is designed to U.S. standards.

Lamp   Light bulb, any type, including fluorescent tube.

Lampholder
1. A very basic lighting fixture.
2. The part of a lighting fixture that actually holds the lamp.

Lath   The material nailed or screwed to studs or joists, on which a plaster surface is applied.

LED   Light-Emitting Diode. A very rugged, reliable, efficient type of indicator light.

Line
1. (Slang) Synonym for Line side.
2. (Slang) Circuit or cable.

Line side   The side of a control or protective device, such as a fuse, into which uncontrolled electricity enters. Compare Load Side.

Listed   Investigated for a specific purpose.

Live   Synonym for Hot.

Load
1. Something that uses up electricity, such as a light or appliance.
2. (Slang) Synonym for Load Side.

Load side   The controlled or protected side of a control or protective device, such as a GFCI. Compare with Line side.

Loadcenter   An enclosure with fuses or circuit breakers used to originate residential branch circuits; Panelboard, in the NEC.

Locknut   A ring that is threaded on the inside and faceted on the outside, used for screwing over cable and raceway connectors, or over threaded conduit, to secure it to openings in enclosures.

Lug   A wire terminal. Usually, the wire's inserted and then setscrews are tightened or a compression tool crimps the lug.

Luminaire   A lighting fixture. In the past, the type of fixture to which the term was applied was usually fluorescent, most often part of a suspended ceiling.

Madison clip   An add-on device for holding a box inside a wall by pressing against the inside surface of the wall.

Main breaker   A panel designed so that power enters the busbars after first passing through a permanently installed breaker; also that breaker itself.

Main lug panel   A panel designed for power to be fed directly to the busbars.

Male-male   A generally dangerous, illegal means of supplying power from a generator into a house's wiring system. It involves throwing off your Main breaker, plugging an extension cord into a receptacle on the generator, fitting the other end of the cord with a second male cord connector, and plugging it into a receptacle in your house's wiring system. A safer, if not necessarily safe, approach involves using a generator that has an "Inlet" male connector, to which you would attach the female end of a normal extension cord.

Master   See Electrician.

Medallion   A large, round, elaborate fitting that is attached to the ceiling for elegant looks or to cover damage.

Meter can   Meter pan.

Meter pan, Meter base   An enclosure for an electric meter: applied to the housing plus the guts.

Meter socket   Loosely, meter pan.

MLO   Main Lug Only. Main Lug panel.

Mogul   Oversize.

Monument   An enclosure that protrudes up from a flat, horizontal surface, with a device—most commonly a receptacle—facing out of it at right angles to the surface to which the monument is mounted.

Mounting rail   The part of a circuit breaker panel onto which circuit breakers are clipped to provide stability and support, positioning them correctly with respect to the busbar stabs they clip on to to receive power.

Mud   (Slang) Plaster or cement; often, also drywall joint compound or spackle.

Mud box   See Box.

Mud ring   (Slang) A device ring, usually for extending a square metal box to attach a light fixture.

Multigang   More than single gang. See Gang.

Multimeter   A meter designed to test voltage, current, resistance, and perhaps other electrical quantities.

Nail plate   A steel plate nailed or screwed to the surface of a structural member such as a wood stud to protect wiring. The wiring may have been pulled through a hole that was dangerously close to the edge of the stud, or laid in a groove chiseled in its edge.

Nalox   (Slang, Trademark) Antioxidant for aluminum terminations.

Neon   Lights and signs based on the noble gas neon, which generally are not legal for use in homes.

Neon tester   (Slang) Any two-lead voltage tester with a small, glowing indicator.

Neutral
  1. A conductor, normally grounded at its source, that is equidistant in voltage between two hot conductors.
  2. (Slang) The return conductor in a circuit.

NFPA National Fire Protection Association.

Nipple A short length of pipe.

> **Chase nipple** A very short nipple, with a flanged (collared) end that bears against the inside of a knockout, generally used to connect enclosures.

> **Close nipple** A very short nipple, all threaded, generally used with two bushings and two or four locknuts to hold two enclosures an inch or so apart.

Nose A connection between conductors and whatever is used to pull them through a raceway; it is smoothly tapered in order to minimize resistance to pulling.

NRTL Nationally Recognized Testing Laboratory; recognized by the U.S. Department of Labor.

Nuisance trip See Trip.

Octopus (Slang) A receptacle adapter permitting the attachment of additional cord connectors beyond the number the receptacle is designed to accommodate.

Ohm Measure of resistance to the flow of electricity.

Ohmmeter A resistance tester.

Open When electrical connection is not made or is unmade, as when:
A switch opens a connection intentionally, for instance to turn off a light.
A fuse opens a circuit when there is a problem that draws too much current.
An "open" in a conductor is created when the conductor breaks. Compare Short.

Open wiring See Wiring.

Outlet Any place where electricity can be drawn—a receptacle or light or appliance.

Overfuse To feed equipment or wiring via an overcurrent device whose rating is such that it will not protect the equipment. Dangerous.

Overlamp To insert a hotter light bulb than a fixture is designed to handle. Dangerous.

Pancake box See Box.

Panel blank A press-in plate to close an unused opening in a panel cover.

Parts counter Synonym for Counter.

Penetrox (Slang, Trademark) Antioxidant for aluminum terminations.

Piggyback (Slang) Synonym for Tandem.

Pigtail
1. Wire spliced to an existing wire or wires.
2. The act of adding a pigtail.
3. A rubber lampholder with two leads of rubber-covered flexible wire, insulated except at their ends.

Pipe Rigid or intermediate metal conduit.

Plaster ring A device-mounting ring for a square box. Technically, a synonym for "mud ring."

Plastic pipe (Slang) Rigid, nonmetallic conduit. Electrical PVC, not to be confused with plumbing pipe or fittings.

Plug This can refer to a switch, receptacle, or attachment plug. (Lay language, thus too nonspecific) Also see Fuse, plug.

Polarity:

Correct When the shell of a light or the neutral terminal of a 120-volt receptacle is connected to the neutral conductor, and the center contact of a light or hot terminal of a receptacle is connected to the hot conductor.

Reversed When the shell of a light or the neutral terminal of a 120-volt receptacle is connected to the hot conductor, and the center contact of a light or hot terminal of a receptacle is connected to the neutral conductor.

Polarized Incorporating a distinction between that part which is to be connected to the hot conductor and that which is to be connected to the neutral conductor. A polarized attachment plug has a narrower blade connecting to the hot, a broader blade connecting to the neutral.

Porcelain
1. The ceramic material itself.
2. A plain lampholder designed to screw directly to an 8B box. It may incorporate a single receptacle on the side, and it may or may not have a pull chain.

Posi-ground (Trademark) A line of self-grounding devices.

Potential With respect to shock risk, "step-" potential concerns current coming through a foot, and "touch" potential refers to current flowing through a hand, or another part of your body that might brush against a surface above the floor.

Potential difference Synonym for voltage.

Pullout A fuseholder, usually for two cartridge fuses, that has a handle enabling you to pull it out of the fuse box, disconnecting the circuit or circuits and gaining safe access to the fuses for examination or replacement.

Punch list A list of items needing to be addressed in order to complete a project.

Pushmatic A circuit breaker that is punched on and off, rather than having a handle flipped.

Pyrolize To char.

Quad (Slang), **Quadplex** (Trademark) For the purposes of this book, a receptacle outlet that permits you to plug in four items at once.

Quickwire (noun, verb) To connect wires by pushing them into holes, generally in devices, to be grabbed by spring metal. Not considered the best connection, and now forbidden as a means of connecting anything but #14 solid copper wire to devices.

Raceway

1. Metal or plastic tube, such as pipe, tubing, or surface metal raceway (e.g., Wiremold), used to run conductors from one enclosure to another.

2. A fluorescent or other lengthy light fixture also may serve as a raceway, in that wires may travel through it from one end to the other.

Rag wire, wiring   See Wire.

Receptacle   An outlet where you can plug in one or more appliances.

Red hat   (Slang) Synonym for Bushing, Antishort.

Red head   (Slang) Synonym for Bushing, Antishort.

Red-tag   (Slang) A notice of violation, failed inspection.

Reducing bushing   See Bushing.

Relamp   To replace a light bulb.

Return, return conductor   See Conductor.

Reverse polarity, reversed polarity   See Polarity.

Ring, device or mud   See Device Ring or Mud Ring.

Ripper   A tool for removing the sheath from romex and, perhaps, UF cable.

Romex   (Slang, trademark) Nonmetallic sheathed cable; presently, type NM-B.

Rosette   A fitting, often somewhat elaborate, covering a round ceiling outlet, with a hole in the center for a pendant.

Rough in   (Slang)

1. (Verb) To install pipe or cable plus enclosures, without devices, or at least without covers.

2. (Noun, adjective) Synonym for Close-in.

Run

1. (Verb) To install.

2. (Noun) A complete section of raceway or cable, at least between one enclosure or fixture and the next; more often a larger subsection of a circuit.

Screws:

Allen head   A screw whose head has a hexagonally shaped indentation.

Combination head   A screw that can be turned by more than one type of screwdriver.

Drywall   A relatively slim, mostly nontapered, flat-head/Phillips screw unthreaded by the head.

Fillister head   A screw whose head is extra thick, nearly flat above and below.

Flat head   A screw, the top of whose head is flat and the bottom curved or conical.

Hex head   A screw whose head is hexagonal, requiring a wrench or a socket driver.

Machine   A screw with parallel, even threads that mates with a nut or a threaded hole.

Oval head   A screw, both the top and the bottom of whose head are gently curved.

Phillips head   A screw with tapering, crossed indents, deepest at the center.

Round head   A screw, the top of whose head is round and the bottom flat.

Setscrew   a screw designed to hold something in place because the end of the screw presses against it.

Sheet metal   A screw with tapering threads, designed to hold pieces of metal together. Generally not suitable for electrical connections, including grounding.

Slot head   A screw designed to be turned by a flat blade.

Self-grounding   A device having a clip on its mounting yoke that ensures adequate electrical contact between a mounting screw going through it and the yoke itself. When (and only when) screwed to an adequately grounded metal box, this grounds the device without the need to run a bonding conductor back to the box.

Service   The point where power makes the transition from being the utility's responsibility to being yours; where you have your service disconnect.

Sectional box   See Box.

Service disconnect   See Disconnect.

Sheath
1. In test instruments, the insulation covering the leads.
2. In cables, the covering over the conductors' insulation.

Short   An accidental connection bypassing utilization equipment, either bridging from conductor (or live part of equipment) to conductor or conductor (or other live part) to ground or other surface.

Showroom   The part of an electrical supply house where you can pick out light fixtures and paddle fans.

Sign off   (Verb, noun)
1. Successful final inspection.
2. An inspector's notification to the utility to proceed with their work.

Single-gang   See Gang.

Size, trade   A formal designation, not the same as the equivalent ruler measurement. For instance, the diameter known as "three-quarter inch trade size" measures about 1-1/4."

Skin   Remove insulation from wire; often used when insulation is to be removed only partly.

Slo-blo, slow-blow   Synonym for Fuse, Dual Element.

Splice   (Verb, noun)
1. To connect wires.
2. A point where wires are connected to each other.

Splicing device   A wirenut, bug, or other means specifically designed for connecting wires.

Split-bus   A panel whose power is fed into one Main section of busbars, from which another section or sections are fed by one of several breakers or fuse-holders attached to the Main section.

Split-wire   To feed a receptacle so that its outlets are independent.

Stab   The extension from a busbar to which a circuit breaker is attached.

Stacked   Combining two or three devices, other than simply two receptacle outlets, on one yoke.

Star drill   A steel bit for hand-drilling through masonry.

Stem   A short, threaded hollow metal cylinder used to connect and support parts of lighting fixtures.

Sticker
1. The record that inspection has been passed.
2. The Label from a Listing lab.

Strap
1. A fitting designed to hold a specific cable or raceway to a surface.
2. A fitting that holds a fixture to its outlet box.

Strip   To remove insulation.

Strip gage   A marking that indicates how much insulation should be removed from the end of a conductor.

Stub-up   The empty pipe sticking out of a floor, or occasionally a wall, during rough-in.

Stubby   A short screwdriver.

Subpanel   Any panel, containing fuses or breakers, other than one that contains your service disconnect.

Switch

Four-way   A switch that works in conjunction with at least two other switches, and has no ON and OFF markings.

Knife   A definitely outdated switch with exposed parts and no spring loading, such as you would see in a 1940s movie showing execution by electric chair.

Mercury   A very long-lasting and silent switch, whose contacts don't arc because they make and break connection by dipping in or out of a tiny pool of liquid mercury.

Three-way   A switch that works in conjunction with at least one other switch and has no ON and OFF markings.

Two-way   British for three-way.

Switch box   See Box.

Tag   See Red-Tag, Green Sticker.

Tandem   A circuit breaker that feeds two 120-volt circuits from the same busbar location, or two 240-volt circuits from the same adjacent busbar locations.

Tap
1. A tool for threading or rethreading the inside of a hole.
2. To draw power for a secondary, often under-protected (Danger) use.

Terminal
1. A screw or other mechanism for connecting wires.
2. The part of a fluorescent or similar fixture the lamp plugs in to.

Terminate   To attach a wire or wires to equipment.

Thinwall   Synonym for EMT.

Time Delay   See Fuse.

Trade Size   See Size, Trade.

Transformer   A magnetic device that converts from one voltage level to another.

Traveller   In three-way switching, where an outlet (usually lighting) or outlets are controlled from two locations, the Travellers are the conductors that are alternately connected to the Common when the switch is operated.

Trim kit   The part of a recessed light fixture that hooks into the frame-in kit from below, that determines both the appearance of the fixture and the beam pattern.

Trip   For a mechanical device to operate so as to interrupt the flow of power.

Nuisance trip   To trip in response to something that is not a genuine hazard.

Upstream   Closer to the source of power.

Utility box   See Box.

Volt, Voltage   Measure of electrical "pressure," and thus risk of arcing and shock.

Wall washer   A fixture whose light is directed along a wall, rather than along the ceiling or out into the room.

Watt   The basic measure of electric power, essentially combining volts and amps.

Weatherhead   In an overhead service, the equipment where the service conductors running up the side of your house are separated, and water kept out. The service drop, the line from the utility's pole, often is attached to it for support. The conductors of the service drop are spliced to the conductors emerging from the weatherhead within a foot or two of where they emerge.

Wiggy   (Trademark, slang) A solenoidal voltage tester, one that vibrates and buzzes as part of its signal when it detects voltage within the range it is designed to register.

Will fit   Unless the component is Listed for your specific use, this means, "We don't stock what you need." There is unfortunately a good chance that applying it as suggested is illegal.

Will-call   The part of the counter at electrical supply houses where phoned-in orders are picked up.

Wire:

> **Building**   Insulated conductors run in cables and raceways; the cables themselves; pretty much everything but low-voltage wiring, extension cords, and cords used as part of tools and appliances is called building wire.
>
> **Fixture**   Insulated stranded conductors, normally smaller than #14, that are part of a light fixture, appliance, or other equipment as it comes from the factory, and are to be spliced to the circuit conductors.
>
> **Knob-and-tube**   The in-the-walls complement to open wiring.
>
> **Open wiring**   An antiquated, once-legitimate wiring system, not requiring cables, raceways, or enclosures. Leave it to a pro.
>
> **Rag**   Wiring utilizing an old type of conductor insulation, consisting of cloth-covered rubber.
>
> Warning: Low-voltage cables such as speaker wire, stereo wire, bell wire, and phone wire are neither suitable for use as building wire nor as fixture wire.

Wire connector   Synonym for Splicing Device.

Wiremold   (Trademark, slang) Surface metal raceway.

Wirenut   (Trademark, slang) A twist-on solderless connector.

Wiring device   See Device.

Working clearance   See Clearance, Working.

Yoke   The part of a wiring device that supports it to the enclosure; usually but not always grounded metal.

AFTERWORD

What I Left Out

This book does not begin to teach you how to do all your own wiring, nor even how to troubleshoot your wiring. For instance, you now may be able to replace a switch. I explained how to determine if a switch is defective and therefore needs replacement. However, I did not go into the details of switching theory, so as to help you determine whether your switches—working or not—were installed as the NEC requires. For this you need to read a more-advanced text, such as my *Old Electrical Wiring*. Until you have absorbed that material, you will not have the information you need to answer such questions. This means you will reach your limits, and, if wise, will seek help.

Sources of Help

I wish I could hand you a key to finding reliable help. Throughout this book, I periodically came to places where I advised, "Call in a pro." But what's a real pro? What's a trustworthy source of advice? There are a number of home repair referral services, which recommend contractors to homeowners who call or e-mail, and charge contractors in return for recommending them. I have not been greatly impressed with the screening performed or the guidance provided by the services I have checked out.

If you simply have questions about how to interpret the electrical code, and your town's or county's electrical inspection department has someone who is willing to talk to you over the phone, you're in luck. That sort of service is less available than it used to be. If you want to know why this is so, and are interested in finding out how electrical inspection services can be restored to offer better protection to the public, contact The Inspection Initiative, 1-800-647-3156, ext. 521.

There are also various resources available via the World Wide Web. Some are listed in Chapter 14, Resources, of my more-advanced text, *Old Electrical Wiring*. You may also get useful answers to questions involving genuinely confusing areas of the NEC by signing on to electronic bulletin boards that serve the electrical trade. One of them can be accessed by going to www.electrician.com. One manufacturer-sponsored Web site that will give you some theoretical and even practical information on electricity and on the components of your electrical system is http://www.ch.cutler-hammer.com/training/slfstudy/navigate/webmanualmenu.htm. www.lightolier.com is another manufacturer site, which offers "Lessons in Lighting."

If you need to hire someone to do electrical work, I can suggest a few indicators that may help you find someone who is relatively competent. First, if you have a choice of getting your work done by someone licensed as a master electrician, he or she is likely to be more knowledgeable than someone with a journeyman's or other lower-level license. Second, someone who specializes in commercial work and rarely does residential wiring may not be your best bet. Third, someone who has chosen to take (and passed) a national certification test to become an electrical inspector may well be more knowledgeable than the average electrician. Fourth, someone who carries a copy of the electrical code in his or her truck probably is relatively conscientious. Fifth, someone who partakes of continuing education takes the work seriously; it makes a difference. Keeping current means periodically attending seminars on electrical code changes and other industry developments, and subscribing to the trade magazines. Sixth and finally, members of groups such as the International Association of Electrical Inspectors and the National Fire Protection Association may have a greater-than-average commitment to electrical safety.

Repair: the Broad View

Maintaining old wiring, when you can do so safely, is a good thing to do in many senses. "Global sustainability," or "deep ecology," means acting responsibly towards future generations and towards the have-nots by not squandering resources. Some of the behaviors this leads to are recycling and not throwing out the still functional for the shinier. On the other hand, repeated, unreliable repair of equipment that is no longer viable serves no great good, other than that of a learning experience. For example, I bought an inexpensive—actually, cheap—paddle fan whose light kit was, while attractive, quite poorly designed. Within a few years, the multiple lampholders were sagging, and the insulation on the wiring feeding them chafed through. I rewired them, and this was a mistake. If the wear had been the result of reasonable age, sure; however, poor design meant that its other wiring wore through within another year or so.

What about . . . ?

It is quite possible that you will run into a problem that is not within the scope of either this or my other book. Perhaps it should be added to future editions of one or the other, or perhaps it can find a place in a book I have yet to write. If you suspect that novice readers really would benefit from a topic that I do not cover here, I'm interested in your thinking. If something in this book simply confuses you, I certainly would like to know. I can't promise to respond to all feedback, but I encourage you to email me (plain text with no attachments, please) at writer@davidelishapiro.com or write to me in care of McGraw-Hill Professional Books, #2 Penn Plaza, NYC 10121-0101.

INDEX

Boldface items indicate illustrations.

ABOUT THE AUTHOR

David E. Shapiro is a master electrician and renowned authority on home wiring. The author of *Old Electrical Wiring*, also from McGraw-Hill, and the residential columnist for *Electrical Contractor* magazine, he has written for the *Washington Post, Practical Homeowner, New England Builder*, and many other magazines. He is also an electrical consultant and a certified inspector and plan reviewer. He lives in Colmar Manor, Maryland, near Washington, D.C., where he does business as Safety First Electrical Contracting, Consulting, and Safety Education.